Recent volumes in MSJ Memoirs

1 **Ivan Cherednik, Peter J. Forrester and Denis Uglov**, Quantum many-body problems and representation theory, 1998 .... JPY4400
2 **Masako Takahashi, Mitsuhiro Okada and Mariangiola Dezani-Ciancaglini (Eds.)**, Theories of types and proofs, 1998 .... JPY6380
3 **Tomotada Ohtsuki**, Combinatorial quantum method in 3-dimensional topology, 1999 .... JPY2200
4 **Atsushi Matsuo and Kiyokazu Nagatomo**, Axioms for a vertex algebra and the locality of quantum fields, 1999 .... JPY2200
5 **Daryl Cooper, Craig D. Hodgson and Steven P. Kerckhoff**, Three-dimensional orbifolds and cone-manifolds, 2000 .... JPY3080
6 **Kong De-xing**, Cauchy problem for quasilinear hyperbolic systems, 2000 .... JPY4400
7 **Vladimir Georgiev**, Semilinear hyperbolic equations, 2000 .... JPY4400
8 **Eric M. Opdam**, Lecture notes on Dunkl operators for real and complex reflection groups, 2000 .... JPY1650
9 **Peter Orlik and Hiroaki Terao**, Arrangements and hyper-geometric integrals, 2001 .... JPY2090
10 **Yuri G. Prokhorov**, Lectures on complements on log surfaces, 2001 .... JPY2750
11 **John R. Stembridge, Jean-Yves Thibon and Marc A. A. van Leeuwen**, Interaction of combinatorics and representation theory, 2001 .... JPY2640
12 **Yves André**, Period mappings and differential equations. From $\mathbb{C}$ to $\mathbb{C}_p$, 2003 .... JPY4950
13 **Shigeaki Koike**, A beginner's guide to the theory of viscosity solutions, 2004 .... JPY2410
14 **Noboru Nakayama**, Zariski-decomposition and Abundance, 2004 ..... JPY4714
15 **Valery Alexeev and Viacheslav V. Nikulin**, Del Pezzo and K3 surfaces, 2006 .... JPY3142
16 **Brian H. Bowditch**, A course on geometric group theory, 2006 .... JPY1676
17 **Arkady Berenstein, David Kazhdan, Cédric Lecouvey, Masato Okado, Anne Schilling, Taichiro Takagi and Alexander Veselov**, Combinatorial Aspect of Integrable Systems, 2007 .... JPY2514
18 **Yasutaka Ihara**, On Congruence Monodromy Problems, 2008 .... JPY2933
19 **Joseph Najnudel, Bernard Roynette and Marc Yor**, A Global View of Brownian Penalisations, 2009 .... JPY1906
20 **Danny Calegari**, scl (Stable Commutator Length), 2009 .... JPY3404
21 **Gautami Bhowmik, Kohji Matsumoto and Hirofumi Tsumura (Eds.)**, Algebraic and Analytic Aspects of Zeta Functions and L-functions, 2010 .... JPY3059
22 **Michael Ruzhansky and James Smith**, Dispersive and Strichartz estimates for hyperbolic equations with constant coefficients, 2010 .... JPY2587

23 **Armen Sergeev**, Kähler geometry of loop spaces, 2010 ... JPY3436

24 **Taro Asuke**, Godbillon–Vey class of transversely holomorphic foliations, 2010 ... JPY2325

25 **Hiroshi Sugita**, Monte Carlo method, random number, and pseudorandom number, 2011 ... JPY2566

26 **Shinya Nishibata and Masahiro Suzuki**, Hierarchy of semiconductor equations: relaxation limits with initial layers for large initial data, 2011 ... JPY2147

27 **Suhyoung Choi**, Geometric structures on 2-orbifolds: Exploration of discrete symmetry, 2012 ... JPY3038

28 **Toshio Oshima**, Fractional calculus of Weyl algebra and Fuchsian differential equations, 2012 ... JPY3614

29 **Takeshi Hirai, Akihito Hora and Etsuko Hirai**, Projective representations and spin characters of complex reflection groups $G(m,p,n)$ and $G(m,p,\infty)$, 2013 ... JPY4431

30 **Tatsuo Nishitani**, Cauchy Problem for Noneffectively Hyperbolic Operators, 2013 ... JPY3048

31 **Satoshi Takanobu**, Bohr–Jessen Limit Theorem, Revisited, 2013 ... JPY3698

32 **Hiroshi Isozaki and Yaroslav Kurylev**, Introduction to spectral theory and inverse problem on asymptotically hyperbolic manifolds, 2014 ... JPY4298

33 **Masaki Maruyama**, Moduli spaces of stable sheaves on schemes, 2016 ... JPY2570

34 **Martin T. Barlow, Tibor Jordán and Andrzej Zuk**, Discrete Geometric Analysis, 2016 ... JPY3150

35 **Osamu Fujino**, Foundations of the minimal model program, 2017 ... JPY5130

36 **Soichiro Katayama**, Global solutions and the asymptotic behavior for nonlinear wave equations with small initial data, 2018 ... JPY5100

37 **Kazuki Hiroe, Hiroshi Kawakami, Akane Nakamura and Hidetaka Sakai**, 4-dimensional Painlevé-type equations, 2018 ... JPY3350

38 **Alex Casella, Dominic Tate and Stephan Tillmann**, Moduli spaces of real projective structures on surfaces, 2020 ... JPY2540

39 **Masaharu Taniguchi**, Traveling Front Solutions in Reaction-Diffusion Equations, 2021 ... JPY3370

40 **Yuji Odaka and Yoshiki Oshima**, Collapsing K3 Surfaces, Tropical Geometry and Moduli Compactifications of Satake, Morgan–Shalen Type, 2021 ... JPY3280

41 **Tomoki Nakanishi**, Cluster Algebras and Scattering Diagrams, 2023 ... JPY5130

Cluster Algebras and Scattering Diagrams

Tomoki Nakanishi

MSJ Memoirs

Mathematical Society of Japan

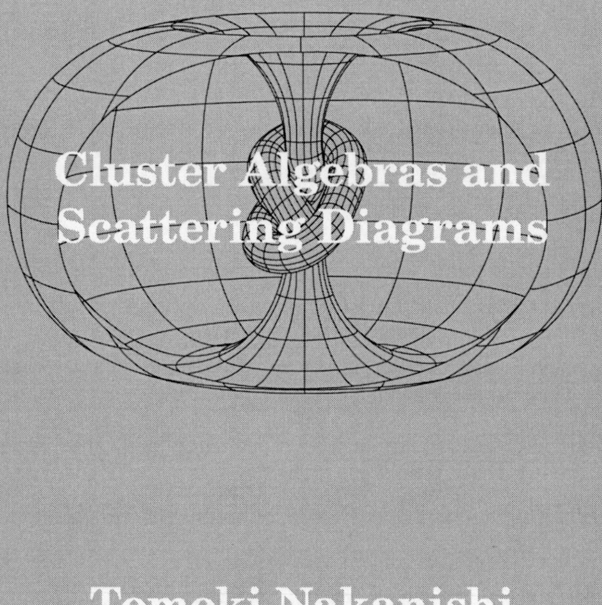

Cluster Algebras and Scattering Diagrams

Tomoki Nakanishi

Volume 41 2023

Author
Tomoki Nakanishi
Nagoya University
Graduate School of Mathematics
Chikusa-ku, Nagoya, 464-8602, Japan.
nakanisi@math.nagoya-u.ac.jp

Keywords: cluster algebra, scattering diagram, dilogarithm, pentagon relation
2020 Mathematics Subject Classifications: Primary 13F60.

MSJ Memoirs

This monograph series is intended to publish lecture notes, graduate textbooks and long research papers* in pure and applied mathematics. Each volume should be an integrated monograph. Proceedings of conferences or collections of independent papers are not accepted.

Articles for the series can be submitted to one of the editors in the form of hard copy. When the article is accepted, the author(s) is (are) requested to send a camera-ready manuscript.

* long research papers are limited to contributions by MSJ members

Editorial Board

The series *MSJ Memoirs* is published occasionally (3–5 volumes each year) by the Mathematical Society of Japan. The titles and prices of recent volumes are listed on page i of this volume. Each volume of the series can be separately purchased by placing an order with:

- The Mathematical Society of Japan, 1-34-8 Taito, Taito-ku, 110-0016 Tokyo, Japan (from inside Japan)
- World Scientific Publishing, 5 Thoh Tuck Link, Singapore 596224 (from outside Japan)

To my family

Preface

As the title suggests, the theme of this monograph is the relation between cluster algebras and scattering diagrams. The former was introduced by Fomin and Zelevinsky as an algebraic and combinatorial structure originated in Lie theory, while the latter was introduced by Kontsevich and Soibelman, and also by Gross and Siebert in the study of homological mirror symmetry.

The text consists of three parts.

- Part I is a first step guide to the theory of cluster algebras for readers without any knowledge on cluster algebras. We especially focus on basic notions, techniques, and results concerning seeds, cluster patterns, and cluster algebras.
- Part II is considered as the main part of the monograph, where we focus on the column sign-coherence of C-matrices and the Laurent positivity for cluster patterns, both of which were conjectured by Fomin and Zelevinsky and proved by Gross, Hacking, Keel, and Kontsevich based on the scattering diagram method. We also give a detailed account of the correspondence between the notions of cluster patterns and scattering diagrams.
- Part III is a self-contained exposition of several fundamental properties of cluster scattering diagrams which are admitted and used in Part II without proof. In particular, detailed proofs are presented for the construction, the mutation invariance, and the positivity of theta functions of cluster scattering diagrams with emphasis on the roles of the dilogarithm elements and the pentagon relation. This is regarded as a supplement to Part II, but also it may be read as an introductory text to cluster scattering diagrams.

More detailed introductions will be found in each part.

As a specific feature of this monograph, each part is written without explicitly relying on the other parts. Thus, readers can start reading from any part depending on their interest and knowledge as suggested below.

- If the reader is new to cluster algebras, simply start from Part I.
- If the reader is already familiar with basic notions in cluster algebras, skip Part I and start from Part II or III depending on the reader's interest.

Both cluster algebras and scattering diagrams are still young subjects. Therefore, it is likely that their notions and formulations would be drastically altered in near future. In fact, the proofs of the sign-coherence and the Laurent positivity in Part II suggest that a cluster pattern and a cluster scattering diagram are regarded as one inseparable object, so that they will be eventually integrated. On the other hand, the usefulness of the current formulation of cluster algebras has been already established in various applications; therefore, the current formulation should also remain to be used especially in practical applications to various subjects due to its efficiency and simplicity. It is my hope that this monograph serves as a useful guide in both perspectives.

We thank Ryota Akagi, Riku Fushimi, Mark Gross, Yasuaki Gyoda, Kodai Matsushita, Nathan Reading, and Takamasa Yoshida for careful reading of the draft, useful comments, and discussions. We also thank anonymous referees for numerous comments and suggestions. This work is supported in part by JSPS Grant No. JP16H03922, JP22H01114.

The errata and update will be posted on arXiv and the following webpage:
`http://www.math.nagoya-u.ac.jp/~nakanisi/index.html`

Tomoki Nakanishi
December, 2022, Nagoya

Contents

Part I
Basics in Cluster Algebras

0 Introduction to Part I

This is a first step guide to the theory of cluster algebras.

Cluster algebra theory was founded by the series of papers entitled as "Cluster Algebras I–IV" (commonly referred to as CA1–4) by Fomin and Zelevinsky, together with Berenstein [FZ02a, FZ03a, BFZ05, FZ07]. These fundamental papers are still the best sources to learn the basic notions, formulations, techniques, important examples, and the perspective on cluster algebras. However, even for someone who seriously wishes to learn cluster algebra theory, reading through all of them, even selectively, at the beginning is not an easy task. This is partly because each of CA1–4 has multiple goals, and some of them are rather advanced and/or specific for the beginners.

Fortunately, several excellent texts/monographs/reviews (including preliminary drafts) are already available to fill the gap. (e.g., [Car06], [Kel10], [GSV10], [Kel13], [Mar13], [Wil14], [GR17], [FZW16], [FZW17], [FZW20], [FZW21]). As another text of the same kind, we especially focus on *basic aspects* in cluster algebra theory. To be more explicit, we paid attention in the following points while preparing the manuscript.

- Most importantly, the text is aimed for anyone who starts to learn cluster algebras seriously without any preliminary knowledge.
- The text should be read in a self-contained way. Also, it should be reasonably concise.
- Consequently, we focus on (most but not all of) the basic and fundamental aspects of cluster algebra theory, namely, basic notions, techniques, and results concerning seeds, cluster patterns, and cluster algebras in CA1–4.
- We especially employ the formulation of seeds and mutations in CA4 throughout the text, because the structure of mutations is most transparent in our point of view. By the same reason, in some part of proofs, we use the notion of *free coefficients* (coefficients in a universal semifield) that replaces the role of coefficients of geometric type in CA1–4.
- We try to present elementary manipulations of mutations in the proofs of basic results explicitly, even if many of them are easy or straightforward. This is based on our belief that the essence of cluster algebras is found in these details of mutation mechanism.
- Naturally, we do not try to go deep into a particular topic. In particular, we have to omit the proof of the finite type classification of cluster algebras in CA2, because it requires deep analysis in conjunction with root systems of finite type.
- We add a short section on generalized cluster algebras as extra material beyond CA1–4.

All results presented in this text are well-known, and, except for the last section, they are taken from CA1–4 or easily derived from the results therein unless otherwise mentioned.

Bon voyage!

1 Getting Started

1.1 The very first step: pentagon periodicity

Let us play with a prototypical example of the cluster algebraic structure without explaining the background.

Let x_1, x_2 be formal variables. We consider the following recursion, or discrete dynamical system with discrete time $t = 0, 1, 2, \ldots$:

- initial condition: $x_1(0) = x_1,\ \ x_2(0) = x_2$.
- time development: For even t,

$$\begin{cases} x_1(t+1) = \dfrac{1}{x_1(t)}(x_2(t)+1), \\ x_2(t+1) = x_2(t), \end{cases} \tag{1.1}$$

  and, for odd t,

$$\begin{cases} x_1(t+1) = x_1(t), \\ x_2(t+1) = \dfrac{1}{x_2(t)}(x_1(t)+1). \end{cases} \tag{1.2}$$

Let us calculate $x_1(t)$ and $x_2(t)$ up to $t = 5$. Anyone who wishes to learn cluster algebras seriously has to work on the following calculations by oneself.

$$\begin{cases} x_1(0) = x_1, \\ x_2(0) = x_2. \end{cases} \tag{1.3}$$

$$\begin{cases} x_1(1) = \dfrac{x_2+1}{x_1}, \\ x_2(1) = x_2. \end{cases} \tag{1.4}$$

$$\begin{cases} x_1(2) = \dfrac{x_2+1}{x_1}, \\ x_2(2) = \dfrac{1}{x_2}\dfrac{x_1+x_2+1}{x_1} = \dfrac{x_1+x_2+1}{x_1x_2}. \end{cases} \tag{1.5}$$

So far, nothing special happened. Let us continue.

$$\begin{cases} x_1(3) = \dfrac{x_1}{x_2+1}\dfrac{x_1x_2+x_1+x_2+1}{x_1x_2} \overset{!}{=} \dfrac{x_1+1}{x_2}, \\ x_2(3) = \dfrac{x_1+x_2+1}{x_1x_2}. \end{cases} \tag{1.6}$$

Now, at the equality with symbol "!" in (1.6), the first nontrivial reduction happens. Let us continue, and we mark the equality with symbol "!" whenever a similar reduction occurs.

$$\begin{cases} x_1(4) = \dfrac{x_1+1}{x_2}, \\ x_2(4) = \dfrac{x_1x_2}{x_1+x_2+1}\dfrac{x_1+x_2+1}{x_2} \overset{!}{=} x_1, \end{cases} \tag{1.7}$$

$$\begin{cases} x_1(5) = \dfrac{x_2}{x_1+1}(x_1+1) \overset{!}{=} x_2, \\ x_2(5) = x_1. \end{cases} \tag{1.8}$$

After $t = 3$, reductions occur *systematically*, so that at $t = 5$ the result is extremely reduced.

Let us observe the above result more closely.

Observation 1.1. (a). *Periodicity.* We have a *half* periodicity at $t = 5$. If we continue calculation, we have a *full* periodicity at $t = 10$.

(b). *Laurent phenomenon.* Thanks to the reduction, each $x_i(t)$ is expressed, not as a general rational function, but as a *Laurent polynomial* in the initial variables x_1, x_2 with integer coefficients.

(c). *Laurent positivity.* Moreover, every nonzero coefficient of the above Laurent polynomial is *positive.*

Remark 1.2. The property (c) is nontrivial even though the rules (1.1) and (1.2) do not contain any negative coefficients. For example, we have

$$\frac{x^3+1}{x+1} = x^2 - x + 1. \tag{1.9}$$

Next, let us consider a similar but different system.

Let y_1, y_2 be another formal variables. We consider the following recursion, or discrete dynamical system with discrete time $t = 0, 1, 2, \dots$:

- initial condition: $y_1(0) = y_1,\ y_2(0) = y_2$.
- time development: For even t,

$$\begin{cases} y_1(t+1) = y_1(t)^{-1}, \\ y_2(t+1) = y_2(t)(1+y_1(t)), \end{cases} \tag{1.10}$$

  and, for odd t,

$$\begin{cases} y_1(t+1) = y_1(t)(1+y_2(t)), \\ y_2(t+1) = y_2(t)^{-1}. \end{cases} \tag{1.11}$$

We calculate up to $t = 5$. As before, we mark the symbol "!" whenever some reduction occurs. Again, it is important to do it oneself.

$$\begin{cases} y_1(0) = y_1, \\ y_2(0) = y_2, \end{cases} \tag{1.12}$$

$$\begin{cases} y_1(1) = y_1^{-1}, \\ y_2(1) = y_2(1+y_1), \end{cases} \tag{1.13}$$

$$\begin{cases} y_1(2) = \dfrac{1+y_2+y_1y_2}{y_1}, \\ y_2(2) = \dfrac{1}{y_2(1+y_1)}, \end{cases} \tag{1.14}$$

$$\begin{cases} y_1(3) = \dfrac{y_1}{1+y_2+y_1y_2}, \\ y_2(3) = \dfrac{1}{y_2(1+y_1)} \dfrac{1+y_1+y_2+y_1y_2}{y_1} \overset{!}{=} \dfrac{1+y_2}{y_1y_2}, \end{cases} \tag{1.15}$$

$$\begin{cases} y_1(4) = \dfrac{y_1}{1+y_2+y_1y_2} \dfrac{1+y_2+y_1y_2}{y_1y_2} \overset{!}{=} \dfrac{1}{y_2}, \\ y_2(4) = \dfrac{y_1y_2}{1+y_2}, \end{cases} \tag{1.16}$$

$$\begin{cases} y_1(5) = y_2, \\ y_2(5) = \dfrac{y_1y_2}{1+y_2} \dfrac{1+y_2}{y_2} \overset{!}{=} y_1. \end{cases} \tag{1.17}$$

Unlike the previous case, the variables $y_i(t)$ are not necessarily Laurent polynomials in the initial variables y_1, y_2. Nevertheless, similar reductions systematically occur after $t = 3$, and the same periodicity is obtained. It is natural to speculate that there is a close relation between two systems.

Indeed this is the simplest example of mutations of *cluster variables* for $x_i(t)$ and *coefficients* for $y_i(t)$, which we are going to study. The periodicity we observed is the celebrated *pentagon periodicity* for a cluster algebra of type A_2.

1.2 Semifields

We introduce the notion of a semifield. There are some variations of the definition depending on the purpose, and we especially use the one in [FZ03a].

Definition 1.3 (Semifield)**.** A multiplicative abelian group $\mathbb{P}$ equipped with a binary operation $\oplus$ is called a *semifield* if the following properties hold: For any $a, b, c, \in \mathbb{P}$,

$$a \oplus b = b \oplus a, \tag{1.18}$$

$$(a \oplus b) \oplus c = a \oplus (b \oplus c), \tag{1.19}$$

$$(a \oplus b)c = ac \oplus bc. \tag{1.20}$$

The operation $\oplus$ is called the *addition* in $\mathbb{P}$. Note that there is no *subtraction* in $\mathbb{P}$.

Example 1.4. The set of all positive rational numbers $\mathbb{Q}_+$ is a semifield by the usual multiplication and addition. On the other hand, the set of all nonnegative rational numbers $\mathbb{Q}_{\geq 0}$ is *not* a semifield by the usual multiplication and addition, because it is not a multiplicative group due to the presence of 0. Similarly, the set of all positive real numbers $\mathbb{R}_+$ is a semifield by the usual multiplication and addition.

The following three examples are especially important for cluster algebras.

Example 1.5. (a). *Universal semifield* $\mathbb{Q}_{\mathrm{sf}}(\mathbf{u})$. Let $\mathbf{u} = (u_1, \dots, u_n)$ be an n-tuple of formal variables. Let $\mathbb{Q}(\mathbf{u})$ be the rational function field of $\mathbf{u}$. We say that a rational function $f(\mathbf{u}) \in \mathbb{Q}(\mathbf{u})$ has a *subtraction-free expression* if it is expressed as $f(\mathbf{u}) = p(\mathbf{u})/q(\mathbf{u})$, where both $p(\mathbf{u})$ and $q(\mathbf{u})$ are *nonzero* polynomials in $\mathbf{u}$ whose coefficients are *nonnegative* integers. For example, $f(\mathbf{u}) = u_1^2 - u_1 + 1 = (u_1^3 + 1)/(u_1 + 1)$ has a subtraction-free expression. Let $\mathbb{Q}_{\mathrm{sf}}(\mathbf{u})$ be the set of all rational functions in $\mathbf{u}$ having subtraction-free expressions. Then, $\mathbb{Q}_{\mathrm{sf}}(\mathbf{u})$ is a semifield by the usual multiplication and addition in $\mathbb{Q}(\mathbf{u})$.

(b). *Tropical semifield* $\mathrm{Trop}(\mathbf{u})$. Let $\mathbf{u} = (u_1, \dots, u_n)$ be an n-tuple of formal variables. Let $\mathrm{Trop}(\mathbf{u})$ be the set of all Laurent monomials in $\mathbf{u}$ with coefficient 1, which is a multiplicative abelian group by the usual multiplication. We define the addition $\oplus$ by

$$\prod_{i=1}^{n} u_i^{a_i} \oplus \prod_{i=1}^{n} u_i^{b_i} := \prod_{i=1}^{n} u_i^{\min(a_i, b_i)}. \tag{1.21}$$

Then, $\mathrm{Trop}(\mathbf{u})$ becomes a semifield. The addition $\oplus$ is called the *tropical sum*.

(c). *Trivial semifield* $\mathbb{1}$. Let $\mathbb{1} = \{1\}$ be the trivial multiplicative group. We define the addition by $1 \oplus 1 = 1$. Then, $\mathbb{1}$ becomes a semifield.

Let $\mathbb{P}$ be any semifield. For any $a \in \mathbb{P}$ and any positive integer m, we write

$$ma := \underbrace{a \oplus \cdots \oplus a}_{m \text{ times}}. \tag{1.22}$$

Also, any positive integer m is identified with an element of $\mathbb{P}$ as $m = m1 \in \mathbb{P}$. For example, in the trivial semifield $\mathbb{1}$, we have $2 = 1 \oplus 1 = 1$, which is a little confusing. (See Remark 1.12 for another problem of this notation.)

Definition 1.6 (Semifield homomorphism)**.** For any semifields $\mathbb{P}$ and $\mathbb{P}'$, a map $\varphi : \mathbb{P} \to \mathbb{P}'$ is called a *semifield homomorphism* if it preserves the multiplication and the addition.

Example 1.7 (Trivial homomorphisms). For any semifield $\mathbb{P}$, the map

$$\varphi_{\text{triv}} : \mathbb{P} \to \mathbb{1},\ a \mapsto 1 \tag{1.23}$$

is a semifield homomorphism.

The following fact justifies the name "the universal semifield" for $\mathbb{Q}_{\text{sf}}(\mathbf{u})$.

Proposition 1.8. *Let $\mathbb{Q}_{\text{sf}}(\mathbf{u})$ be the universal semifield with $\mathbf{u} = (u_1, \dots, u_n)$. Let $\mathbb{P}$ be any semifield, and let $\mathbf{a} = (a_1, \dots, a_n)$ be any n-tuple of elements in $\mathbb{P}$. Then, there is a unique semifield homomorphism $\pi : \mathbb{Q}_{\text{sf}}(\mathbf{u}) \to \mathbb{P}$ such that $\pi(u_i) = a_i$ for any $i = 1, \dots, n$.*

Proof. First, for any nonzero polynomial $p(\mathbf{u})$ with nonnegative integer coefficients, the image $\pi(p(\mathbf{u}))$ is uniquely determined by replacing u_i with a_i and $+$ with $\oplus$ in $\mathbb{P}$ in the polynomial $p(\mathbf{u})$, where the notation (1.22) is taken into account. The equality $\pi(p(\mathbf{u})q(\mathbf{u})) = \pi(p(\mathbf{u}))\pi(q(\mathbf{u}))$ is guaranteed by the axiom (1.18)–(1.20). Next, for a given $f(\mathbf{u}) \in \mathbb{Q}_{\text{sf}}(\mathbf{u})$, take any subtraction-free expression $f(\mathbf{u}) = p(\mathbf{u})/q(\mathbf{u})$. Then, the image $\pi(f(\mathbf{u}))$ is defined by $\pi(p(\mathbf{u}))/\pi(q(\mathbf{u}))$. To see that it is well-defined, let us take another subtraction-free expression $f(\mathbf{u}) = p'(\mathbf{u})/q'(\mathbf{u})$. Then, we have $p(\mathbf{u})q'(\mathbf{u}) = p'(\mathbf{u})q(\mathbf{u})$. Thus, $\pi(p(\mathbf{u}))\pi(q'(\mathbf{u})) = \pi(p'(\mathbf{u}))\pi(q(\mathbf{u}))$. Therefore, we have $\pi(p(\mathbf{u}))/\pi(q(\mathbf{u})) = \pi(p'(\mathbf{u}))/\pi(q'(\mathbf{u}))$. It is easy to show that π is a semifield homomorphism. □

Definition 1.9 (Specialization). Let $\mathbb{Q}_{\text{sf}}(\mathbf{u})$, $\mathbf{a} \in \mathbb{P}$, and π be the ones in Proposition 1.8. For $f(\mathbf{u}) \in \mathbb{Q}_{\text{sf}}(\mathbf{u})$, the image $\pi(f(\mathbf{u}))$ is called the *specialization* of $f(\mathbf{u})$ at $\mathbf{a}$ in $\mathbb{P}$, and denoted by $f|_{\mathbb{P}}(\mathbf{a})$.

Example 1.10 (Tropicalization homomorphism). Let us consider the semifields $\mathbb{Q}_{\text{sf}}(\mathbf{u})$ and $\text{Trop}(\mathbf{u})$ with common formal variables $\mathbf{u} = (u_1, \dots, u_n)$. By applying Proposition 1.8 with $\mathbb{P} = \text{Trop}(\mathbf{u})$ and $\mathbf{a} = \mathbf{u}$, we have a unique semifield homomorphism

$$\pi_{\text{trop}} : \mathbb{Q}_{\text{sf}}(\mathbf{u}) \to \text{Trop}(\mathbf{u}) \tag{1.24}$$

such that $\pi_{\text{trop}}(u_i) = u_i$ for any $i = 1, \dots, n$. We call it the *tropicalization homomorphism.* For example, for $\mathbf{u} = (u_1, u_2, u_3)$,

$$\pi_{\text{trop}}\left(\frac{3u_1u_2^2u_3^2 + 2u_1^2u_2u_3}{3u_2^2 + u_1^2u_2^2 + u_1u_2^3u_3}\right) = \frac{u_1u_2u_3}{u_2^2} = u_1u_2^{-1}u_3. \tag{1.25}$$

Roughly speaking, it extracts the "leading Laurent monomial" of a function $f(\mathbf{u})$ in $\mathbb{Q}_{\text{sf}}(\mathbf{u})$.

Since a semifield $\mathbb{P}$ is a multiplicative abelian group, we can construct the group ring $\mathbb{ZP}$ of $\mathbb{P}$. The addition in $\mathbb{ZP}$, denoted by $+$ as usual, should be distinguished from the addition $\oplus$ in $\mathbb{P}$.

To construct the field of fractions of the ring $\mathbb{ZP}$, the ring $\mathbb{ZP}$ should be a domain. This is automatically guaranteed.

Proposition 1.11. *For any semifield $\mathbb{P}$, the following facts hold.*

(a). There is no torsion element in $\mathbb{P}$ other than 1. Namely, for any $p \in \mathbb{P}$, if $p^n = 1$ for some positive integer n, then $p = 1$.

(b). The group ring $\mathbb{ZP}$ is a domain. Namely, there is no zero divisor other than 0.

Proof. (a). Suppose that $p^n = 1$ for some positive integer n. Then, we have

$$p = \frac{p \oplus p^2 \oplus \cdots \oplus p^n}{1 \oplus p \oplus \cdots \oplus p^{n-1}} = \frac{p \oplus p^2 \oplus \cdots \oplus 1}{1 \oplus p \oplus \cdots \oplus p^{n-1}} = 1. \tag{1.26}$$

(b). This follows from the known fact that, for any multiplicative abelian group G with no torsion element other than 1, its group ring $\mathbb{Z}G$ is a domain (e.g., [May69]). For completeness, let us give a proof. Take any $a, b \in \mathbb{Z}G$. They are written as finite sums $a = \sum_i m_i g_i$, $b = \sum_i m'_i g'_i$ ($g_i, g'_i \in G$) with nonzero coefficients $m_i, m'_i \in \mathbb{Z}$. Let H be the subgroup of G generated by all g_i, g'_i. Then, H is finitely generated, and we have $a, b, ab \in \mathbb{Z}H$. By the assumption, H has no torsion element other than 1. Therefore, thanks to the fundamental theorem of finitely generated abelian groups, we have $H \simeq \mathbb{Z}^n$ for some n. Thus, $\mathbb{Z}H$ is isomorphic to the Laurent polynomial ring $\mathbb{Z}[x_1^{\pm 1}, \dots, x_n^{\pm 1}]$, which is a domain. Therefore, if $ab = 0$, then $a = 0$ or $b = 0$. □

Thanks to Proposition 1.11, the field of fractions of $\mathbb{ZP}$ is well-defined, and it is denoted by $\mathbb{QP}$.

Remark 1.12. There is some notational conflict between $\mathbb{P}$ and $\mathbb{ZP}$ (and also $\mathbb{QP}$). Namely, for $a \in \mathbb{P}$, $2a$, for example, may stand for $a \oplus a$ in $\mathbb{P}$, or $a + a$ in $\mathbb{ZP}$ depending on the context, and they are distinct. Fortunately, we do not have serious difficulty from this conflict, because usually this notation appears together with $\oplus$ or $+$, which clarifies the context. Anyway, we have to be careful.

1.3 Matrix and quiver mutations

Below we fix a positive integer n, which is called the *rank* of the forthcoming seeds, cluster patterns, cluster algebras, etc.

Definition 1.13 (Skew-symmetrizable matrix)**.** An $n \times n$ integer matrix $B = (b_{ij})_{i,j=1}^n$ is said to be *skew-symmetrizable* if there is a diagonal matrix $D = (d_i \delta_{ij})_{i,j=1}^n$ whose diagonal entries d_i are positive rational numbers such that DB is skew-symmetric, i.e,

$$d_i b_{ij} = -d_j b_{ji}. \tag{1.27}$$

The matrix D is called a *(left) skew-symmetrizer* of B, which is not unique to B. In particular, any skew-symmetric matrix is skew-symmetrizable with a skew-symmetrizer $D = I$.

The condition (1.27) can be rephrased in the matrix notation as

$$DB = -B^T D, \text{ or } DBD^{-1} = -B^T, \tag{1.28}$$

where B^T is the transpose of B. By (1.27), we have

$$b_{ii} = 0, \tag{1.29}$$
$$b_{ij} = 0 \Longleftrightarrow b_{ji} = 0, \tag{1.30}$$
$$b_{ij} > 0 \Longleftrightarrow b_{ji} < 0. \tag{1.31}$$

Example 1.14. The following matrices exhaust all 2×2 skew-symmetrizable matrices:

$$\begin{pmatrix} 0 & 0 \\ 0 & 0 \end{pmatrix}, \quad \begin{pmatrix} 0 & \mp b \\ \pm a & 0 \end{pmatrix} \quad (a, b \in \mathbb{Z}_{>0}). \tag{1.32}$$

In the former case, any diagonal matrix whose diagonals are positive integers is a skew-symmetrizer. In the latter case, a skew-symmetrizer is given by

$$D = \begin{pmatrix} a & 0 \\ 0 & b \end{pmatrix}. \tag{1.33}$$

For any integer a, we define

$$[a]_+ := \max(a, 0). \tag{1.34}$$

We have the following useful equalities:

$$a = [a]_+ - [-a]_+, \tag{1.35}$$
$$a[b]_+ + [-a]_+ b = a[-b]_+ + [a]_+ b, \tag{1.36}$$

where (1.36) follows from (1.35).

Definition 1.15 (Matrix mutation)**.** For any $n \times n$ skew-symmetrizable matrix $B = (b_{ij})$ and $k \in \{1, \dots, n\}$, we define a new matrix $B' = \mu_k(B) = (b'_{ij})$ by the following rule:

$$b'_{ij} = \begin{cases} -b_{ij} & i = k \text{ or } j = k, \\ b_{ij} + b_{ik}[b_{kj}]_+ + [-b_{ik}]_+ b_{kj} & i, j \neq k. \end{cases} \tag{1.37}$$

The matrix $\mu_k(B)$ is called the *mutation of B in direction k*.

Remark 1.16. By the identity (1.36), the second case of (1.37) is also written as

$$b'_{ij} = b_{ij} + b_{ik}[-b_{kj}]_+ + [b_{ik}]_+ b_{kj} \quad (i, j \neq k). \tag{1.38}$$

Remark 1.17. The formula (1.37) is verbally rephrased that B' is obtained from B by the following elementary transformations of matrices:

- For each $j \neq k$, add the kth column multiplied by $[b_{kj}]_+$ to the jth column. Also, for each $i \neq k$, add the kth row multiplied by $[-b_{ik}]_+$ to the ith row. (By Remark 1.16, one can simultaneous change the sign of b_{kj} and b_{ik} above.)
- Then, multiply -1 to the kth column and the kth row. (This is well-defined because $b_{kk} = 0$.)

Proposition 1.18. *Let B a skew-symmetrizable matrix, and let $B' = \mu_k(B)$. Then, the following facts hold:*

(a). Any skew-symmetrizer of D is also a skew-symmetrizer of B'. Therefore, B' is also skew-symmetrizable.

(b). We have $\det B = \det B'$. In particular, if B is nonsingular, B' is also nonsingular.

(c). We have $B = \mu_k(B')$. Namely, each mutation μ_k is involutive.

Proof. (a). For $i = k$ or $j = k$, we have

$$d_i b'_{ij} = -d_i b_{ij} = d_j b_{ji} = -d_j b'_{ji}. \tag{1.39}$$

For $i, j \neq k$, we have

$$\begin{aligned} d_i b'_{ij} &= d_i(b_{ij} + b_{ik}[b_{kj}]_+ + [-b_{ik}]_+ b_{kj}) \\ &= -d_j b_{ji} - d_k b_{ki}[b_{kj}]_+ + [d_k b_{ki}]_+ b_{kj} \\ &= -d_j(b_{ji} + b_{ki}[-b_{jk}]_+ + [b_{ki}]_+ b_{jk}) = -d_j b'_{ji}. \end{aligned} \tag{1.40}$$

(b). The verbal version of the mutation in Remark 1.17 preserves the determinant.

(c). Let $B'' = \mu_k(B')$. For $i = k$ or $j = k$, we have

$$b''_{ij} = -b'_{ij} = b_{ij}. \tag{1.41}$$

For $i, j \neq k$,

$$\begin{aligned} b''_{ij} &= b'_{ij} + b'_{ik}[b'_{kj}]_+ + [-b'_{ik}]_+ b'_{kj} \\ &= (b_{ij} + b_{ik}[b_{kj}]_+ + [-b_{ik}]_+ b_{kj}) - b_{ik}[-b_{kj}]_+ - [b_{ik}]_+ b_{kj} = b_{ij}, \end{aligned} \tag{1.42}$$

where in the last equality we used (1.35) or (1.36). □

Example 1.19. Here is an example of a matrix mutation.

$$B = \begin{pmatrix} 0 & 6 & -3 \\ -12 & 0 & 6 \\ 2 & -2 & 0 \end{pmatrix}, \quad B' = \mu_1(B) = \begin{pmatrix} 0 & -6 & 3 \\ 12 & 0 & -30 \\ -2 & 10 & 0 \end{pmatrix}, \tag{1.43}$$

where a common skew-symmetrizer of B and B' is given by

$$D = \begin{pmatrix} 2 & 0 & 0 \\ 0 & 1 & 0 \\ 0 & 0 & 3 \end{pmatrix}. \tag{1.44}$$

The matrix mutation is compatible with the matrix decomposition.

Proposition 1.20. *Suppose that an $n \times n$ skew-symmetrizable matrix B is decomposed as*

$$B = \begin{pmatrix} B_1 & O \\ O & B_2 \end{pmatrix}, \tag{1.45}$$

where B_1 and B_2 are $n_1 \times n_1$ and $n_2 \times n_2$ matrices, respectively. Then, we have

$$\mu_k(B) = \begin{cases} \begin{pmatrix} \mu_k(B_1) & O \\ O & B_2 \end{pmatrix} & k = 1, \dots, n_1, \\ \begin{pmatrix} B_1 & O \\ O & \mu_k(B_2) \end{pmatrix} & k = n_1 + 1, \dots, n. \end{cases} \tag{1.46}$$

Proof. This is clear from the verbal version of the mutation in Remark 1.17. □

Any *skew-symmetric* matrix can be represented by a *quiver*.

Definition 1.21 (Quiver). A (finite) *quiver* is a finite directed graph. Namely, it consists of a finite set of vertices and a finite set of arrows between the vertices. The following arrows are called a *loop* and a *2-cycle*, respectively.

Example 1.22. Here is an example of a quiver with a loop and a 2-cycle.

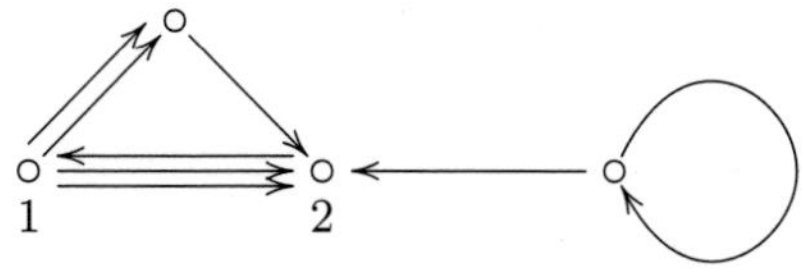

To be more precise, there are two ways to count 2-cycles in this quiver. If we do not distinguish two arrows from the vertex 1 to the vertex 2, there are only *one* 2-cycle. On the other hand, if we distinguish them, there are *two* 2-cycles. Here, we employ the former viewpoint. Namely, we only care about the multiplicity of arrows.

For a quiver with n vertices, we assume that its vertices are labeled with $1, \ldots, n$ without duplication. For a skew-symmetric matrix $B = (b_{ij})$, one can associate a quiver $Q(B)$ without loops and 2-cycles by the following rule:

- For each matrix element $b_{ij} > 0$, we assign b_{ij} arrows from the vertex i to the vertex j.

Since $b_{ii} = 0$, there are no loops. Also, there are no 2-cycles, because $b_{ji} = -b_{ij} < 0$ if $b_{ij} > 0$.

Conversely, one can recover a skew-symmetric matrix B from a quiver without loops and 2-cycles by applying the above rule in the opposite direction. It is clear that this correspondence is one-to-one.

One can translate the matrix mutation into the quiver mutation as follows.

Definition 1.23 (Quiver mutation). For any quiver Q without loops and 2-cycles and $k = 1, \ldots, n$, we define a new quiver $Q' = \mu_k(Q)$ by the following operation:

- For each pair i, j $(i \neq j)$ such that $i, j \neq k$, if there are $p > 0$ arrows from the vertex i to the vertex k, and $q > 0$ arrows from the vertex k to the vertex j, then add pq arrows from the vertex i to the vertex j.
- Remove the resulting 2-cycles as many as possible.
- Invert all arrows into and out of the vertex k.

The quiver Q' is called the *mutation of* Q *at the vertex* k.

To see the equivalence to the matrix mutation (1.37), we rewrite the second case of (1.37) as follows: For $i, j \neq k$,

$$b'_{ij} = \begin{cases} b_{ij} + b_{ik}b_{kj} & b_{ik}, b_{kj} > 0, \\ b_{ij} - b_{ik}b_{kj} & b_{ik}, b_{kj} < 0, \\ b_{ij} & \text{otherwise.} \end{cases} \tag{1.47}$$

It is easy to see that the first two operations in Definition 1.23 correspond to (1.47), while the last operation corresponds to the first case of (1.37).

Example 1.24. Here is an example of a skew-symmetric matrix B and the corresponding quiver $Q(B)$, where the number attached to each arrow represents the multiplicity of the arrows of the same kind:

$$B = \begin{pmatrix} 0 & 3 & -2 & 2 \\ -3 & 0 & 4 & 0 \\ 2 & -4 & 0 & 1 \\ -2 & 0 & -1 & 0 \end{pmatrix}, \quad Q(B) =$$

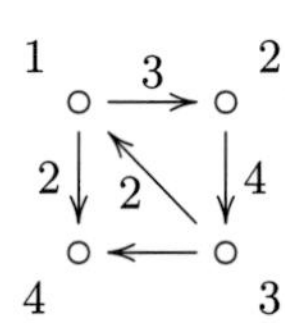

.

The mutation of $Q(B)$ at the vertex 1 is done as follows:

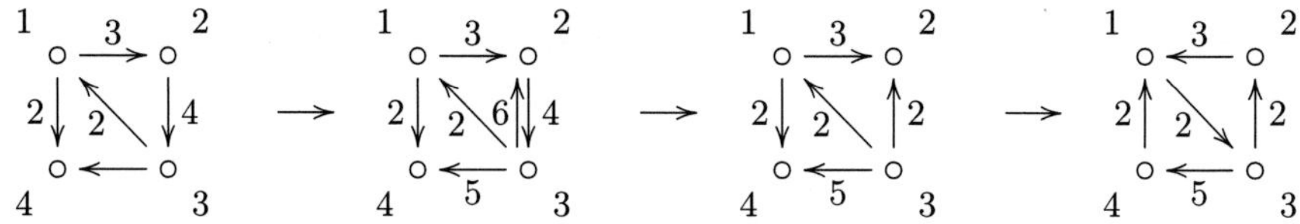

Putting it back to the matrix form, we obtain

$$\mu_1(B) = \begin{pmatrix} 0 & -3 & 2 & -2 \\ 3 & 0 & -2 & 0 \\ -2 & 2 & 0 & 5 \\ 2 & 0 & -5 & 0 \end{pmatrix}. \tag{1.48}$$

The above correspondence between skew-symmetric matrices and quivers is the basis of the connection between cluster algebras and quiver representations. See [Pla18] for a concise survey on the subject.

2 Basic Notions

In this section we introduce basic notions for cluster algebras.

2.1 Seeds and mutations

Let us introduce the most fundamental notions in cluster algebra theory, namely, *seeds* and *mutations.*

Recall that, for a given semifield $\mathbb{P}$, $\mathbb{QP}$ is the field of fractions of the group algebra $\mathbb{ZP}$ in Section 1.2.

Definition 2.1 (Seed/Cluster variable/Coefficient). Let n be any positive integer. Let $\mathbb{P}$ be any semifield, and let $\mathcal{F}$ be a field that is isomorphic to the rational function field of n-variables with coefficients in the field $\mathbb{QP}$.

- A *(labeled) seed with coefficients in* $\mathbb{P}$ *(or a seed in* $\mathcal{F}$*) of rank* n is a triplet $\Sigma = (\mathbf{x}, \mathbf{y}, B)$ such that $\mathbf{x} = (x_1, \dots, x_n)$ is an n-tuple of algebraically independent and generating elements in $\mathcal{F}$ (i.e., a transcendence basis of $\mathcal{F}$), $\mathbf{y} = (y_1, \dots, y_n)$ is an n-tuple of any elements in $\mathbb{P}$, and $B = (b_{ij})_{i,j=1}^n$ is an $n \times n$ skew-symmetrizable (integer) matrix.
- We call $\mathbf{x}$, $\mathbf{y}$, and B, respectively, the *cluster*, the *coefficient tuple*, and the *exchange matrix* of Σ. The elements x_i and y_i, respectively, are called the *cluster variables* and the *coefficients.*
- We call $\mathbb{P}$ and $\mathcal{F}$, respectively, the *coefficient semifield* and the *ambient field* of a seed Σ, and also, of the forthcoming cluster patterns, cluster algebras, etc.

Remark 2.2. Sometimes, cluster variables x_i and coefficients y_i are casually called *x-variables* and *y-variables*, respectively. Or, we may simply say *variable x_i, variable y_i*, etc. Alternatively, they are also denoted by A_i and X_i, and called *A-coordinates (A-variables)* and *X-coordinates (X-variables)*, respectively, following the convention of another pioneering works on cluster algebra theory by Fock and Goncharov [FG09a, FG09b].

There are some related notions to Definition 2.1.

Definition 2.3 (Seed without coefficients/Y-seed).

- When the coefficient semifield $\mathbb{P}$ is taken to be the trivial semifield $\mathbb{1}$, all coefficients y_i are 1. Then, we can reduce a triplet $(\mathbf{x}, \mathbf{y}, B)$ in Definition 2.1 to a pair $(\mathbf{x}, B)$, which is called a *seed without coefficients.*
- For any coefficient semifield $\mathbb{P}$, we have an option to ignore cluster variables and concentrate on a pair $Y = (\mathbf{y}, B)$ in Definition 2.1, which is called a *Y-seed* in $\mathbb{P}$.

For any seed $\Sigma = (\mathbf{x}, \mathbf{y}, B)$, we attach an n-tuple $\hat{\mathbf{y}} = (\hat{y}_1, \dots, \hat{y}_n)$ of elements in $\mathcal{F}$,

$$\hat{y}_i = y_i \prod_{j=1}^{n} x_j^{b_{ji}}. \tag{2.1}$$

They play an important role in the cluster algebra theory. We call them *$\hat{y}$-variables.*

Definition 2.4 (Seed mutation). For any seed $\Sigma = (\mathbf{x}, \mathbf{y}, B)$ in $\mathcal{F}$ and $k \in \{1, \dots, n\}$, we define a new seed $\Sigma' = (\mathbf{x}', \mathbf{y}', B')$ in $\mathcal{F}$ by the following rule:

$$x'_i = \begin{cases} x_k^{-1}\left(\prod_{j=1}^{n} x_j^{[-b_{jk}]_+}\right)\dfrac{1+\hat{y}_k}{1 \oplus y_k} & i = k, \\ x_i & i \neq k, \end{cases} \tag{2.2}$$

$$y'_i = \begin{cases} y_k^{-1} & i = k, \\ y_i y_k^{[b_{ki}]_+}(1 \oplus y_k)^{-b_{ki}} & i \neq k, \end{cases} \tag{2.3}$$

$$b'_{ij} = \begin{cases} -b_{ij} & i = k \text{ or } j = k, \\ b_{ij} + b_{ik}[b_{kj}]_+ + [-b_{ik}]_+ b_{kj} & i, j \neq k, \end{cases} \tag{2.4}$$

where $\hat{y}_k$ in (2.2) is defined by (2.1). The seed Σ' is called the *mutation of Σ in direction k*, and denoted by $\mu_k(\Sigma) = \mu_k(\mathbf{x}, \mathbf{y}, B)$. The mutations of a seed without coefficients $(\mathbf{x}, B)$ and a Y-seed $(\mathbf{x}, B)$ are also defined by the same formulas.

We will soon show that Σ' in the above is indeed a seed in $\mathcal{F}$.

The mutation of B in (2.4) is the matrix mutation already introduced in Definition 1.15. Thanks to the identity (1.35), the first case of (2.2) is also written as the following more standard form:

$$x'_k = x_k^{-1}\left(\frac{y_k}{1 \oplus y_k}\prod_{j=1}^{n} x_j^{[b_{jk}]_+} + \frac{1}{1 \oplus y_k}\prod_{j=1}^{n} x_j^{[-b_{jk}]_+}\right). \tag{2.5}$$

Related with the mutation (2.2), we note the following useful identity:

$$\frac{1+\hat{y}_k^{-1}}{1 \oplus y_k^{-1}} = \frac{1+\hat{y}_k}{1 \oplus y_k}\prod_{j=1}^{n} x_j^{-b_{jk}}. \tag{2.6}$$

The following fact, together with Proposition 1.18, ensures that $(\mathbf{x}', \mathbf{y}', B')$ is indeed a seed in $\mathcal{F}$.

Proposition 2.5. *(a). The mutation μ_k is involutive. Namely, $\mu_k(\Sigma') = \Sigma$ for $\Sigma' = \mu_k(\Sigma)$.*

(b). The elements in $\mathbf{x}'$ are algebraically independent and generating elements in $\mathcal{F}$.

Proof. (a). Let $(\mathbf{x}'', \mathbf{y}'', B'') = \mu_k(\mathbf{x}', \mathbf{y}', B')$ for $(\mathbf{x}', \mathbf{y}', B')$ in Definition 2.4. $B'' = B$ was already shown in Proposition 1.18 (b). To show $\mathbf{x}'' = \mathbf{x}$, it is enough to show that $x''_k = x_k$. We first note that

$$\hat{y}'_k = y'_k \prod_{j=1}^{n} {x'_j}^{b'_{jk}} = y_k^{-1} \prod_{j=1}^{n} {x_j}^{-b_{jk}} = \hat{y}_k^{-1}, \tag{2.7}$$

where we used the fact $b_{kk} = 0$. Then, we have

$$\begin{aligned} x''_k &= {x'_k}^{-1} \left(\prod_{j=1}^{n} {x_j}^{[-b'_{jk}]_+} \right) \frac{1 + \hat{y}'_k}{1 \oplus y'_k} = {x'_k}^{-1} \left(\prod_{j=1}^{n} {x_j}^{[b_{jk}]_+} \right) \frac{1 + \hat{y}_k^{-1}}{1 \oplus y_k^{-1}} \\ &= {x'_k}^{-1} \left(\prod_{j=1}^{n} {x_j}^{[-b_{jk}]_+} \right) \frac{1 + \hat{y}_k}{1 \oplus y_k} = x_k, \end{aligned} \tag{2.8}$$

where we used (2.6) and (1.35). Let us show $\mathbf{y}'' = \mathbf{y}$. We have $y''_k = {y'_k}^{-1} = y_k$. For $i \neq k$,

$$\begin{aligned} y''_i &= y'_i {y'_k}^{[b'_{ki}]_+} (1 \oplus y'_k)^{-b'_{ki}} \\ &= (y_i y_k^{[b_{ki}]_+} (1 \oplus y_k)^{-b_{ki}}) y_k^{-[-b_{ki}]_+} (1 \oplus y_k^{-1})^{b_{ki}} = y_i. \end{aligned} \tag{2.9}$$

(b). By (a), $x_1, \dots, x_n$ are expressed as rational functions of $x'_1, \dots, x'_n$ over $\mathbb{QP}$. It follows that $x'_1, \dots, x'_n$ generate $\mathcal{F}$; moreover, they are algebraically independent, because, if not, the transcendence degree of $\mathcal{F}$ over $\mathbb{QP}$ becomes less than n, which is a contradiction. □

The following fact is the first manifestation of the close relationship (duality) between the mutations of $\mathbf{x}$ and $\mathbf{y}$.

Proposition 2.6. *The $\hat{y}$-variables in* (2.1) *mutate in the ambient field $\mathcal{F}$ as*

$$\hat{y}'_i = \begin{cases} \hat{y}_k^{-1} & i = k, \\ \hat{y}_i \hat{y}_k^{[b_{ki}]_+} (1 + \hat{y}_k)^{-b_{ki}} & i \neq k, \end{cases} \tag{2.10}$$

which is the same rule for the coefficients in (2.3).

Proof. Below (and elsewhere) we use the fact $b_{kk} = 0$ effectively. For $i = k$, it was already shown in (2.7). For $i \neq k$,

$$\begin{aligned}
\hat{y}'_i &= y'_i \prod_{j=1}^{n} x'^{\,b'_{ji}}_j \\
&= y_i y_k^{[b_{ki}]_+}(1 \oplus y_k)^{-b_{ki}} \left(\prod_{\substack{j=1 \\ j\neq k}}^{n} x_j^{b_{ji}+b_{jk}[b_{ki}]_+ + [-b_{jk}]_+ b_{ki}} \right) \\
&\quad \times \left(x_k^{-1} \left(\prod_{j=1}^{n} x_j^{[-b_{jk}]_+} \right) \frac{1+\hat{y}_k}{1 \oplus y_k} \right)^{-b_{ki}} \\
&= \hat{y}_i \hat{y}_k^{[b_{ki}]_+}(1+\hat{y}_k)^{-b_{ki}}.
\end{aligned} \tag{2.11}$$

□

There is some flexibility to write the mutation formulas in Definition 2.4 and Proposition 2.6.

Proposition 2.7 (ε-expressions). *The right hand sides of the following formulas do not depend on the choice of* $\varepsilon \in \{1, -1\}$.

$$x'_i = \begin{cases} x_k^{-1} \left(\prod_{j=1}^{n} x_j^{[-\varepsilon b_{jk}]_+} \right) \dfrac{1+\hat{y}_k^{\varepsilon}}{1 \oplus y_k^{\varepsilon}} & i = k, \\ x_i & i \neq k, \end{cases} \tag{2.12}$$

$$y'_i = \begin{cases} y_k^{-1} & i = k, \\ y_i y_k^{[\varepsilon b_{ki}]_+}(1 \oplus y_k^{\varepsilon})^{-b_{ki}} & i \neq k, \end{cases} \tag{2.13}$$

$$b'_{ij} = \begin{cases} -b_{ij} & i = k \text{ or } j = k, \\ b_{ij} + b_{ik}[\varepsilon b_{kj}]_+ + [-\varepsilon b_{ik}]_+ b_{kj} & i, j \neq k, \end{cases} \tag{2.14}$$

$$\hat{y}'_i = \begin{cases} \hat{y}_k^{-1} & i = k, \\ \hat{y}_i \hat{y}_k^{[\varepsilon b_{ki}]_+}(1 + \hat{y}_k^{\varepsilon})^{-b_{ki}} & i \neq k. \end{cases} \tag{2.15}$$

Proof. For the matrix mutation, this was already pointed out in Remark 1.16. The other cases can be also shown by (1.35) as follows:

$$\left(\prod_{j=1}^{n} \frac{x_j^{[-b_{jk}]_+}}{x_j^{[b_{jk}]_+}} \right) \frac{1+\hat{y}_k}{1+\hat{y}_k^{-1}} \frac{1 \oplus y_k^{-1}}{1 \oplus y_k} = \left(\prod_{j=1}^{n} x_j^{-b_{jk}} \right) \hat{y}_k y_k^{-1} = 1, \tag{2.16}$$

$$\frac{y_k^{[b_{ki}]_+}}{y_k^{[-b_{ki}]_+}} \frac{(1 \oplus y_k)^{-b_{ki}}}{(1 \oplus y_k^{-1})^{-b_{ki}}} = y_k^{b_{ki}} y_k^{-b_{ki}} = 1. \tag{2.17}$$

The case (2.15) is similar. □

We continue to give related notions for seeds.

Definition 2.8 (S_n-action/Unlabeled seed)**.**

- For a seed $\Sigma = (\mathbf{x}, \mathbf{y}, B)$ in $\mathcal{F}$ and a permutation σ of $\{1, \dots, n\}$, we define the action of σ on Σ by

$$\sigma\Sigma = (\sigma\mathbf{x}, \sigma\mathbf{y}, \sigma B), \tag{2.18}$$

where $\sigma\mathbf{x} = \mathbf{x}'$, $\sigma\mathbf{y} = \mathbf{y}'$, $\sigma B = B'$ are defined by

$$x'_i = x_{\sigma^{-1}(i)}, \quad y'_i = y_{\sigma^{-1}(i)}, \quad b'_{ij} = b_{\sigma^{-1}(i)\sigma^{-1}(j)}. \tag{2.19}$$

Clearly, $\sigma\Sigma$ is a seed in $\mathcal{F}$, and this yields a left action of the symmetric group S_n of degree n on the set of seeds in $\mathcal{F}$; namely, we have $\tau(\sigma\Sigma) = \tau\sigma(\Sigma)$ for $\sigma, \tau \in S_n$. This also induces the action of σ on $\hat{y}$-variables $\sigma\hat{\mathbf{y}} = \hat{\mathbf{y}}'$ as

$$\hat{y}'_i := y'_i \prod_{j=1}^{n} x'^{\,b'_{ji}}_j = y_{\sigma^{-1}(i)} \prod_{j=1}^{n} x_{\sigma^{-1}(j)}^{b_{\sigma^{-1}(j)\sigma^{-1}(i)}} = \hat{y}_{\sigma^{-1}(i)}. \tag{2.20}$$

- We introduce an equivalence condition for (labeled) seeds in $\mathcal{F}$,

$$\Sigma' \sim \Sigma \tag{2.21}$$

if there is some permutation $\sigma \in S_n$ such that $\Sigma' = \sigma\Sigma$. Then, each equivalence class $[(\mathbf{x}, \mathbf{y}, B)]$ is called an *unlabeled seed* in $\mathcal{F}$.

The mutations and the action of σ is compatible in the following sense.

Proposition 2.9. *The following equality holds:*

$$\mu_{\sigma(k)}(\sigma\Sigma) = \sigma(\mu_k(\Sigma)). \tag{2.22}$$

Proof. We set the left and right hand sides as $(\mathbf{x}', \mathbf{y}', B')$ and $(\mathbf{x}'', \mathbf{y}'', B'')$, respectively. We calculate $(\mathbf{x}', \mathbf{y}', B')$ below, which turns out to coincide with $(\mathbf{x}'', \mathbf{y}'', B'')$.

$$x'_i = \begin{cases} x_k^{-1}\left(\prod_{j=1}^{n} x_{\sigma^{-1}(j)}^{[-b_{\sigma^{-1}(j)k}]_+}\right)\dfrac{1+\hat{y}_k}{1 \oplus y_k} & i = \sigma(k), \\ x_{\sigma^{-1}(i)} & i \neq \sigma(k), \end{cases} \tag{2.23}$$

$$y'_i = \begin{cases} y_k^{-1} & i = \sigma(k), \\ y_{\sigma^{-1}(i)} y_k^{[b_{k\sigma^{-1}(i)}]_+}(1 \oplus y_k)^{-b_{k\sigma^{-1}(i)}} & i \neq \sigma(k), \end{cases} \tag{2.24}$$

$$b'_{ij} = \begin{cases} -b_{\sigma^{-1}(i)\sigma^{-1}(j)} & i = \sigma(k) \text{ or } j = \sigma(k), \\ b_{\sigma^{-1}(i)\sigma^{-1}(j)} + b_{\sigma^{-1}(i)k}[b_{k\sigma^{-1}(j)}]_+ & i, j \neq \sigma(k). \\ +[-b_{\sigma^{-1}(i)k}]_+ b_{k\sigma^{-1}(j)} & \end{cases} \tag{2.25}$$

□

Two mutations are not commutative, in general. However, under some simple condition, they are commutative.

Proposition 2.10. *For a seed $\Sigma = (\mathbf{x}, \mathbf{y}, B)$ and a pair k, ℓ $(k \neq \ell)$, suppose that*

$$b_{k\ell} = b_{\ell k} = 0 \tag{2.26}$$

holds. Then, we have

$$\mu_k \mu_\ell(\Sigma) = \mu_\ell \mu_k(\Sigma), \tag{2.27}$$

or equivalently,

$$\mu_\ell \mu_k \mu_\ell \mu_k(\Sigma) = \Sigma. \tag{2.28}$$

Proof. We set $\Sigma' = \mu_k(\Sigma)$ and $\Sigma'' = \mu_\ell(\Sigma')$, and we show that Σ'' is symmetric with respect to k and ℓ. By (2.26), we have

$$x'_i = \begin{cases} x_k^{-1} \left(\prod_{j=1}^{n} x_j^{[-b_{jk}]_+} \right) \dfrac{1 + \hat{y}_k}{1 \oplus y_k} & i = k, \\ x_i & i \neq k, \end{cases} \tag{2.29}$$

$$y'_i = \begin{cases} y_k^{-1} & i = k, \\ y_\ell & i = \ell, \\ y_i y_k^{[b_{ki}]_+} (1 \oplus y_k)^{-b_{ki}} & i \neq k, \ell, \end{cases} \tag{2.30}$$

$$b'_{ij} = \begin{cases} -b_{ij} & i = k \text{ or } j = k, \\ b_{ij} & i = \ell \text{ or } j = \ell, \\ b_{ij} + b_{ik}[b_{kj}]_+ + [-b_{ik}]_+ b_{kj} & \text{otherwise.} \end{cases} \tag{2.31}$$

These also imply that

$$\hat{y}'_\ell = \hat{y}_\ell. \tag{2.32}$$

Then, we have

$$x''_i = \begin{cases} x_k^{-1} \left(\prod_{j=1}^{n} x_j^{[-b_{jk}]_+} \right) \dfrac{1 + \hat{y}_k}{1 \oplus y_k} & i = k, \\ x_\ell^{-1} \left(\prod_{j=1}^{n} x_j^{[-b_{j\ell}]_+} \right) \dfrac{1 + \hat{y}_\ell}{1 \oplus y_\ell} & i = \ell, \\ x_i & i \neq k, \ell, \end{cases} \tag{2.33}$$

$$y''_i = \begin{cases} y_k^{-1} & i = k, \\ y_\ell^{-1} & i = \ell, \\ y_i y_k^{[b_{ki}]_+} y_\ell^{[b_{\ell i}]_+} (1 \oplus y_k)^{-b_{ki}} (1 \oplus y_\ell)^{-b_{\ell i}} & i \neq k, \ell, \end{cases} \tag{2.34}$$

$$b''_{ij} = \begin{cases} -b_{ij} & i=k \text{ or } j=k, \\ -b_{ij} & i=\ell \text{ or } j=\ell, \\ b_{ij} + b_{ik}[b_{kj}]_+ + [-b_{ik}]_+ b_{kj} & \text{otherwise.} \\ \quad + b_{i\ell}[b_{\ell j}]_+ + [-b_{i\ell}]_+ b_{\ell j} & \end{cases} \tag{2.35}$$

□

2.2 Cluster patterns and cluster algebras

We introduce cluster patterns and cluster algebras, which are the main objects to be studied in cluster algebra theory.

Definition 2.11 (n-Regular tree).

- Let $\mathbb{T}_n$ denote the *n-regular tree*; namely, it is a tree graph such that each vertex has exactly n edges attached to it. Moreover, the edges are labeled by $1,\dots,n$ so that the edges attached to each vertex are labeled without duplication. By abusing the notation, the set of vertices of $\mathbb{T}_n$ is also denoted by $\mathbb{T}_n$.
- We say that a pair of vertices t and t' in $\mathbb{T}_n$ are *k-adjacent*, or t' is *k-adjacent to* t, if they are connected with an edge labeled by k.

The graph $\mathbb{T}_n$ is finite for $n=1$ and infinite otherwise.

Example 2.12. Here are the n-regular trees $\mathbb{T}_n$ for $n=1,2,3$.

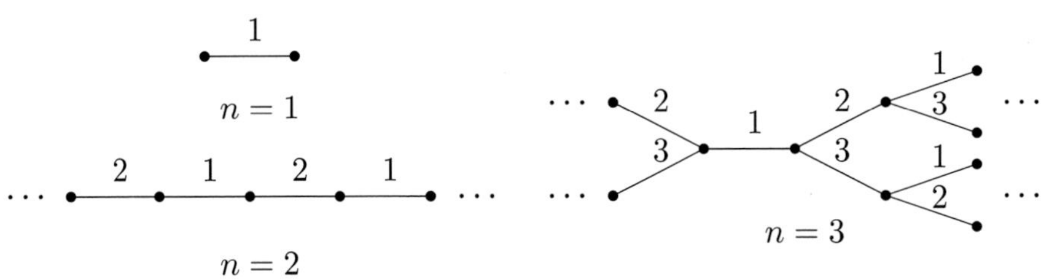

Definition 2.13 (Cluster pattern/Y-pattern/B-pattern).

- A collection of seeds $\mathbf{\Sigma} = \{\Sigma_t = (\mathbf{x}_t, \mathbf{y}_t, B_t)\}_{t\in\mathbb{T}_n}$ with coefficients in $\mathbb{P}$ indexed by $\mathbb{T}_n$ is called a *cluster pattern with coefficients in* $\mathbb{P}$ if, for any pair $t,t' \in \mathbb{T}_n$ that are k-adjacent, the equality $\Sigma_{t'} = \mu_k(\Sigma_t)$ holds. A cluster pattern is also called a *seed pattern*.
- We replace the above $\mathbf{\Sigma}$ with a collection of seeds $\mathbf{\Sigma} = \{\Sigma_t = (\mathbf{x}_t, B_t)\}_{t\in\mathbb{T}_n}$ without coefficients. Then, it is called a *cluster pattern without coefficients*.
- We replace the above $\mathbf{\Sigma}$ with a collection of Y-seeds $\mathbf{Y} = \{Y_t = (\mathbf{y}_t, B_t)\}_{t\in\mathbb{T}_n}$ in $\mathbb{P}$. Then, it is called a *Y-pattern in* $\mathbb{P}$.

- From the above $\mathbf{\Sigma}$, we extract a collection of the exchange matrices $\mathbf{B} = \{B_t\}_{t\in\mathbb{T}_n}$. We call it the *B-pattern of* $\mathbf{\Sigma}$.

Often, it is convenient to choose arbitrarily a distinguished vertex $t_0 \in \mathbb{T}_n$ called the *initial vertex*. Since any cluster pattern $\mathbf{\Sigma} = \{\Sigma_t = (\mathbf{x}_t, \mathbf{y}_t, B_t) \mid t \in \mathbb{T}_n\}$ is uniquely determined from the *initial seed* Σ_{t_0} at t_0 by repeating mutations, we may write $\mathbf{\Sigma} = \mathbf{\Sigma}(\Sigma_{t_0})$.

For a seed $\Sigma_t = (\mathbf{x}_t, \mathbf{y}_t, B_t)$ in a cluster pattern $\mathbf{\Sigma}$, we use the notation

$$\mathbf{x}_t = (x_{1;t}, \dots, x_{n;t}), \quad \mathbf{y}_t = (y_{1;t}, \dots, y_{n;t}), \quad B_t = (b_{ij;t})_{i,j=1}^n. \tag{2.36}$$

Often, we omit the index t_0 for the initial seed as

$$\mathbf{x}_{t_0} = \mathbf{x} = (x_1, \dots, x_n), \quad \mathbf{y}_{t_0} = \mathbf{y} = (y_1, \dots, y_n), \quad B_{t_0} = B = (b_{ij})_{i,j=1}^n. \tag{2.37}$$

We use similar notations for $\hat{y}$-variables as well:

$$\hat{\mathbf{y}}_t = (\hat{y}_{1;t}, \dots, \hat{y}_{n;t}), \quad \hat{\mathbf{y}}_{t_0} = \hat{\mathbf{y}} = (\hat{y}_1, \dots, \hat{y}_n). \tag{2.38}$$

By fixing the initial vertex, one may regard each coefficient $y_{i;t}$ as a rational function with a subtraction-free expression in the initial coefficients $\mathbf{y}$, since the mutations of coefficients in (2.3) are subtraction-free. Similarly, one may regard each cluster variable $x_{i;t}$ as a rational function with a subtraction-free expression in the initial cluster variables $\mathbf{x}$; moreover, its coefficients are rational functions with subtraction-free expressions in $\mathbf{y}$.

Now we give the definition of a cluster algebra.

Definition 2.14 (Cluster algebra). For any cluster pattern $\mathbf{\Sigma}$, the *cluster algebra* $\mathcal{A} = \mathcal{A}(\mathbf{\Sigma})$ associated with $\mathbf{\Sigma}$ is the $\mathbb{ZP}$-subalgebra of the ambient field $\mathcal{F}$ generated by all cluster variables $x_{i;t}$ $(i = 1, \dots, n; t \in \mathbb{T}_n)$ of $\mathbf{\Sigma}$.

If $\mathbf{\Sigma}$ has only finitely many distinct cluster variables, $\mathcal{A}(\mathbf{\Sigma})$ is clearly finitely generated. On the other hand, if $\mathbf{\Sigma}$ has infinitely many distinct cluster variables, $\mathcal{A}(\mathbf{\Sigma})$ may be finitely generated or not, depending on $\mathbf{\Sigma}$.

Example 2.15 (Type A_1). Let $n = 1$. We consider the following arrangement of a cluster pattern $\mathbf{\Sigma}$ on $\mathbb{T}_1$.

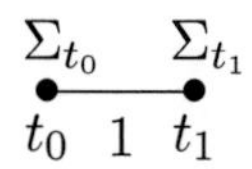

We have the unique choice of the exchange matrices

$$B_{t_0} = B_{t_1} = (0). \tag{2.39}$$

Let t_0 be the initial vertex, and we set $x_{1;t_0} = x_1$ and $y_{1;t_0} = y_1$. We have $\hat{y}_{1;t_0} = \hat{y}_1 = y_1$. Accordingly, we have

$$x_{1;t_1} = x_1^{-1}\frac{1+y_1}{1 \oplus y_1}, \quad y_{1;t_1} = y_1^{-1}. \tag{2.40}$$

One can directly confirm the involution property $\mu_1(\Sigma_{t_1}) = \Sigma_{t_0}$ as follows:

$$x_{1;t_1}^{-1} \frac{1+y_1^{-1}}{1 \oplus y_1^{-1}} = x_1 \frac{1 \oplus y_1}{1+y_1} \frac{1+y_1^{-1}}{1 \oplus y_1^{-1}} = x_1, \quad y_{1;t_1}^{-1} = y_1. \tag{2.41}$$

The associated cluster algebra $\mathcal{A}(\mathbf{\Sigma})$ is the $\mathbb{ZP}$-algebra generated by

$$x_1, \ x_1^{-1} \frac{1+y_1}{1 \oplus y_1}. \tag{2.42}$$

This is called a cluster algebra of *type* A_1, which depends on the choice of a semifield $\mathbb{P}$ and also the initial coefficient y_1.

Example 2.16 (Type $A_1 \times A_1$). Let $n = 2$. We consider the following arrangement of a cluster pattern on $\mathbb{T}_2$.

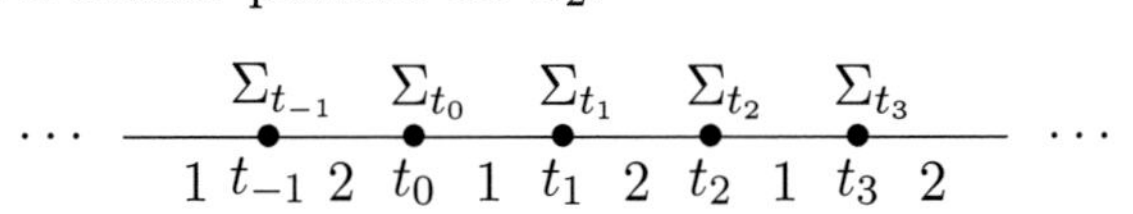

Below we use the simplified notations such as $\Sigma_{t_s} = \Sigma_s$, $B_{t_s} = B_s$, $\mathbf{x}_{t_s} = \mathbf{x}_s$, $\mathbf{y}_{t_s} = \mathbf{y}_s$, etc. Let t_0 be the initial vertex, and we set $\mathbf{x}_0 = \mathbf{x}$, $\mathbf{y}_0 = \mathbf{y}$. Again, we consider the simplest case

$$B_0 = B = O, \tag{2.43}$$

so that the initial $\hat{y}$ variables are given by

$$\hat{y}_1 = y_1, \quad \hat{y}_2 = y_2. \tag{2.44}$$

For any $s \in \mathbb{Z}$, we have

$$B_s = O. \tag{2.45}$$

Accordingly, we have

$$\begin{cases} x_{1;1} = x_1^{-1} \dfrac{1+y_1}{1 \oplus y_1}, \\ x_{2;1} = x_2, \end{cases} \qquad \begin{cases} y_{1;1} = y_1^{-1}, \\ y_{2;1} = y_2, \end{cases} \tag{2.46}$$

$$\begin{cases} x_{1;2} = x_1^{-1} \dfrac{1+y_1}{1 \oplus y_1}, \\ x_{1;2} = x_2^{-1} \dfrac{1+y_2}{1 \oplus y_2}, \end{cases} \qquad \begin{cases} y_{1;2} = y_1^{-1}, \\ y_{2;2} = y_2^{-1}, \end{cases} \tag{2.47}$$

$$\begin{cases} x_{1;3} = x_1, \\ x_{2;3} = x_2^{-1} \dfrac{1+y_2}{1 \oplus y_2}, \end{cases} \qquad \begin{cases} y_{1;3} = y_1, \\ y_{2;3} = y_2^{-1}, \end{cases} \tag{2.48}$$

$$\begin{cases} x_{1;4} = x_1, \\ x_{1;4} = x_2, \end{cases} \qquad \begin{cases} y_{1;4} = y_1, \\ y_{2;4} = y_2, \end{cases} \tag{2.49}$$

where we did the same calculation as (2.41) in the last two steps. We observe the periodicity with period 4. Namely,

$$\mathbf{x}_{s+4} = \mathbf{x}_s, \quad \mathbf{y}_{s+4} = \mathbf{y}_s, \quad B_{s+4} = B_s \quad (s \in \mathbb{Z}). \tag{2.50}$$

In other words, $\Sigma_{s+4} = \Sigma_s$. This is regarded as a special case of (2.28). The cluster algebra $\mathcal{A}(\mathbf{\Sigma})$ is the $\mathbb{ZP}$-algebra generated by

$$x_1,\ x_2,\ x_1^{-1}\frac{1+y_1}{1 \oplus y_1},\ x_2^{-1}\frac{1+y_2}{1 \oplus y_2}. \tag{2.51}$$

This is called a cluster algebra of *type* $A_1 \times A_1$.

Observe that a cluster algebra of type $A_1 \times A_1$ obtained above is isomorphic to the tensor product of two cluster algebras of type A_1 as a $\mathbb{ZP}$-algebra. One can extend this result to a more general situation.

Proposition 2.17. *Let $\mathbf{\Sigma} = \mathbf{\Sigma}(\Sigma_{t_0})$ be a cluster pattern of rank n with the initial seed $\Sigma_{t_0} = (\mathbf{x}, \mathbf{y}, B)$. Suppose that the initial exchange matrix $B_{t_0} = B$ is decomposed as*

$$B = \begin{pmatrix} B' & O \\ O & B'' \end{pmatrix}, \tag{2.52}$$

where the size of B' and B'' are n' and n'' ($n' + n'' = n$), respectively. Accordingly, consider cluster patterns $\mathbf{\Sigma}'$ and $\mathbf{\Sigma}''$ of rank n' and n'' whose initial seeds are given by

$$\Sigma'_{t_0} = ((x_i)_{i=1}^{n'}, (y_i)_{i=1}^{n'}, B'), \quad \Sigma''_{t_0} = ((x_i)_{i=n'+1}^{n}, (y_i)_{i=n'+1}^{n}, B''), \tag{2.53}$$

respectively. Then, as $\mathbb{ZP}$-algebras, we have

$$\mathcal{A}(\mathbf{\Sigma}) \simeq \mathcal{A}(\mathbf{\Sigma}') \otimes_{\mathbb{ZP}} \mathcal{A}(\mathbf{\Sigma}''). \tag{2.54}$$

Proof. We first note that, by Proposition 1.20, for any $t \in \mathbb{T}_n$, the exchange matrix B_t of $\mathbf{\Sigma}$ is decomposed in the same form as

$$B_t = \begin{pmatrix} B'_t & O \\ O & B''_t \end{pmatrix}. \tag{2.55}$$

For a seed $\Sigma_t = (\mathbf{x}_t, \mathbf{y}_t, B_t)$ of $\mathbf{\Sigma}$, let us consider the mutation at $k \leq n'$. Then, the mutation of B_t is given by the first case in (1.46). Also, for $i \leq n'$, $x_{i;t}$ and $y_{i;t}$ mutate effectively by the sub-exchange matrix B'_t, while, for $i > n'$, they are stable. The other case $k > n'$ is similar. It follows that the set of the cluster variables of $\mathbf{\Sigma}$ is the disjoint union of those of $\mathbf{\Sigma}'$ or $\mathbf{\Sigma}''$. Also, there is no nontrivial algebraic relation between cluster variables of $\mathbf{\Sigma}'$ and those of $\mathbf{\Sigma}''$. Therefore, $\mathcal{A}(\mathbf{\Sigma})$ is naturally isomorphic to $\mathcal{A}(\mathbf{\Sigma}') \otimes_{\mathbb{ZP}} \mathcal{A}(\mathbf{\Sigma}'')$ under the identification $zz' \mapsto z \otimes_{\mathbb{ZP}} z'$ ($z \in \mathcal{A}(\mathbf{\Sigma})$, $z' \in \mathcal{A}(\mathbf{\Sigma}')$). $\square$

We say that a square matrix M is *decomposable* (resp. *indecomposable*) if, after some simultaneous permutation of the column and row indices of M, M is a direct sum of two square matrices (resp. otherwise).

Thanks to Proposition 2.17, in many situations it is enough to concentrate on cluster patterns and cluster algebras with *indecomposable* exchange matrices.

2.3 Rank 2 periodicities

We continue to use the parametrization and the notation for cluster patterns of rank 2 in Example 2.16. Let us take the following initial exchange matrix

$$B_0 = B = \begin{pmatrix} 0 & -b \\ a & 0 \end{pmatrix}, \quad (a, b > 0). \tag{2.56}$$

By the mutations (2.4), the exchange matrices are given by

$$B_s = \begin{cases} B & s\text{: even}, \\ -B & s\text{: odd}. \end{cases} \tag{2.57}$$

Accordingly, we have, for even s,

$$\hat{y}_{1;s} = y_{1;s} x_{2;s}^{a}, \quad \hat{y}_{2;s} = y_{2;s} x_{1;s}^{-b}, \tag{2.58}$$

$$\begin{cases} x_{1;s+1} = x_{1;s}^{-1} \dfrac{1 + \hat{y}_{1;s}}{1 \oplus y_{1;s}}, \\ x_{2;s+1} = x_{2;s}, \end{cases} \quad \begin{cases} y_{1;s+1} = y_{1;s}^{-1}, \\ y_{2;s+1} = y_{2;s}(1 \oplus y_{1;s})^{b}, \end{cases} \tag{2.59}$$

and, for odd s,

$$\hat{y}_{1;s} = y_{1;s} x_{2;s}^{-a}, \quad \hat{y}_{2;s} = y_{2;s} x_{1;s}^{b}, \tag{2.60}$$

$$\begin{cases} x_{1;s+1} = x_{1;s}, \\ x_{2;s+1} = x_{2;s}^{-1} \dfrac{1 + \hat{y}_{2;s}}{1 \oplus y_{2;s}}, \end{cases} \quad \begin{cases} y_{1;s+1} = y_{1;s}(1 \oplus y_{2;s})^{a}, \\ y_{2;s+1} = y_{2;s}^{-1}. \end{cases} \tag{2.61}$$

Below we concentrate on the case $ab \leq 3$. Without losing generality we may assume that $b = 1$ and $a = 1, 2, 3$, because the opposite case is obtained by exchanging the indices 1 and 2. We strongly recommend the readers to carry out the following calculations to appreciate the systematic occurrences of reductions ("miracles"). It may require a few hours. So, be prepared!

Example 2.18 (Type A_2). Consider the case $a = 1$, where

$$B_0 = B = \begin{pmatrix} 0 & -1 \\ 1 & 0 \end{pmatrix}, \tag{2.62}$$

so that the initial $\hat{y}$-variables are given by

$$\hat{y}_1 = y_1 x_2, \quad \hat{y}_2 = y_2 x_1^{-1}. \tag{2.63}$$

In fact, this is the case already calculated in Section 1.1, where cluster variables are without coefficients. Here, we effectively use Proposition 2.6 so that we do not have to look into the contents of $\hat{y}$-variables during the calculation except for the initial one (2.63). We have the following result:

$$\begin{cases} x_{1;1} = x_1^{-1}\dfrac{1+\hat{y}_1}{1\oplus y_1}, \\ x_{2;1} = x_2, \end{cases} \qquad \begin{cases} y_{1;1} = y_1^{-1}, \\ y_{2;1} = y_2(1\oplus y_1), \end{cases} \tag{2.64}$$

$$\begin{cases} x_{1;2} = x_1^{-1}\dfrac{1+\hat{y}_1}{1\oplus y_1}, \\ x_{2;2} = x_2^{-1}\dfrac{1+\hat{y}_2+\hat{y}_1\hat{y}_2}{1\oplus y_2\oplus y_1y_2}, \end{cases} \qquad \begin{cases} y_{1;2} = y_1^{-1}(1\oplus y_2\oplus y_1y_2), \\ y_{2;2} = y_2^{-1}(1\oplus y_1)^{-1}, \end{cases} \tag{2.65}$$

$$\begin{cases} x_{1;3} = x_1x_2^{-1}\dfrac{1+\hat{y}_2}{1\oplus y_2}, \\ x_{2;3} = x_2^{-1}\dfrac{1+\hat{y}_2+\hat{y}_1\hat{y}_2}{1\oplus y_2\oplus y_1y_2}, \end{cases} \qquad \begin{cases} y_{1;3} = y_1(1\oplus y_2\oplus y_1y_2)^{-1}, \\ y_{2;3} = y_1^{-1}y_2^{-1}(1\oplus y_2), \end{cases} \tag{2.66}$$

$$\begin{cases} x_{1;4} = x_1x_2^{-1}\dfrac{1+\hat{y}_2}{1\oplus y_2}, \\ x_{2;4} = x_1, \end{cases} \qquad \begin{cases} y_{1;4} = y_2^{-1}, \\ y_{2;4} = y_1y_2(1\oplus y_2)^{-1}, \end{cases} \tag{2.67}$$

$$\begin{cases} x_{1;5} = x_2, \\ x_{2;5} = x_1, \end{cases} \qquad \begin{cases} y_{1;5} = y_2, \\ y_{2;5} = y_1. \end{cases} \tag{2.68}$$

Let us extend Observation 1.1 in more details.

(a). *Periodicity/Finiteness.* Even with the presence of coefficients for cluster variables, we still have the same pentagon periodicity

$$\Sigma_{s+5} = \tau_{12}(\Sigma_s), \tag{2.69}$$

where τ_{12} is the transpose of 1 and 2, and its action was defined in (2.18). In particular, the cluster variables of $\mathbf{\Sigma}$ are exhausted by

$$x_1, \quad x_2, \quad x_1^{-1}\frac{1+\hat{y}_1}{1\oplus y_1}, \quad x_2^{-1}\frac{1+\hat{y}_2+\hat{y}_1\hat{y}_2}{1\oplus y_2\oplus y_1y_2}, \quad x_1x_2^{-1}\frac{1+\hat{y}_2}{1\oplus y_2}. \tag{2.70}$$

By (2.63), as Laurent polynomials in x_1 and x_2 with coefficients in $\mathbb{ZP}$, they are written as

$$x_1, \quad x_2, \quad x_1^{-1}\frac{1+y_1x_2}{1\oplus y_1}, \quad x_1^{-1}x_2^{-1}\frac{x_1+y_2+y_1y_2x_2}{1\oplus y_2\oplus y_1y_2}, \quad x_2^{-1}\frac{x_1+y_2}{1\oplus y_2}. \tag{2.71}$$

(b). *Laurent phenomenon.* In (2.71), each cluster variable $x_{i;t}$ is expressed as a Laurent polynomial in the initial cluster variables $\mathbf{x}$ with coefficients in $\mathbb{ZP}$.

(c). *Laurent positivity.* Every coefficient of the above Laurent polynomial is nonnegative in $\mathbb{ZP}$.

(d). *F-polynomials.* Let us rephrase the properties (b) and (c) in a more specific way. Each cluster variable $x_{i;t}$ is expressed in a unified way as follows:

$$x_{i;t} = \left(\prod_{j=1}^{n} x_j^{g_{ji;t}} \right) \frac{F_{i;t}(\hat{\mathbf{y}})}{F_{i;t}|_{\mathbb{P}}(\mathbf{y})}, \tag{2.72}$$

where $F_{i;t}(\mathbf{y})$ is a polynomial in formal variables $\mathbf{y} = (y_1, y_2)$ with nonnegative integer coefficients, and $F_{i;t}|_{\mathbb{P}}(\mathbf{y})$ is the specialization in $\mathbb{P}$ at the initial coefficients $\mathbf{y}_{t_0} = \mathbf{y}$ in Definition 1.9. (Here, we conveniently abuse the symbol $\mathbf{y}$ for two different usages.) Later, $F_{i;t}(\mathbf{y})$ is called an *F-polynomial.*

(e). *Unit constant term.* Every F-polynomial $F_{i;t}(\mathbf{y})$ has constant term 1.

(f). *Duality.* As a new observation, we see that the cluster variables and coefficients share some common/parallel structure. In particular, the above F-polynomials also appear for coefficients.

Example 2.19 (Type B_2)**.** Consider the case $a = 2$, where

$$B_0 = B = \begin{pmatrix} 0 & -1 \\ 2 & 0 \end{pmatrix}, \quad D = \begin{pmatrix} 2 & 0 \\ 0 & 1 \end{pmatrix}, \tag{2.73}$$

so that the initial $\hat{y}$-variables are given by

$$\hat{y}_1 = y_1 x_2^2, \quad \hat{y}_2 = y_2 x_1^{-1}. \tag{2.74}$$

We have the following result:

$$\begin{cases} x_{1;1} = x_1^{-1} \dfrac{1+\hat{y}_1}{1 \oplus y_1}, \\ x_{2;1} = x_2, \end{cases} \qquad \begin{cases} y_{1;1} = y_1^{-1}, \\ y_{2;1} = y_2(1 \oplus y_1), \end{cases} \tag{2.75}$$

$$\begin{cases} x_{1;2} = x_1^{-1} \dfrac{1+\hat{y}_1}{1 \oplus y_1}, \\ x_{2;2} = x_2^{-1} \dfrac{1+\hat{y}_2+\hat{y}_1\hat{y}_2}{1 \oplus y_2 \oplus y_1 y_2}, \end{cases} \qquad \begin{cases} y_{1;2} = y_1^{-1}(1 \oplus y_2 \oplus y_1 y_2)^2, \\ y_{2;2} = y_2^{-1}(1 \oplus y_1)^{-1}, \end{cases} \tag{2.76}$$

$$\begin{cases} x_{1;3} = x_1 x_2^{-2} \dfrac{1+2\hat{y}_2+\hat{y}_2^2+\hat{y}_1\hat{y}_2^2}{1 \oplus 2y_2 \oplus y_2^2 \oplus y_1 y_2^2}, \\ x_{2;3} = x_2^{-1} \dfrac{1+\hat{y}_2+\hat{y}_1\hat{y}_2}{1 \oplus y_2 \oplus y_1 y_2}, \end{cases} \tag{2.77}$$

$$\begin{cases} y_{1;3} = y_1(1 \oplus y_2 \oplus y_1 y_2)^{-2}, \\ y_{2;3} = y_1^{-1} y_2^{-1}(1 \oplus 2y_2 \oplus y_2^2 \oplus y_1 y_2^2), \end{cases}$$

$$\begin{cases} x_{1;4} = x_1 x_2^{-2} \dfrac{1 + 2\hat{y}_2 + \hat{y}_2^2 + \hat{y}_1 \hat{y}_2^2}{1 \oplus 2y_2 \oplus y_2^2 \oplus y_1 y_2^2}, \\ x_{2;4} = x_1 x_2^{-1} \dfrac{1 + \hat{y}_2}{1 \oplus y_2}, \end{cases} \tag{2.78}$$

$$\begin{cases} y_{1;4} = y_1^{-1} y_2^{-2} (1 \oplus y_2)^2, \\ y_{2;4} = y_1 y_2 (1 \oplus 2y_2 \oplus y_2^2 \oplus y_1 y_2^2)^{-1}, \end{cases}$$

$$\begin{cases} x_{1;5} = x_1, \\ x_{2;5} = x_1 x_2^{-1} \dfrac{1 + \hat{y}_2}{1 \oplus y_2}, \end{cases} \quad \begin{cases} y_{1;5} = y_1 y_2^2 (1 \oplus y_2)^{-2}, \\ y_{2;5} = y_2^{-1}, \end{cases} \tag{2.79}$$

$$\begin{cases} x_{1;6} = x_1, \\ x_{2;6} = x_2, \end{cases} \quad \begin{cases} y_{1;6} = y_1, \\ y_{2;6} = y_2. \end{cases} \tag{2.80}$$

During the calculation we used the following polynomial identities:

$$y_1 + (1 + y_2 + y_1 y_2)^2 = (1 + y_1)(1 + 2y_2 + y_2^2 + y_1 y_2^2), \tag{2.81}$$

$$y_1 y_2 + (1 + 2y_2 + y_2^2 + y_1 y_2^2) = (1 + y_2 + y_1 y_2)(1 + y_2). \tag{2.82}$$

We observe the periodicity

$$\Sigma_{s+6} = \Sigma_s. \tag{2.83}$$

In particular, the cluster variables of $\mathbf{\Sigma}$ are exhausted by

$$\begin{gathered} x_1, \quad x_2, \quad x_1^{-1} \frac{1 + \hat{y}_1}{1 \oplus y_1}, \quad x_2^{-1} \frac{1 + \hat{y}_2 + \hat{y}_1 \hat{y}_2}{1 \oplus y_2 \oplus y_1 y_2}, \\ x_1 x_2^{-2} \frac{1 + 2\hat{y}_2 + \hat{y}_2^2 + \hat{y}_1 \hat{y}_2^2}{1 \oplus 2y_2 \oplus y_2^2 \oplus y_1 y_2^2}, \quad x_1 x_2^{-1} \frac{1 + \hat{y}_2}{1 \oplus y_2}. \end{gathered} \tag{2.84}$$

As Laurent polynomials in x_1 and x_2, they are written as

$$\begin{gathered} x_1, \quad x_2, \quad x_1^{-1} \frac{1 + y_1 x_2^2}{1 \oplus y_1}, \quad x_1^{-1} x_2^{-1} \frac{x_1 + y_2 + y_1 y_2 x_2^2}{1 \oplus y_2 \oplus y_1 y_2}, \\ x_1^{-1} x_2^{-2} \frac{x_1^2 + 2y_2 x_1 + y_2^2 + y_1 y_2^2 x_2^2}{1 \oplus 2y_2 \oplus y_2^2 \oplus y_1 y_2^2}, \quad x_2^{-1} \frac{x_1 + y_2}{1 \oplus y_2}. \end{gathered} \tag{2.85}$$

All other observations for type A_2 are commonly applied.

Example 2.20 (Type G_2). Consider the case $a = 3$, where

$$B_0 = B = \begin{pmatrix} 0 & -1 \\ 3 & 0 \end{pmatrix}, \quad D = \begin{pmatrix} 3 & 0 \\ 0 & 1 \end{pmatrix}, \tag{2.86}$$

so that the initial $\hat{y}$-variables are given by

$$\hat{y}_1 = y_1 x_2^3, \quad \hat{y}_2 = y_2 x_1^{-1}. \tag{2.87}$$

We have the following result:

$$\begin{cases} x_{1;1} = x_1^{-1}\dfrac{1+\hat{y}_1}{1\oplus y_1}, \\ x_{2;1} = x_2, \end{cases} \qquad \begin{cases} y_{1;1} = y_1^{-1}, \\ y_{2;1} = y_2(1\oplus y_1), \end{cases} \tag{2.88}$$

$$\begin{cases} x_{1;2} = x_1^{-1}\dfrac{1+\hat{y}_1}{1\oplus y_1}, \\ x_{2;2} = x_2^{-1}\dfrac{1+\hat{y}_2+\hat{y}_1\hat{y}_2}{1\oplus y_2\oplus y_1y_2}, \end{cases} \qquad \begin{cases} y_{1;2} = y_1^{-1}(1\oplus y_2\oplus y_1y_2)^3, \\ y_{2;2} = y_2^{-1}(1\oplus y_1)^{-1}, \end{cases} \tag{2.89}$$

$$\begin{aligned} &\begin{cases} x_{1;3} = x_1x_2^{-3}\dfrac{1+3\hat{y}_2+3\hat{y}_2^2+\hat{y}_2^3+3\hat{y}_1\hat{y}_2^2+2\hat{y}_1\hat{y}_2^3+\hat{y}_1^2\hat{y}_2^3}{1\oplus 3y_2\oplus 3y_2^2\oplus y_2^3\oplus 3y_1y_2^2\oplus 2y_1y_2^3\oplus y_1^2y_2^3}, \\ x_{2;3} = x_2^{-1}\dfrac{1+\hat{y}_2+\hat{y}_1\hat{y}_2}{1\oplus y_2\oplus y_1y_2}, \end{cases} \\ &\begin{cases} y_{1;3} = y_1(1\oplus y_2\oplus y_1y_2)^{-3}, \\ y_{2;3} = y_1^{-1}y_2^{-1}(1\oplus 3y_2\oplus 3y_2^2\oplus y_2^3\oplus 3y_1y_2^2\oplus 2y_1y_2^3\oplus y_1^2y_2^3), \end{cases} \end{aligned} \tag{2.90}$$

$$\begin{aligned} &\begin{cases} x_{1;4} = x_1x_2^{-3}\dfrac{1+3\hat{y}_2+3\hat{y}_2^2+\hat{y}_2^3+3\hat{y}_1\hat{y}_2^2+2\hat{y}_1\hat{y}_2^3+\hat{y}_1^2\hat{y}_2^3}{1\oplus 3y_2\oplus 3y_2^2\oplus y_2^3\oplus 3y_1y_2^2\oplus 2y_1y_2^3\oplus y_1^2y_2^3}, \\ x_{2;4} = x_1x_2^{-2}\dfrac{1+2\hat{y}_2+\hat{y}_2^2+\hat{y}_1\hat{y}_2}{1\oplus 2y_2\oplus y_2^2\oplus y_1y_2^2}, \end{cases} \\ &\begin{cases} y_{1;4} = y_1^{-2}y_2^{-3}(1\oplus 2y_2\oplus y_2^2\oplus y_1y_2^2)^3, \\ y_{2;4} = y_1y_2(1\oplus 3y_2\oplus 3y_2^2\oplus y_2^3\oplus 3y_1y_2^2\oplus 2y_1y_2^3\oplus y_1^2y_2^3)^{-1}, \end{cases} \end{aligned} \tag{2.91}$$

$$\begin{aligned} &\begin{cases} x_{1;5} = x_1^2x_2^{-3}\dfrac{1+3\hat{y}_2+3\hat{y}_2^2+\hat{y}_2^3+\hat{y}_1\hat{y}_2^3}{1\oplus 3y_2\oplus 3y_2^2\oplus y_2^3\oplus y_1y_2^3}, \\ x_{2;5} = x_1x_2^{-2}\dfrac{1+2\hat{y}_2+\hat{y}_2^2+\hat{y}_1\hat{y}_2}{1\oplus 2y_2\oplus y_2^2\oplus y_1y_2^2}, \end{cases} \\ &\begin{cases} y_{1;5} = y_1^2y_2^3(1\oplus 2y_2\oplus y_2^2\oplus y_1y_2^2)^{-3}, \\ y_{2;5} = y_1^{-1}y_2^{-2}(1\oplus 3y_2\oplus 3y_2^2\oplus y_2^3\oplus y_1y_2^3), \end{cases} \end{aligned} \tag{2.92}$$

$$\begin{aligned} &\begin{cases} x_{1;6} = x_1^2x_2^{-3}\dfrac{1+3\hat{y}_2+3\hat{y}_2^2+\hat{y}_2^3+\hat{y}_1\hat{y}_2^3}{1\oplus 3y_2\oplus 3y_2^2\oplus y_2^3\oplus y_1y_2^3}, \\ x_{2;6} = x_1x_2^{-1}\dfrac{1+\hat{y}_2}{1\oplus y_2}, \end{cases} \\ &\begin{cases} y_{1;6} = y_1^{-1}y_2^{-3}(1\oplus y_2)^3, \\ y_{2;6} = y_1y_2^2(1\oplus 3y_2\oplus 3y_2^2\oplus y_2^3\oplus y_1y_2^3)^{-1}, \end{cases} \end{aligned} \tag{2.93}$$

$$\begin{cases} x_{1;7} = x_1, \\ x_{2;7} = x_1x_2^{-1}\dfrac{1+\hat{y}_2}{1\oplus y_2}, \end{cases} \qquad \begin{cases} y_{1;7} = y_1y_2^3(1\oplus y_2)^{-3}, \\ y_{2;7} = y_2^{-1}, \end{cases} \tag{2.94}$$

$$\begin{cases} x_{1;8} = x_1, \\ x_{2;8} = x_2, \end{cases} \qquad \begin{cases} y_{1;8} = y_1, \\ y_{2;8} = y_2. \end{cases} \tag{2.95}$$

During the calculation we used the following polynomial identities:

$$\begin{aligned}&y_1+(1+y_2+y_1y_2)^3\\&=(1+y_1)(1+3y_2+3y_2^2+y_2^3+3y_1y_2^2+2y_1y_2^3+y_1^2y_2^3),\end{aligned}\tag{2.96}$$

$$\begin{aligned}&y_1y_2+(1+3y_2+3y_2^2+y_2^3+3y_1y_2^2+2y_1y_2^3+y_1^2y_2^3)\\&=(1+y_2+y_1y_2)(1+2y_2+y_2^2+y_1y_2^2),\end{aligned}\tag{2.97}$$

$$\begin{aligned}&y_1^2y_2^3+(1+2y_2+y_2^2+y_1y_2^2)^3\\&=(1+3y_2+3y_2^2+y_2^3+3y_1y_2^2+2y_1y_2^3+y_1^2y_2^3)\\&\qquad\times(1+3y_2+3y_2^2+y_2^3+y_1y_2^3),\end{aligned}\tag{2.98}$$

$$\begin{aligned}&y_1y_2^2+(1+3y_2+3y_2^2+y_2^3+y_1y_2^3)\\&=(1+2y_2+y_2^2+y_1y_2^2)(1+y_2),\end{aligned}\tag{2.99}$$

where the most complicated identity (2.98) may be quickly checked by computer. We observe the periodicity

$$\Sigma_{s+8}=\Sigma_s.\tag{2.100}$$

In particular, the cluster variables of $\mathbf{\Sigma}$ are exhausted by

$$\begin{aligned}&x_1,\quad x_2,\quad x_1^{-1}\frac{1+\hat{y}_1}{1\oplus y_1},\quad x_2^{-1}\frac{1+\hat{y}_2+\hat{y}_1\hat{y}_2}{1\oplus y_2\oplus y_1y_2},\\&x_1x_2^{-3}\frac{1+3\hat{y}_2+3\hat{y}_2^2+\hat{y}_2^3+3\hat{y}_1\hat{y}_2^2+2\hat{y}_1\hat{y}_2^3+\hat{y}_1^2\hat{y}_2^3}{1\oplus 3y_2\oplus 3y_2^2\oplus y_2^3\oplus 3y_1y_2^2\oplus 2y_1y_2^3\oplus y_1^2y_2^3},\\&x_1x_2^{-2}\frac{1+2\hat{y}_2+\hat{y}_2^2+\hat{y}_1\hat{y}_2^2}{1\oplus 2y_2\oplus y_2^2\oplus y_1y_2^2},\\&x_1^2x_2^{-3}\frac{1+3\hat{y}_2+3\hat{y}_2^2+\hat{y}_2^3+\hat{y}_1\hat{y}_2^3}{1\oplus 3y_2\oplus 3y_2^2\oplus y_2^3\oplus y_1y_2^3},\quad x_1x_2^{-1}\frac{1+\hat{y}_2}{1\oplus y_2}.\end{aligned}\tag{2.101}$$

As Laurent polynomials in x_1 and x_2, they are written as

$$\begin{aligned}&x_1,\quad x_2,\quad x_1^{-1}\frac{1+y_1x_2^3}{1\oplus y_1},\quad x_1^{-1}x_2^{-1}\frac{x_1+y_2+y_1y_2x_2^3}{1\oplus y_2\oplus y_1y_2},\\&x_1^{-2}x_2^{-3}\frac{x_1^3+3y_2x_1^2+3y_2^2x_1+y_2^3+3y_1y_2^2x_1x_2^3+2y_1y_2^3x_2^3+y_1^2y_2^3x_2^6}{1\oplus 3y_2\oplus 3y_2^2\oplus y_2^3\oplus 3y_1y_2^2\oplus 2y_1y_2^3\oplus y_1^2y_2^3},\\&x_1^{-1}x_2^{-2}\frac{x_1^2+2y_2x_1+y_2^2+y_1y_2^2x_2^3}{1\oplus 2y_2\oplus y_2^2\oplus y_1y_2^2},\\&x_1^{-1}x_2^{-3}\frac{x_1^3+3y_2x_1^2+3y_2^2x_1+y_2^3+y_1y_2^3x_2^3}{1\oplus 3y_2\oplus 3y_2^2\oplus y_2^3\oplus y_1y_2^3},\quad x_2^{-1}\frac{x_1+y_2}{1\oplus y_2}.\end{aligned}\tag{2.102}$$

All other observations for type A_2 are commonly applied.

Congratulations! You have successfully gone through the famous “ordeal” in cluster algebra theory.

Remark 2.21. (a). For the matrices in (2.56) with $ab \geq 4$, the periodicity of seeds does not occur. Accordingly, we have infinitely many distinct cluster variables for those cluster patterns.

(b). The periods 5, 6, 8 in the above examples are interpreted as $h+2$, where $h = 3, 4, 6$ are the *Coxeter numbers* of the root systems of type A_2, B_2, G_2, respectively. A more account on the connection to the root systems will be given in Section 3.2.

2.4 Free coefficients

Let us introduce a notion, which is not defined in CA1-4 explicitly.

Definition 2.22. We say that a cluster pattern $\mathbf{\Sigma}$ is *with free coefficients at* $t_0 \in \mathbb{T}_n$ if the following conditions are satisfied:

- The coefficient semifield of $\mathbf{\Sigma}$ is the universal semifield $\mathbb{Q}_{\mathrm{sf}}(\mathbf{y})$ with generators $\mathbf{y} = (y_1, \dots, y_n)$, where n is the rank of $\mathbf{\Sigma}$.
- The coefficient tuple $\mathbf{y}_{t_0}$ at t_0 coincides with $\mathbf{y}$.

Note that, for each t, $y_{1;t}$, $\dots$, $y_{n;t}$ are algebraically independent.

Remark 2.23. It might be natural to call such coefficients *universal coefficients.* However, in CA4, a related but different notion is defined and called so. Therefore, we call it differently to avoid confusion.

Remark 2.24. Though it is not necessary, we often identify the above base vertex t_0 with the initial vertex for $\mathbf{\Sigma}$. In that case the notation $\mathbf{y}$ above is compatible with the convention for the initial coefficients (2.37). Otherwise, we may dismiss the notation (2.37) to avoid the conflict.

For any $t \in \mathbb{T}_n$, the coefficients $y_{1;t_0}$, $\dots$, $y_{n;t_0}$ in Definition 2.22 are expressed as rational functions in $\mathbf{y}_t$ with subtraction-free expressions by doing mutations from t to t_0. They induce a canonical isomorphism between the semifields $\mathbb{Q}_{\mathrm{sf}}(\mathbf{y}_{t_0})$ and $\mathbb{Q}_{\mathrm{sf}}(\mathbf{y}_t)$. Therefore, under this identification the base vertex can be shifted arbitrarily, so that the choice of the base point t_0 is superficial.

Recall the universality of $\mathbb{Q}_{\mathrm{sf}}(\mathbf{y})$ in Proposition 1.8. We are going to extend this homomorphism $\pi : \mathbb{Q}_{\mathrm{sf}}(\mathbf{y}) \to \mathbb{P}$ to a map between cluster variables with coefficients in $\mathbb{Q}_{\mathrm{sf}}(\mathbf{y})$ and $\mathbb{P}$ sharing a common B-pattern. However, there is a pitfall to avoid. Let $\varphi : \mathbb{P} \to \mathbb{P}'$ be a semifield homomorphism. Then, it is uniquely extended to a ring homomorphism $\varphi_1 : \mathbb{ZP} \to \mathbb{ZP}'$. However, it can be extended to a field homomorphism $\varphi_2 : \mathbb{QP} \to \mathbb{QP}'$ only if φ is injective. Indeed, if φ is not injective, φ_1 is not injective. Then, a fraction of $\mathbb{ZP}$ whose denominator is in $\operatorname{Ker} \varphi_1$ does not have a well-defined image of φ_2.

Below, for $\mathbb{P} = \mathbb{Q}_{\mathrm{sf}}(\mathbf{y})$, we write $\mathbb{QP}$ as $\mathbb{Q}\mathbb{Q}_{\mathrm{sf}}(\mathbf{y})$ according to our convention, though it looks a little cumbersome.

Proposition 2.25. *Let $\mathbf{\Sigma}$ be a cluster pattern with free coefficients at t_0. Let $\mathbf{\Sigma}'$ be any cluster pattern with coefficients in any semifield $\mathbb{P}$ sharing the common B-pattern with $\mathbf{\Sigma}$.*

(a). Let π be the semifield homomorphism defined by

$$\begin{array}{rccc} \pi: & \mathbb{Q}_{\mathrm{sf}}(\mathbf{y}) & \rightarrow & \mathbb{P} \\ & y_i = y_{i;t_0} & \mapsto & y'_{i;t_0}. \end{array} \tag{2.103}$$

Then, we have

$$\pi(y_{i;t}) = y'_{i;t}. \tag{2.104}$$

(b). Let $\mathcal{X}(\mathbf{\Sigma})$ be the set of the cluster variables of $\mathbf{\Sigma}$. Let

$$\varphi: \ \mathcal{X}(\mathbf{\Sigma}) \ \rightarrow \ (\mathbb{QP})(\mathbf{x}'_{t_0}) \tag{2.105}$$

be the map such that, in each element $x_{i;t} \in (\mathbb{Q}\mathbb{Q}_{\mathrm{sf}}(\mathbf{y}))(\mathbf{x}_{t_0})$, $x_{i;t_0}$ is replaced with $x'_{i;t_0}$, and $f(\mathbf{y}) \in \mathbb{Q}_{\mathrm{sf}}(\mathbf{y})$ is replaced with $f(\mathbf{y}') = \pi(f(\mathbf{y}))$. Then, the map is well defined; moreover, we have

$$\varphi(x_{i;t}) = x'_{i;t}. \tag{2.106}$$

Proof. (a). Any coefficient $y_{i;t}$ is expressed as a rational function $Y_{i;t}(\mathbf{y}) \in \mathbb{Q}_{\mathrm{sf}}(\mathbf{y})$. Applying π yields the expression for $y'_{i;t}$ in $\mathbf{y}'_{t_0}$, because both $y_{i;t}$ and $y'_{i;t}$ obey formally the same mutation formula (2.3).

(b). Any cluster variable $x_{i;t}$ is expressed as a rational function $X_{i;t}(\mathbf{x}_{t_0}) \in (\mathbb{Q}\mathbb{Q}_{\mathrm{sf}}(\mathbf{y}))(\mathbf{x}_{t_0})$. Since the mutation (2.2) does not involve any subtraction, $X_{i;t}(\mathbf{x}_{t_0})$ has a subtraction-free expression in $(\mathbb{Q}\mathbb{Q}_{\mathrm{sf}}(\mathbf{y}))(\mathbf{x}_{t_0})$. In particular, for the denominator of any coefficient in $X_{i;t}(\mathbf{x}_{t_0})$, its image by π does not vanish. Thus, the map φ is well defined. Applying φ yields the expression for $x'_{i;t}$ in $\mathbf{x}'_{t_0}$, because both $x_{i;t}$ and $x'_{i;t}$ obey formally the same mutation formula (2.2). □

The result is plainly rephrased that cluster variables with *any specific choice of coefficients* can be obtained from cluster variables with *free coefficients* by *specializing the coefficients.* To see how it works more concretely, let us look at the formulas in Section 2.3 (e.g., (2.64)–(2.68)). They can be viewed as expressions of $x_{i;t}$ in $(\mathbb{Q}\mathbb{Q}_{\mathrm{sf}}(\mathbf{y}))(\mathbf{x}_{t_0})$ and $y_{i;t}$ in $\mathbb{Q}_{\mathrm{sf}}(\mathbf{y})$, because we did not use any specific property of a given semifield $\mathbb{P}$ and initial coefficients $\mathbf{y}$ therein. Meanwhile, they can be viewed also as expressions for any specific choice of coefficients. (Viewing so is nothing but the specialization by φ and π.)

The following is an immediate consequence of Proposition 2.25.

Proposition 2.26. *Let* $\mathbf{\Sigma}$ *be a cluster pattern with free coefficients at* t_0. *Let* $\mathbf{\Sigma}'$ *be a cluster pattern with coefficients in any semifield* $\mathbb{P}$ *sharing the common* B*-pattern with* $\mathbf{\Sigma}$. *Then, for any* $t, t' \in \mathbb{T}_n$ *and a permutation* $\sigma \in S_n$, *the following fact holds:*

$$\Sigma_t = \sigma \Sigma_{t'} \quad \Longrightarrow \quad \Sigma'_t = \sigma \Sigma'_{t'}. \tag{2.107}$$

In other words, any periodicity of $\mathbf{\Sigma}$ *implies the same periodicity of* $\mathbf{\Sigma}'$.

Proof. The equality $\Sigma_t = \sigma\Sigma_{t'}$ reduces to the equality $\Sigma'_t = \sigma\Sigma'_{t'}$ under the specialization in Proposition 2.25. □

Whether the opposite implication of (2.107) holds or not is an important issue posed in CA4. Namely, there is a possibility that additional periodicities of $\mathbf{\Sigma}'$ occur under some specialization of coefficients, though it does not happen in the rank 2 examples in Section 2.3. We will discuss more about the problem later in Section 4.4.

3 Fundamental Results

In this section we present some of the most fundamental results in cluster algebra theory.

3.1 Laurent phenomenon

We prove the Laurent phenomenon observed in rank 2 examples. This is the main result in CA1 and also the most fundamental fact on cluster algebras.

Theorem 3.1 (Laurent phenomenon [FZ02a]). *Let $\mathbf{\Sigma}$ be any cluster pattern with coefficients in any semifield $\mathbb{P}$. Let $t_0, t \in \mathbb{T}_n$ be any vertices. Then, any cluster variable $x_{i;t}$ is expressed as a Laurent polynomial in $\mathbf{x}_{t_0}$ with coefficients in $\mathbb{ZP}$.*

Here we present the "classic proof" in CA1. In view of Proposition 2.25, it is enough to prove Theorem 3.1 for any cluster pattern $\mathbf{\Sigma}$ with *free coefficients at* t_0. This is because the Laurent polynomial expression of $x_{i;t}$ in $\mathbf{x}_{t_0}$ for $\mathbf{\Sigma}$ reduces to the one for any choice of coefficients under the specialization in Proposition 2.25.

Definition 3.2 (Coprime). We say that two Laurent polynomials in $\mathbb{ZP}[\mathbf{x}_{t_0}^{\pm 1}]$ are *coprime* if there is no common factor except for Laurent monomials in $\mathbf{x}_{t_0}$ with coefficients in $\mathbb{ZP}^\times = \{\pm 1\}\mathbb{P}$.

From now on, we assume that $\mathbf{\Sigma}$ has free coefficients at t_0, that is, $\mathbb{P} = \mathbb{Q}_{\mathrm{sf}}(\mathbf{y})$ and $\mathbf{y}_{t_0} = \mathbf{y}$. The proof in CA1 relies on the following lemma. (The assumption of coefficients is necessary only for the claim (b).)

Lemma 3.3. *Let $t_1, t_2, t_3 \in \mathbb{T}_n$ be vertices that are sequentially adjacent to t_0 in the following way, where $k \neq \ell$:*

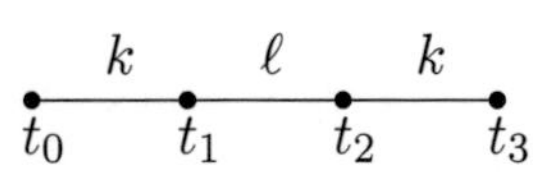

(3.1)

Then, the following facts hold:

(a). The cluster variable $x_{k;t_3}$ can be expressed as a Laurent polynomial in $\mathbf{x}_{t_0}$ with coefficients in $\mathbb{ZP}$.

(b). As elements in $\mathbb{ZP}[\mathbf{x}_{t_0}^{\pm 1}]$, $x_{k;t_1}$ is coprime with $x_{k;t_3}$ and $x_{\ell;t_3}$.

Let us temporarily assume Lemma 3.3 and prove Theorem 3.1.

Let $d(t, t')$ denote the distance between t and t' in $\mathbb{T}_n$, that is, the number of edges of $\mathbb{T}_n$ between t and t'.

Proof of Theorem 3.1. Let $d = d(t_0, t)$. Let t_1, t_2, t_3 be the ones in (3.1).

Recall that

$$x_{i;t_1} = \begin{cases} x_{k;t_0}^{-1} \left(\prod_{j=1}^{n} x_{j;t_0}^{[-b_{jk;t_0}]_+} \right) \dfrac{1+\hat{y}_{k;t_0}}{1 \oplus y_{k;t_0}} & i = k, \\ x_{i;t_0} & i \neq k. \end{cases} \tag{3.2}$$

They are certainly Laurent polynomials in $\mathbf{x}_{t_0}$. Thus, the claim holds for $d = 1$. The case $d = 2$ is similar, where only $x_{\ell;t_2}$ is concerned. For $d = 3$, the problem arises for $x_{k;t_3}$, because the term $x_{k;t_2} = x_{k;t_1}$ in the denominator is no longer a monomial in $\mathbf{x}_{t_0}$. However, this case is covered by Lemma 3.3 (a).

We prove the claim by the induction on $d = d(t_0, t) \geq 3$, *where we fix t and vary t_0*. Assume that the claim hold up to d. We consider the situation in the following graph, where $d(t_1, t) = d(t_3, t) = d$ and $d(t_0, t) = d + 1$:

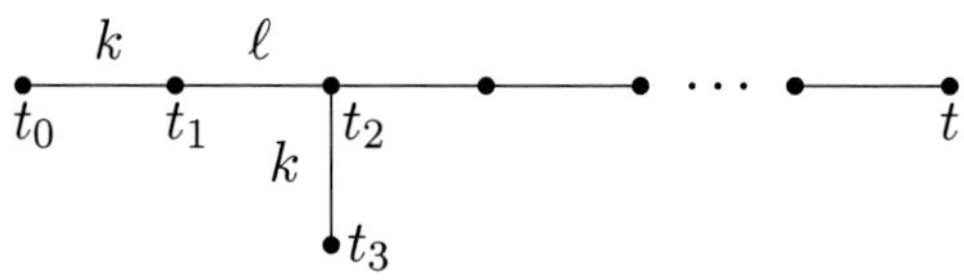

(3.3)

By the induction assumption, $x_{i;t}$ is expressed as a Laurent polynomial in $\mathbf{x}_{t_1}$. We write

$$x_{i;t} = x_{k;t_1}^{-a} f(\mathbf{x}_{t_1}), \quad f(\mathbf{x}_{t_1}) \in \mathbb{ZP}[\mathbf{x}_{t_1}^{\pm 1}], \tag{3.4}$$

where $a \geq 0$ is a sufficiently large integer such that $f(\mathbf{x}_{t_1})$ does not contain any negative power of $x_{k;t_1}$. Substituting the expression (3.2) for $x_{k;t_1}$ in $f(\mathbf{x}_{t_1})$, we have

$$x_{i;t} = x_{k;t_1}^{-a} \tilde{f}(\mathbf{x}_{t_0}), \quad \tilde{f}(\mathbf{x}_{t_0}) \in \mathbb{ZP}[\mathbf{x}_{t_0}^{\pm 1}]. \tag{3.5}$$

Meanwhile, $x_{i;t}$ is also expressed as a Laurent polynomial in $\mathbf{x}_{t_3}$. Similarly, we write

$$x_{i;t} = x_{k;t_3}^{-b} x_{\ell;t_3}^{-c} g(\mathbf{x}_{t_3}), \quad g(\mathbf{x}_{t_3}) \in \mathbb{ZP}[\mathbf{x}_{t_3}^{\pm 1}], \tag{3.6}$$

where $b, c \geq 0$ are sufficiently large integers such that $g(\mathbf{x}_{t_3})$ does not contain any negative powers of $x_{k;t_3}$ and $x_{\ell;t_3}$. Recall that $x_{k;t_3}$ and $x_{\ell;t_3}$ can be expressed as Laurent polynomials in $\mathbf{x}_{t_0}$. Substituting these expressions for $x_{k;t_3}$ and $x_{\ell;t_3}$ in $g(\mathbf{x}_{t_3})$, we have

$$x_{i;t} = x_{k;t_3}^{-b} x_{\ell;t_3}^{-c} \tilde{g}(\mathbf{x}_{t_0}), \quad \tilde{g}(\mathbf{x}_{t_0}) \in \mathbb{ZP}[\mathbf{x}_{t_0}^{\pm 1}]. \tag{3.7}$$

Comparing (3.4) and (3.7), we obtain the equality

$$x_{k;t_3}^{b} x_{\ell;t_3}^{c} \tilde{f}(\mathbf{x}_{t_0}) = x_{k;t_1}^{a} \tilde{g}(\mathbf{x}_{t_0}) \quad \text{in } \mathbb{ZP}[\mathbf{x}_{t_0}^{\pm 1}]. \tag{3.8}$$

Then, by Lemma 3.3 (b), $\tilde{f}(\mathbf{x}_{t_0})$ is divisible by $x_{k;t_1}^{a}$ in $\mathbb{ZP}[\mathbf{x}_{t_0}^{\pm 1}]$. Thus, by (3.5), we conclude that $x_{i;t} \in \mathbb{ZP}[\mathbf{x}_{t_0}^{\pm 1}]$. □

Remark 3.4. The above method is commonly applicable to various systems showing the Laurent phenomenon beyond cluster algebras [FZ02b].

Let us go back and prove Lemma 3.3.

Proof of Lemma 3.3. Below we temporarily view any cluster variable as an element in $\mathbb{QP}(\mathbf{x}_{t_0})$. For $a, b \in \mathbb{QP}(\mathbf{x}_{t_0})$, we write $a \sim b$ if there is a Laurent monomial m in $\mathbf{x}_{t_0}$ with coefficients in $\mathbb{ZP}^\times$ such that $a = mb$.

(a). We need to compare two cluster variables

$$x_{k;t_1} = x_{k;t_0}^{-1} \left(\prod_{j=1}^{n} x_{j;t_0}^{[-b_{jk;t_0}]_+} \right) \frac{1 + \hat{y}_{k;t_0}}{1 \oplus y_{k;t_0}}, \tag{3.9}$$

$$x_{k;t_3} = x_{k;t_2}^{-1} \left(\prod_{j=1}^{n} x_{j;t_2}^{[-\varepsilon b_{jk;t_2}]_+} \right) \frac{1 + \hat{y}_{k;t_2}^{\varepsilon}}{1 \oplus y_{k;t_2}^{\varepsilon}}, \tag{3.10}$$

where for $x_{k;t_3}$ we employ the ε-expression in Proposition 2.7. Recall that $x_{k;t_2} = x_{k;t_1}$ and $x_{i;t_2} = x_{i;t_0}$ for $i \neq k, \ell$. Thus, we have

$$x_{k;t_1} \sim 1 + \hat{y}_{k;t_0}. \tag{3.11}$$

If $b_{\ell k;t_0} \neq 0$, we set

$$\varepsilon = \operatorname{sign}(b_{\ell k;t_0}) = -\operatorname{sign}(b_{\ell k;t_1}) = \operatorname{sign}(b_{\ell k;t_2}) \in \{1, -1\}, \tag{3.12}$$

where $\operatorname{sign}(a) = 1$ if $a > 0$, and -1 if $a < 0$. If $b_{\ell k;t_0} = 0$, we choose $\varepsilon = \pm 1$ arbitrarily. Then, $[-\varepsilon b_{\ell k;t_2}]_+ = 0$, so that we have

$$x_{k;t_3} \sim x_{\ell;t_2}^{[-\varepsilon b_{\ell k;t_2}]_+} \frac{1 + \hat{y}_{k;t_2}^{\varepsilon}}{x_{k;t_1}} \sim \frac{1 + \hat{y}_{k;t_2}^{\varepsilon}}{1 + \hat{y}_{k;t_0}}. \tag{3.13}$$

Meanwhile, by Proposition 2.7, with the same ε in (3.12), we have

$$\hat{y}_{\ell;t_1} = \hat{y}_{\ell;t_0} \hat{y}_{k;t_0}^{[\varepsilon b_{k\ell;t_0}]_+} (1 + \hat{y}_{k;t_0}^{\varepsilon})^{-b_{k\ell;t_0}} = \hat{y}_{\ell;t_0} (1 + \hat{y}_{k;t_0}^{\varepsilon})^{-b_{k\ell;t_0}}, \tag{3.14}$$

$$\hat{y}_{k;t_2} = \hat{y}_{k;t_1} \hat{y}_{\ell;t_1}^{[\varepsilon b_{\ell k;t_1}]_+} (1 + \hat{y}_{\ell;t_1}^{\varepsilon})^{-b_{\ell k;t_1}} = \hat{y}_{k;t_0}^{-1} (1 + \hat{y}_{\ell;t_1}^{\varepsilon})^{-b_{\ell k;t_1}}. \tag{3.15}$$

We note that $-\varepsilon b_{k\ell;t_0}$ and $-\varepsilon b_{\ell k;t_1}$ are both nonnegative. Thus, $1 + \hat{y}_{k;t_2}^{\varepsilon} \in \mathbb{ZP}[\mathbf{x}_{t_0}^{\pm 1}]$. Therefore, it is enough to prove that $1 + \hat{y}_{k;t_2}^{\varepsilon}$ is divisible by $1 + \hat{y}_{k;t_0}$ in $\mathbb{ZP}[\mathbf{x}_{t_0}^{\pm 1}]$. If $b_{\ell k;t_0} = 0$, we have $\hat{y}_{k;t_2}^{\varepsilon} = \hat{y}_{k;t_0}^{-\varepsilon}$ by (3.15). Therefore, the claim holds. Suppose that $b_{\ell k;t_0} \neq 0$. Let I be the ideal of $\mathbb{ZP}[\mathbf{x}_{t_0}^{\pm 1}]$ generated by $1 + \hat{y}_{k;t_0}$. By (3.14), we have $\hat{y}_{\ell;t_1}^{\varepsilon} \equiv 0 \mod I$. Thus, by (3.15), we have $\hat{y}_{k;t_2}^{\varepsilon} \equiv \hat{y}_{k;t_0}^{-\varepsilon} \mod I$. Therefore, we have

$$1 + \hat{y}_{k;t_2}^{\varepsilon} \equiv 1 + \hat{y}_{k;t_0}^{-\varepsilon} = \hat{y}_{k;t_0}^{-\varepsilon} (1 + \hat{y}_{k;t_0}^{\varepsilon}) \equiv 0 \mod I. \tag{3.16}$$

(b). First we note that

$$x_{\ell;t_3} = x_{\ell;t_2} \sim x_{k;t_1}^{[-\varepsilon b_{k\ell;t_1}]_+}(1+\hat{y}^{\varepsilon}_{\ell;t_1}) = 1+\hat{y}^{\varepsilon}_{\ell;t_1}, \tag{3.17}$$

where ε is the same sign in (3.12). Recall that by the assumption of free coefficients, $y_{\ell;t_0} = y_\ell$ and $y_{k;t_0} = y_k$ are algebraically independent. By (3.14) and (3.15), the following facts hold:

- $1+\hat{y}_{k;t_0}$ is a constant with respect to y_ℓ^ε.
- $1+\hat{y}^{\varepsilon}_{\ell;t_1}$ is a binomial with respect to y_ℓ^ε whose constant term is 1.
- $1+\hat{y}^{\varepsilon}_{k;t_2}$ is a polynomial with respect to y_ℓ^ε whose constant term is $1+\hat{y}^{-\varepsilon}_{k;t_0}$

It follows that $x_{k;t_1} \sim 1+\hat{y}_{k;t_0}$ and $x_{\ell;t_3} \sim 1+\hat{y}^{\varepsilon}_{\ell;t_1}$ are coprime. Also, it follows that $(1+\hat{y}^{\varepsilon}_{k;t_2})/(1+\hat{y}_{k;t_0})$, which is in $\mathbb{ZP}[\mathbf{x}_{t_0}^{\pm1}]$ by (a), is a polynomial with respect to y_ℓ^ε whose constant term is $\hat{y}^{-1}_{k;t_0}$ if $\varepsilon = 1$ and 1 if $\varepsilon = -1$. Therefore, $x_{k;t_3} \sim (1+\hat{y}^{\varepsilon}_{k;t_2})/(1+\hat{y}_{k;t_0})$ and $x_{k;t_1} \sim 1+\hat{y}_{k;t_0}$ are coprime. □

3.2 Finite type classification

We present the finite type classification of cluster algebras and cluster patterns without proofs. This is the main result in CA2.

To state the result, we briefly explain the background in Lie theory.

The following definition is the counterpart of Definition 1.13.

Definition 3.5 (Symmetrizable matrix). An integer square matrix $A = (a_{ij})_{i,j=1}^n$ is said to be *symmetrizable* if there is a diagonal matrix $D = (d_i\delta_{ij})_{i,j=1}^n$ whose diagonal entries d_i are positive integers such that DA is symmetric, i.e,

$$d_i a_{ij} = d_j a_{ji}. \tag{3.18}$$

The matrix D is called a *(left) symmetrizer* of A. In particular, any symmetric matrix is symmetrizable.

Definition 3.6 (Cartan matrix). An $n \times n$ (integer) square matrix $A = (a_{ij})_{i,j=1}^n$ is called a *(generalized) Cartan matrix* if it satisfies the following conditions:

- For any i, we have $a_{ii} = 2$.
- For any i, j $(i \neq j)$, we have $a_{ij} \leq 0$; moreover, $a_{ij} < 0$ if and only if $a_{ji} < 0$.

For each symmetrizable Cartan matrix X, one can define the associated Kac-Moody algebra, root system, and Weyl group [Kac90].

There is a natural (many-to-one) correspondence from a skew-symmetrizable matrix to a symmetrizable Cartan matrix.

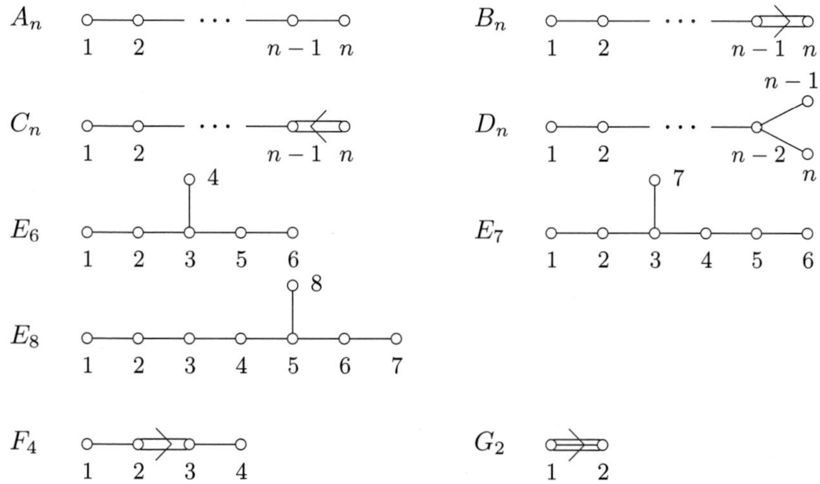

Figure 1: Dynkin diagrams of finite type

Definition 3.7 (Cartan counterpart). With any skew-symmetrizable matrix $B = (b_{ij})_{i,j=1}^{n}$, we associate a symmetrizable Cartan matrix $A = A(B)$ as

$$a_{ij} = \begin{cases} 2 & i = j, \\ -|b_{ij}| & i \neq j. \end{cases} \tag{3.19}$$

The matrix $A(B)$ is called the *Cartan counterpart* of B.

Example 3.8. For the initial exchange matrices B_0 of type A_2, B_2, G_2 in Section 2.3, the associated Cartan matrices $A(B_0)$ are given as follows:

$$\begin{pmatrix} 2 & -1 \\ -1 & 2 \end{pmatrix}, \quad \begin{pmatrix} 2 & -1 \\ -2 & 2 \end{pmatrix}, \quad \begin{pmatrix} 2 & -1 \\ -3 & 2 \end{pmatrix}. \tag{3.20}$$

They are the Cartan matrices of type A_2, B_2, G_2, respectively, as explained below.

We recall the Dynkin diagrams of finite type, which is well-known in the context of semisimple Lie algebras and crystallographic root systems [Bou68, Hum90].

Definition 3.9. We call the graphs given in the list in Figure 1 the *Dynkin diagrams of finite type*. Here, A_n $(n \geq 1)$, B_n $(n \geq 2)$, C_n $(n \geq 2)$, D_n $(n \geq 4)$, and we identify B_2 and C_2 (up to the relabeling of the vertices).

For each Dynkin diagram of type X_n in Figure 1, we define the $n \times n$ matrix $A = A(X_n)$ such that $a_{ii} = 2$, and nondiagonal entries are given as follows, where we follow the convention in [Kac90], which is the transpose to the one in [Bou68] :

$$\begin{cases} a_{ij} = a_{ji} = 0 \\ a_{ij} = a_{ji} = -1 \\ a_{ij} = -1,\ a_{ji} = -2 \\ a_{ij} = -1,\ a_{ji} = -3 \end{cases} \tag{3.21}$$

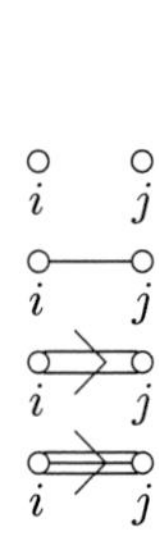

The resulting matrix A is a symmetrizable Cartan matrix. These matrices (up to the simultaneous permutations of row and column indices) are called the *Cartan matrices of finite type (of type X_n).*

Example 3.10. The Cartan matrices of type A_3, B_3, C_3 (up to the simultaneous permutations of row and column indices) are given, respectively, by

$$\begin{pmatrix} 2 & -1 & 0 \\ -1 & 2 & -1 \\ 0 & -1 & 2 \end{pmatrix}, \quad \begin{pmatrix} 2 & -1 & 0 \\ -1 & 2 & -1 \\ 0 & -2 & 2 \end{pmatrix}, \quad \begin{pmatrix} 2 & -1 & 0 \\ -1 & 2 & -2 \\ 0 & -1 & 2 \end{pmatrix}. \tag{3.22}$$

For an indecomposable symmetrizable Cartan matrix A, it is known that the following finiteness conditions are equivalent [Kac90, Prop. 4.9]:

(a). A is of finite type.

(b). The Kac-Moody algebra associated with A is finite-dimensional.

(c). The root system associated with A is a finite set.

(d). The Weyl group associated with A is a finite group.

Moreover, the isomorphism classes of such Kac-Moody algebras, root systems, and Weyl groups, respectively, are classified by the Dynkin diagrams of finite type. In addition, the above finiteness conditions are equivalent to the following condition [Kac90, Prop. 4.7]:

(e). All principal minors of A are positive.

Now we consider the counterpart in cluster algebras.

Definition 3.11 (Finite type)**.** A cluster pattern $\mathbf{\Sigma}$ is said to be *of finite type* if there are only finitely many distinct seeds of $\mathbf{\Sigma}$. The cluster algebra $\mathcal{A}(\mathbf{\Sigma})$ associated with a cluster pattern $\mathbf{\Sigma}$ is said to be *of finite type* if the cluster pattern $\mathbf{\Sigma}$ is of finite type.

Remark 3.12. It is known [FZ03a] that a cluster pattern $\mathbf{\Sigma}$ is of finite type if and only if there are finitely many distinct *clusters* (equivalently, finitely many distinct *cluster variables*) of $\mathbf{\Sigma}$. The only if part holds trivially, while the if part requires a detailed study.

Definition 3.13 (Strongly isomorphic). Let $\mathbf{\Sigma}$ and $\mathbf{\Sigma}'$ be cluster patterns of a common rank n with a common coefficient semifield $\mathbb{P}$.

- We say that cluster patterns $\mathbf{\Sigma}$ and $\mathbf{\Sigma}'$ are *isomorphic* if there are some $t, t' \in \mathbb{T}_n$ and a permutation $\sigma \in S_n$ such that $(\mathbf{y}'_{t'}, B'_{t'}) = (\sigma(\mathbf{y}_t), \sigma(B_t))$.
- We say that the cluster algebras $\mathcal{A}(\mathbf{\Sigma})$ and $\mathcal{A}(\mathbf{\Sigma}')$ are *strongly isomorphic* if the underlying cluster patterns $\mathbf{\Sigma}$ and $\mathbf{\Sigma}'$ are isomorphic.

Proposition 3.14. *If $\mathcal{A}(\mathbf{\Sigma})$ and $\mathcal{A}(\mathbf{\Sigma}')$ are* strongly isomorphic, *they are isomorphic as $\mathbb{ZP}$-algebras.*

Proof. By the assumption, there are some $t_0, t_1 \in \mathbb{T}_n$ and a permutation $\sigma \in S_n$ such that $(\mathbf{y}'_{t_1}, B'_{t_1}) = (\sigma(\mathbf{y}_{t_0}), \sigma(B_{t_0}))$. Let us consider the cluster pattern $\sigma\mathbf{\Sigma} = \{\sigma\Sigma_t\}_{t\in\mathbb{T}_n}$. Then, the correspondence $\mathbf{x}'_{t_1} \mapsto \sigma\mathbf{x}_{t_0}$ induces a $\mathbb{ZP}$-algebra isomorphism $\varphi : \mathcal{A}(\mathbf{\Sigma}') \xrightarrow{\sim} \mathcal{A}(\sigma\mathbf{\Sigma})$. Meanwhile, thanks to Proposition 2.9, the set of cluster variables of $\sigma\mathbf{\Sigma}$ coincides with the one of $\mathbf{\Sigma}$. Therefore, $\mathcal{A}(\sigma\mathbf{\Sigma}) = \mathcal{A}(\mathbf{\Sigma})$. □

Now we are ready to state the classification of cluster patterns/algebras of finite type. In view of Proposition 2.17, it is enough to concentrate on the cluster patterns whose exchange matrices are indecomposable.

Theorem 3.15 (Finite type classification [FZ03a]). *For cluster patterns with indecomposable exchange matrices, the following facts hold.*

(a). The cluster pattern $\mathbf{\Sigma}$ is of finite type if and only if there is some $t \in \mathbb{T}_n$ such that the Cartan matrix $A(B_t)$ associated with B_t is of finite type.

(b). The Dynkin type of $A(B_t)$ does not depend on the choice of such t.

By Theorem 3.15, any isomorphism class of cluster patterns of finite type with indecomposable exchange matrices is uniquely labeled by the Dynkin type of the corresponding Cartan matrix $A(B_t)$ therein. We have already used this labeling for the rank 2 examples in Section 2.3.

For a cluster pattern of finite type, the root system of the corresponding type naturally parametrizes the cluster variables.

Definition 3.16 (Denominator vector/Non-initial cluster variable). Let $\mathbf{\Sigma}$ be any cluster pattern with a given initial vertex t_0.

- For any cluster variable $x_{i;t}$, we define an integer vector $\mathbf{d}_{i;t} = (d_{ji;t})_{j=1}^n$ such that $-d_{ji;t}$ is the lowest degree of $x_{j;t_0}$ in the Laurent polynomial expression of $x_{i;t}$ in $\mathbf{x}_{t_0}$. The vector $\mathbf{d}_{i;t} = (d_{ji;t})_{j=1}^n$ is called the *denominator vector* (or *d-vector*, for short) of $x_{i;t}$. In particular, for an initial variable $x_{i;t_0}$, we have $\mathbf{d}_{i;t_0} = -\mathbf{e}_i$.
- We say that a cluster variable $x_{i;t}$ is *non-initial* if it does not coincide with any initial cluster variables $x_{1;t_0}, \dots, x_{n;t_0}$.

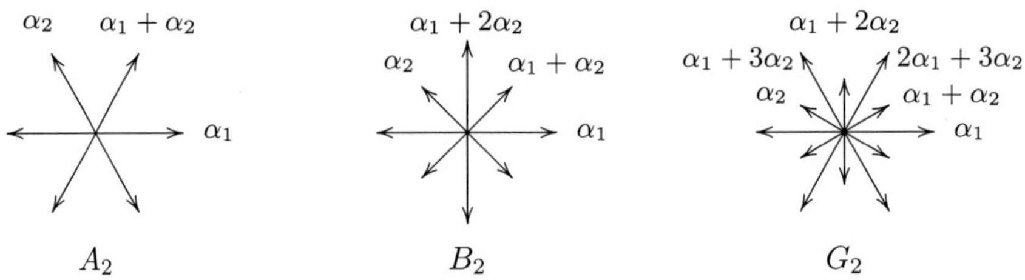

Figure 2: Root systems of rank 2.

Example 3.17. In the examples in Section 2.3, we see the following list of the denominator vectors for the non-initial cluster variables.

$$(A_2) : (1,0), (1,1), (0,1), \tag{3.23}$$
$$(B_2) : (1,0), (1,1), (1,2), (0,1), \tag{3.24}$$
$$(G_2) : (1,0), (1,1), (1,2), (1,3), (2,3), (0,1). \tag{3.25}$$

They are naturally identified with the *positive roots* of the corresponding root systems. See Figure 2.

This phenomenon can be fully generalized to any cluster pattern of finite type if we properly choose the initial vertex t_0.

Theorem 3.18 ([FZ03a]). *Let $\mathbf{\Sigma}$ be any cluster pattern of finite type with a initial vertex t_0 such that $A(B_{t_0})$ is a Cartan matrix of finite type. Then, the denominator vector of any non-initial cluster variable is identified with a positive root of the corresponding root system. Moreover, this correspondence is one-to-one.*

3.3 Cluster algebras of geometric type

Below we focus on a class of cluster patterns whose coefficient semifields are *tropical semifields.* They are important in various applications of cluster algebras.

Let $\mathrm{Trop}(\mathbf{u})$ be the tropical semifield in Example 1.10.

Definition 3.19 (Geometric type)**.** A cluster pattern (resp. cluster algebra) with coefficients in a tropical semifield $\mathrm{Trop}(\mathbf{u})$ is called a cluster pattern (resp. cluster algebra) of *geometric type*, where $\mathbf{u} = (u_1, \dots, u_m)$ and m is taken independently of the rank n of a cluster pattern.

Remark 3.20. The terminology originates in the fact that cluster algebras of this type typically arise as the coordinating rings of certain algebraic varieties, which are called the *geometric realization* of cluster algebras in [FZ03a].

Let $\mathbf{\Sigma}$ be a cluster pattern of geometric type. Recall that an element of $\mathrm{Trop}(\mathbf{u})$ is a Laurent monomial in $\mathbf{u}$ with coefficient 1. Therefore, any coefficient $y_{i;t}$ of $\mathbf{\Sigma}$ at $t \in \mathbb{T}_n$ is represented as

$$y_{i;t} = \prod_{j=1}^{m} u_j^{c_{ji;t}}, \quad i = 1, \dots, n. \tag{3.26}$$

This determines an $m \times n$ integer matrix $C_t = (c_{ij;t})$ for each $t \in \mathbb{T}_n$, and the coefficient tuple $\mathbf{y}_t$ and the matrix C_t are identified. Under this identification, the mutation in (2.3) is translated as follows.

Proposition 3.21. *Let $t, t' \in \mathbb{T}_n$ be k-adjacent vertices. Then, we have the following mutation formula of matrices C_t:*

$$c_{ij;t'} = \begin{cases} -c_{ik;t} & j = k, \\ c_{ij;t} + c_{ik;t}[b_{kj;t}]_+ + [-c_{ik;t}]_+ b_{kj;t} & j \neq k. \end{cases} \tag{3.27}$$

Proof. For any integer a, the following identity holds:

$$\min(0, a) = -[-a]_+. \tag{3.28}$$

Thus, by the definition of the tropical sum in (1.21), we have

$$1 \oplus y_{k;t} = \prod_{i=1}^{m} u_i^{-[-c_{ik;t}]_+}. \tag{3.29}$$

By replacing the index i in (2.3) with j, and comparing the power of u_i, we obtain (3.27). □

Observe that the formula (3.27) is parallel to the mutation (2.4) of the exchange matrix B_t. This motivates us to define the following $(n+m) \times n$ matrix

$$\tilde{B}_t = \begin{pmatrix} B_t \\ C_t \end{pmatrix}, \tag{3.30}$$

which is called an *extended exchange matrix* of $\mathbf{\Sigma}$. Abusing the notation, let us write the (i,j)-entry of $\tilde{B}_t$ as b_{ij}. Then, we can unify the mutations (2.4) and (3.27) as a single matrix mutation of $\tilde{B}_t$ defined by

$$b_{ij;t'} = \begin{cases} -b_{ij;t} & i = k \text{ or } j = k, \\ b_{ij;t} + b_{ik;t}[b_{kj;t}]_+ + [-b_{ik;t}]_+ b_{kj;t} & i, j \neq k, \end{cases} \tag{3.31}$$

where $i = 1, \dots, n+m$, and $j, k = 1, \dots, n$. Next, let us observe that the mutations of cluster variables can be also expressed in terms of the matrix $\tilde{B}_t$. By (3.29), we have

$$\frac{y_{k;t}}{1 \oplus y_{k;t}} = \prod_{j=1}^{m} u_j^{[c_{jk;t}]_+}, \qquad \frac{1}{1 \oplus y_{k;t}} = \prod_{j=1}^{m} u_j^{[-c_{jk;t}]_+}. \tag{3.32}$$

Thus, the mutation in the form of (2.5) is written as

$$x_{i;t'} = \begin{cases} \dfrac{1}{x_{k;t}}\left(\displaystyle\prod_{j=1}^{m} u_j^{[c_{jk;t}]_+} \prod_{j=1}^{n} x_{j;t}^{[b_{jk;t}]_+} + \prod_{j=1}^{m} u_j^{[-c_{jk;t}]_+} \prod_{j=1}^{n} x_{j;t}^{[-b_{jk;t}]_+}\right) & i = k, \\ x_{i;t} & i \neq k. \end{cases} \tag{3.33}$$

To summarize the result so far, the coefficients $\mathbf{y}_t$ of geometric type can be safely replaced with the lower part C_t of the extended exchange matrices $\tilde{B}_t$.

To make the picture complete, we introduce an *extended cluster*,

$$\tilde{\mathbf{x}}_t = (x_{1;t}, \dots, x_{n+m:t}) := (x_{1;t}, \dots, x_{n;t}, u_1, \dots, u_m). \tag{3.34}$$

Then, together with the extended exchange matrix notation, the mutation (3.33) is written as, for $k = 1, \dots, n$,

$$x_{i;t'} = \begin{cases} \dfrac{1}{x_{k;t}}\left(\displaystyle\prod_{j=1}^{n+m} x_{j;t}^{[b_{jk;t}]_+} + \prod_{j=1}^{n+m} x_{j;t}^{[-b_{jk;t}]_+}\right) & i = k, \\ x_{i;t} & i \neq k. \end{cases} \tag{3.35}$$

This formally coincides with the mutation of cluster variables *without coefficients.*

Let us formulate this observation more completely in terms of a cluster pattern of rank $n+m$ *without coefficients.* We consider the extension of $\tilde{B}_{t_0}$ to a full $(n+m) \times (n+m)$ skew-symmetrizable matrix. Such an extension (together with its skew-symmetrizer) is not unique. To given an example, let D be a skew-symmetrizer of B_{t_0}, and let d be the least common multiple of the diagonal entries of D. Then, the following matrix gives a skew-symmetrizable extension of $\tilde{B}_{t_0}$:

$$\hat{B}_{t_0} = \begin{pmatrix} B_{t_0} & -D^{-1}dC_{t_0}^T \\ C_{t_0} & O \end{pmatrix}, \tag{3.36}$$

where a skew-symmetrizer is given by $\hat{D} = D \oplus dI_m$. (The factor d ensures that the matrix $\hat{B}_{t_0}$ is an integer matrix.) For a given initial vertex t_0 in $\mathbb{T}_{n+m}$, we regard $\mathbb{T}_n$ as a subtree of $\mathbb{T}_{n+m}$ such that $\mathbb{T}_n$ contains t_0 and all vertices that are sequentially adjacent to t_0 with edges labeled by $\{1, \dots, n\}$ in $\mathbb{T}_{n+m}$. Then, the above cluster pattern $\mathbf{\Sigma}$ of geometric type is equivalent to consider a cluster pattern of rank $n+m$ *without coefficients*, restricted to the subtree $\mathbb{T}_n \subset \mathbb{T}_{n+m}$, and with the initial seed $(\tilde{\mathbf{x}}_{t_0}, \hat{B}_{t_0})$. The elements $x_{n+1;t} = x_{n+1}$, $\dots$, $x_{n+m;t} = x_{n+m}$ are called the *frozen variables*, because they are not mutated. We note that the right half of $\hat{B}_t$ does not influence neither (3.31) nor (3.35). Thus, we may safely replace seeds $(\tilde{\mathbf{x}}_t, \hat{B}_t)$ with $(\tilde{\mathbf{x}}_t, \tilde{B}_t)$ without losing information.

Let us summarize the results obtained above as a proposition:

Proposition 3.22. *We have the following three equivalent presentations of a cluster pattern of geometric type:*

- *a cluster pattern consisting seeds* $(\mathbf{x}_t, \mathbf{y}_t, B_t)$ *with coefficients in* $\mathrm{Trop}(\mathbf{u})$, *by definition.*
- *a cluster pattern consisting seeds* $(\mathbf{x}_t, \tilde{B}_t)$ *whose coefficients in* $\mathrm{Trop}(\mathbf{u})$ *are encoded in the extend exchange matrices* $\tilde{B}_t$.
- *a cluster pattern consisting extended seeds* $(\tilde{\mathbf{x}}_t, \hat{B}_t)$ *or* $(\tilde{\mathbf{x}}_t, \tilde{B}_t)$ *without coefficients and with frozen variables* $x_{n+1}, \dots, x_{n+m}$ *that are identified with* $u_1, \dots, u_m$. *(In this case we still say that it is of rank* n.*)*

Even if the above three pictures are equivalent as *cluster patterns*, there is some discrepancy for the corresponding *cluster algebras* due to the difference of the coefficient semifields. Note that, for $\mathbb{P} = \mathrm{Trop}(\mathbf{u})$, the group ring $\mathbb{ZP}$ is identified with the Laurent polynomial ring $\mathbb{Z}[\mathbf{u}^{\pm 1}]$. Then, for the first two pictures, the corresponding cluster algebras are given by

$$\mathcal{A} = \mathbb{Z}[\mathbf{u}^{\pm 1}][\mathbf{x}_t]_{t \in \mathbb{T}_n} = \mathbb{Z}[\mathbf{x}_t, \mathbf{u}^{\pm 1}]_{t \in \mathbb{T}_n}, \tag{3.37}$$

while, for the last one, including the frozen variables as honorary members, the corresponding cluster algebra is given by

$$\mathcal{A} = \mathbb{Z}[\tilde{\mathbf{x}}_t]_{t \in \mathbb{T}_n} = \mathbb{Z}[\mathbf{x}_t, \mathbf{u}]_{t \in \mathbb{T}_n}. \tag{3.38}$$

Therefore, one has to be careful in which sense cluster algebra is considered.

For a cluster pattern of geometric type, the Laurent phenomenon in Theorem 3.1 claims

$$x_{i;t} \in \mathbb{Z}[\mathbf{u}^{\pm 1}][\mathbf{x}_{t_0}^{\pm 1}] = \mathbb{Z}[\mathbf{x}_{t_0}^{\pm 1}, \mathbf{u}^{\pm 1}]. \tag{3.39}$$

Actually, the following strong version of the Laurent phenomenon holds.

Theorem 3.23 ([FZ03a]). *For any cluster pattern of geometric type, we have*

$$x_{i;t} \in \mathbb{Z}[\mathbf{x}_{t_0}^{\pm 1}, \mathbf{u}]. \tag{3.40}$$

Proof. We consider the following claim, which is slightly stronger than the desired result.

Claim. For each j, $x_{i;t}$ is a polynomial in u_j whose constant term (with respect to u_j) is a *nonzero* polynomial with a subtraction-free rational expression in $\mathbf{x}_{t_0}^{\pm 1}$ and $\mathbf{u}$ other than u_j. (See Definition 1.5 for a subtraction-free expression.)

We prove the claim by the induction on the distance $d = d(t_0, t)$ in $\mathbb{T}_n$. For $d = 0$, we have $x_{i;t} = x_i$, so that the claim holds. Suppose that the claim holds for $d = d(t_0, t)$. Let $t' \in \mathbb{T}_n$ be k-adjacent to t. It is enough to prove the claim for $x_{k;t'}$. Let us look at the first case of (3.33). Note that, for each j, either $[c_{jk;t}]_+$ or $[-c_{jk;t}]_+$ is zero. Thus, by the induction assumption, the numerator in (3.33) is a polynomial in u_j whose constant term is a nonzero polynomial with a subtraction-free rational expression in $\mathbf{x}_{t_0}^{\pm 1}$ and $\mathbf{u}$ other than u_j. (Here, the condition of having a subtraction-free expression guarantees that, even in the special case $c_{jk;t} = 0$, the constant terms from the two products in (3.33) never cancel accidentally.) Also, the denominator $x_{k;t}$ has the same property. Since we already know that $x_{k;t'}$ is in $\mathbb{Z}[\mathbf{x}_{t_0}^{\pm 1}, \mathbf{u}^{\pm 1}]$, the denominator divides the numerator such that the result is a polynomial in u_j having the same property. □

Remark 3.24. The above claim in the proof does not imply that $x_{i;t}$ has a nonzero constant term in $\mathbf{u}$.

3.4 Grassmannian $\mathrm{Gr}(2,5)$

Let us present a prototypical example of a cluster algebra (of geometric type) appearing in Lie theory.

First, we briefly recall basic definitions/facts on Grassmannian $\mathrm{Gr}(2,5)$. See [Ful97, Section 9] for the background and more information.

Definition 3.25 (Grassmannian $\mathrm{Gr}(2,5)$).

- In short, the *Grassmannian* $\mathrm{Gr}(2,5)$ is the complex projective variety consisting of all 2-dimensional subspaces in a 5-dimensional vector space V over $\mathbb{C}$.
- An element of $\mathrm{Gr}(2,5)$ is identified with its basis, which is represented by a 2×5 complex matrix $M = (m_{ij})$ of rank 2 modulo the left action of $GL(2,\mathbb{C})$. This tells us that $\dim \mathrm{Gr}(2,5) = 10 - 4 = 6$.
- For each pair $1 \leq i < j \leq 5$, let

$$p_{ij} = p_{ij}(M) := \begin{vmatrix} m_{1i} & m_{1j} \\ m_{2i} & m_{2j} \end{vmatrix}. \tag{3.41}$$

  Then, we define a map $i : \mathrm{Gr}(2,5) \to \mathbb{CP}^9$ by

$$i : M \mapsto [p_{12} : p_{13} : \cdots : p_{45}], \tag{3.42}$$

  where $[z_1 : \cdots : z_{10}]$ is the homogeneous coordinates of the complex projective space $\mathbb{CP}^9$. The map is injective, and it is called the *Plücker embedding*, while p_{ij} are called the *Plücker coordinates*.
- For $1 \leq i < j < k < \ell \leq 5$, the Plücker coordinates satisfy the relations

$$R_{ijk\ell} : \ p_{ij}p_{k\ell} - p_{ik}p_{j\ell} + p_{i\ell}p_{jk} = 0. \tag{3.43}$$

  They are called the *Plücker relations.* Note that there are five Plücker relations, R_{1234}, R_{1235}, R_{1245}, R_{1345}, and R_{2345}.

- The homogeneous coordinate ring of Gr(2, 5) is given by

$$\mathbb{C}[\mathrm{Gr}(2,5)] = \mathbb{C}[\mathbf{P}]/I_R, \quad \mathbf{P} = (p_{ij})_{1\le i<j\le 5}, \tag{3.44}$$

  where I_R is the homogeneous ideal generated by the Plücker relations; moreover, I_R is prime. Thus, Gr(2, 5) is a projective variety in $\mathbb{CP}^9$.
- Alternatively, we may consider the *affine cone* $\widehat{\mathrm{Gr}}(2,5)$ over Gr(2, 5), that is, the affine variety in the complex affine space $\mathbb{A}^{10}$ defined by the same Plücker relations (3.43). The coordinate ring of $\widehat{\mathrm{Gr}}(2,5)$ has the same form as (3.44),

$$\mathbb{C}[\widehat{\mathrm{Gr}}(2,5)] = \mathbb{C}[\mathbf{P}]/I_R. \tag{3.45}$$

  However, the ring in (3.44) is regarded as a graded ring, while it is a (not graded) ring here.

We are going to show that $\mathbb{C}[\mathrm{Gr}(2,5)]$ (or $\mathbb{C}[\widehat{\mathrm{Gr}}(2,5)]$ by forgetting the grading) has the cluster algebra structure. To do that, we consider a cluster pattern of geometric type, where there are two cluster variables $x_{i;t}$ $(i = 1, 2)$ and five frozen variables $x_a, \dots, x_e$. The cluster pattern $\mathbf{\Sigma}$ we consider is encoded in the following initial extended exchange matrix $\tilde{B}_0$, or equivalently, the corresponding initial *extended quiver* $\tilde{Q}_0$, where the vertices $a, \dots, e$ are *frozen vertices* and we will mutate only at the vertices 1 and 2.

$$\tilde{B}_0 = \left(\begin{array}{cc} 0 & -1 \\ 1 & 0 \\ \hline -1 & 0 \\ 1 & 0 \\ -1 & 1 \\ 0 & -1 \\ 0 & 1 \end{array}\right), \qquad \tilde{Q}_0 = \tag{3.46}$$

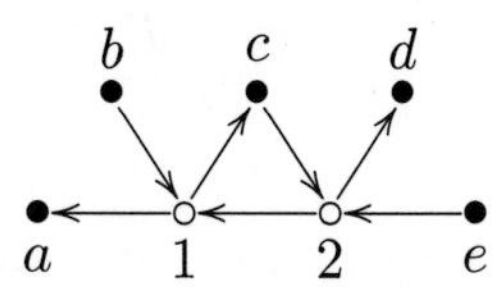

In particular, this is a special case of cluster pattern of type A_2.

In addition, we introduce an alternative presentation of the above data by a *triangulation of a pentagon*, where the frozen vertices of the quiver correspond to the sides of the pentagon, while the unfrozen vertices correspond to two diagonals as follows:

$T_0 =$ (3.47)

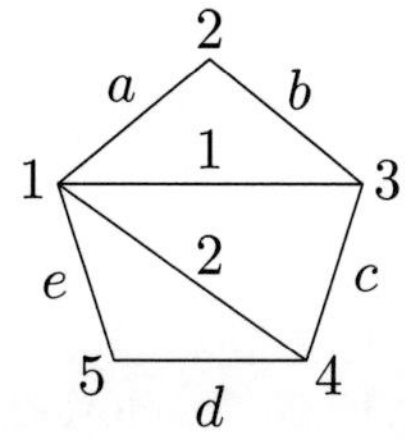

To explain in more detail, we identify a subquiver in $\tilde{Q}_0$ and a triangle in T_0 by the following correspondence:

(3.48)

where we regard that the arrows among frozen vertices (i.e, $a \to b$ and $d \to e$) are omitted in $\tilde{Q}_0$. (This is reasonable, because they do not influence the mutations of seeds $(\tilde{\mathbf{x}}_t, \tilde{B}_t)$ at all.)

The advantage of the presentation by a triangulation is that one can manage a mutation of a seed pictorially as a *flip* of a diagonal. For example, the mutation of the quiver $\tilde{Q}_0$ at the vertex 1 gives the following quiver $\tilde{Q}_1$, where again the arrows among frozen vertices are omitted:

$\tilde{Q}_1 =$ 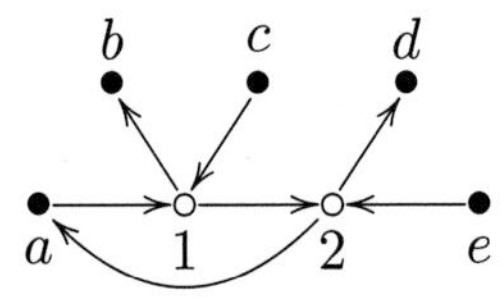

(3.49)

The corresponding triangulation T_1 is given by the flip of the diagonal 1 inside the quadrilateral with sides a, b, c, 2 depicted as below:

$T_1 =$ 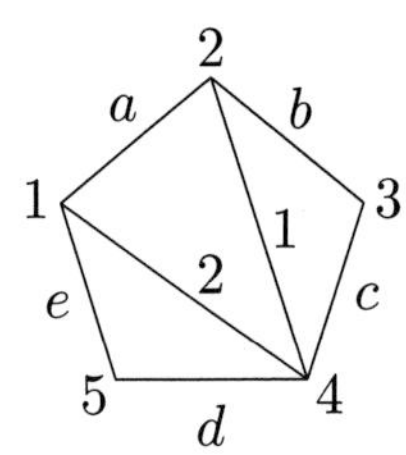

(3.50)

Meanwhile, the mutation of the cluster variable x_1 is given by

$$x_1' = \frac{x_b x_2 + x_a x_c}{x_1}. \tag{3.51}$$

By rewriting it in the form

$$x_1 x_1' = x_b x_2 + x_a x_c, \tag{3.52}$$

we can recognize it as *Ptolemy's theorem* for cyclic quadrilaterals.

The mutations of these data are presented up to $t = 5$ in Table 1, where we observe the familiar pentagon periodicity of type A_2 as the periodicity of the alternative flips in a pentagon. In fact, this is the reason why we call this periodicity so.

Remark 3.26. This is indeed a prototypical example of the *surface realization* of cluster patterns/algebras developed by Fock-Goncharov [FG07] and

t	$\tilde{B}_t$	$\tilde{Q}_t$	T_t	$x_{1;t}, x_{2;t}$
0	$\begin{pmatrix} 0 & -1 \\ 1 & 0 \\ \hline -1 & 0 \\ 1 & 0 \\ -1 & 1 \\ 0 & -1 \\ 0 & 1 \end{pmatrix}$	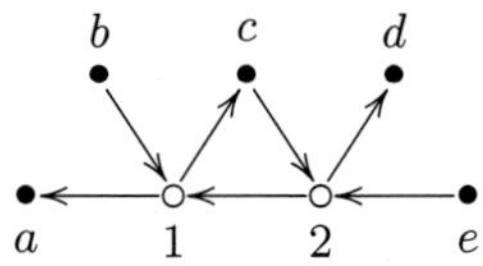	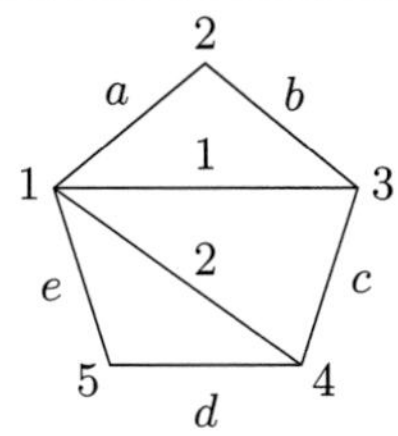	x_{13}, x_{14}
1	$\begin{pmatrix} 0 & 1 \\ -1 & 0 \\ \hline 1 & -1 \\ -1 & 0 \\ 1 & 0 \\ 0 & -1 \\ 0 & 1 \end{pmatrix}$	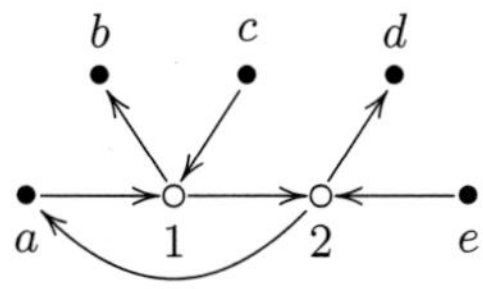	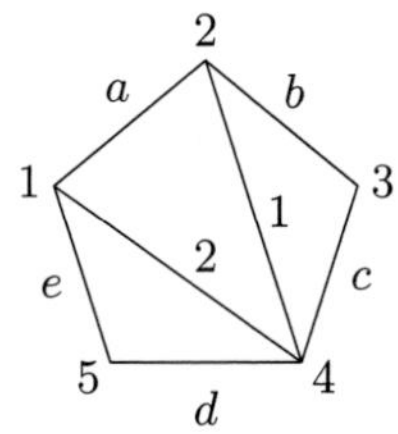	x_{24}, x_{14}
2	$\begin{pmatrix} 0 & -1 \\ 1 & 0 \\ \hline 0 & 1 \\ -1 & 0 \\ 1 & 0 \\ -1 & 1 \\ 0 & -1 \end{pmatrix}$	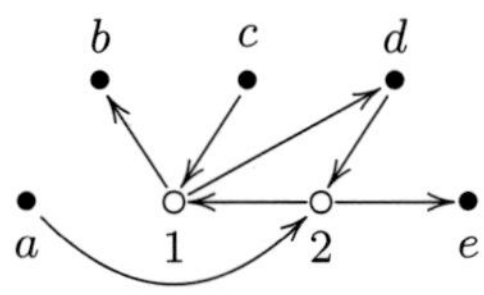	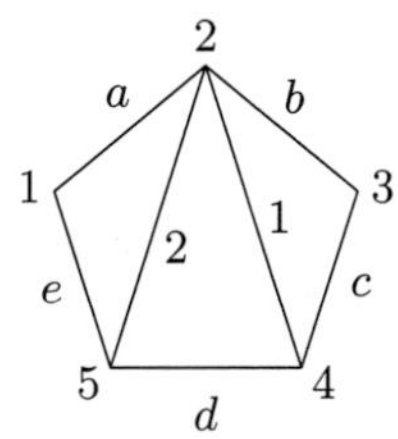	x_{24}, x_{25}
3	$\begin{pmatrix} 0 & 1 \\ -1 & 0 \\ \hline 0 & 1 \\ 1 & -1 \\ -1 & 0 \\ 1 & 0 \\ 0 & -1 \end{pmatrix}$	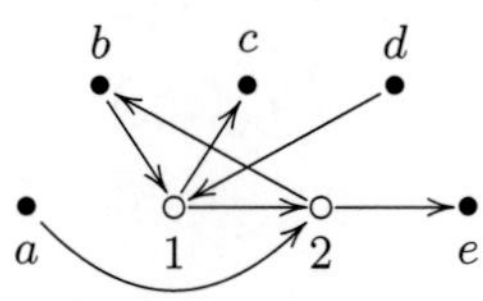	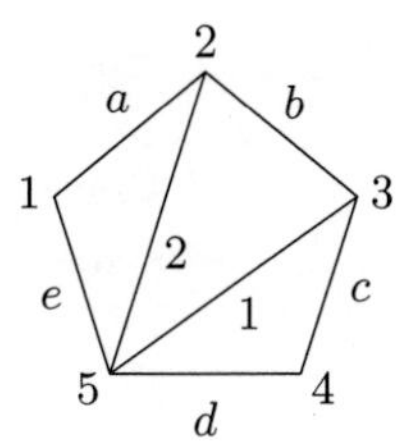	x_{35}, x_{25}
4	$\begin{pmatrix} 0 & -1 \\ 1 & 0 \\ \hline 0 & -1 \\ 0 & 1 \\ -1 & 0 \\ 1 & 0 \\ -1 & 1 \end{pmatrix}$	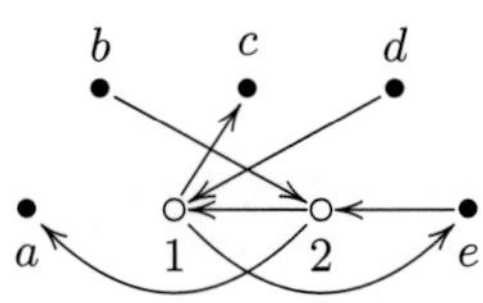	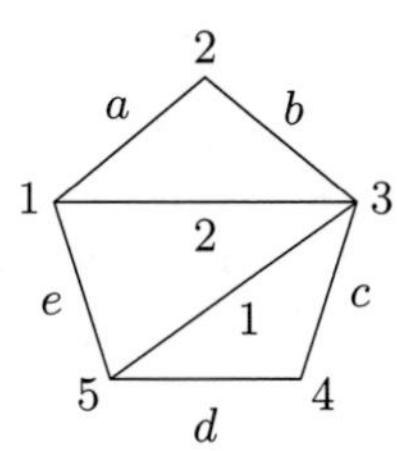	x_{35}, x_{13}
5	$\begin{pmatrix} 0 & 1 \\ -1 & 0 \\ \hline 0 & -1 \\ 0 & 1 \\ 1 & -1 \\ -1 & 0 \\ 1 & 0 \end{pmatrix}$	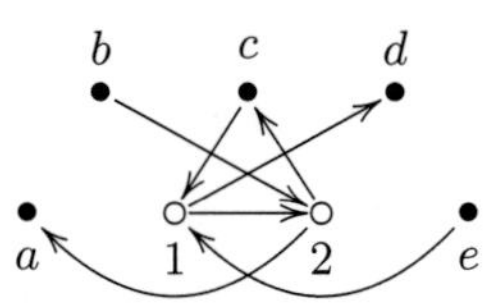	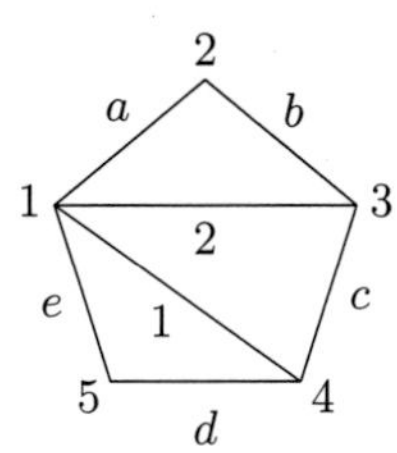	x_{14}, x_{13}

Table 1: Mutations of extended exchange matrices.

Fomin-Shapiro-Thurston [FST08]. Only a limited class of cluster patterns/algebras are realized in this way. In [FST08] a complete classification of the class of cluster algebras that admit the surface realization is given.

Let $\mathcal{A}$ be the cluster algebra associated with the cluster pattern $\mathbf{\Sigma}$ in the sense of (3.38). In order to match $\mathbb{C}[\mathrm{Gr}(2,5)]$ in (3.44), we replace the ground ring $\mathbb{Z}$ with $\mathbb{C}$. In view of the result in Example 2.18, it is given by

$$\mathcal{A} = \mathbb{C}[x_{1;0}, x_{2;0}, x_{1;1}, x_{2;2}, x_{1;3}, x_a, \dots, x_e]. \tag{3.53}$$

We regard it as a graded ring such that every generator is of degree 1.

We have the following conclusion.

Theorem 3.27 ([FZ02a, FZ03a]). *The graded ring* $\mathbb{C}[\mathrm{Gr}(2,5)]$ *is isomorphic to the cluster algebra* $\mathcal{A}$ *in* (3.53). *In particular,* $\mathbb{C}[\mathrm{Gr}(2,5)]$ *has a cluster algebra structure.*

Proof. Let $[ij]$ be the chord (diagonal or side) whose end points are labeled by i and j in the underlying pentagon of T_t. We give the following labeling of cluster/frozen variables of $\mathcal{A}$:

- Each frozen variable x_i $(i = a, \dots, e)$ is labeled as x_{jk} if the side labeled by i is $[jk]$. Explicitly,

$$x_a = x_{12},\ x_b = x_{23},\ x_c = x_{34},\ x_d = x_{45},\ x_e = x_{15}. \tag{3.54}$$

- Each cluster variable $x_{i;t}$ $(i = 1, 2)$ is labeled as x_{jk} if the diagonal labeled by i in T_t is $[jk]$ in Table 1. In particular,

$$x_{1;0} = x_{13},\ x_{2;0} = x_{14},\ x_{1;1} = x_{24},\ x_{2;2} = x_{25},\ x_{1;3} = x_{35}. \tag{3.55}$$

The cluster algebra $\mathcal{A}$ in (3.53) is now written as

$$\mathcal{A} = \mathbb{C}[\mathbf{X}], \quad \mathbf{X} = (x_{ij})_{1 \le i < j \le 5}. \tag{3.56}$$

On the other hand, the mutation from $t = 0$ to 1 in (3.52) is identified with the Plücker relation

$$R_{1234}: \quad x_{13}x_{24} = x_{14}x_{23} + x_{12}x_{34}. \tag{3.57}$$

Similarly, the mutations from $t = 1$ to 5 are identified with the Plücker relations R_{1245}, R_{2345}, R_{1235}, R_{1345}. Therefore, we have a graded ring homomorphism

$$\begin{array}{rccc} \varphi: & \mathbb{C}[\mathrm{Gr}(2,5)] = \mathbb{C}[\mathbf{P}]/I_R & \to & \mathcal{A} = \mathbb{C}[\mathbf{X}] \\ & p_{ij} & \mapsto & x_{ij}. \end{array} \tag{3.58}$$

The map φ is clearly surjective. To show that φ is injective, we rely on the following facts:

- The ideal I_R is prime as stated in Definition 3.25.
- In particular, the ideal I_R contains no monomial in $\mathbf{P}$. (If there is a monomial in I_R, it should be at least quadratic, which contradict the primeness.)

Suppose that a polynomial $F(\mathbf{P})$ in $\mathbf{P} = (p_{ij})$ is in $\operatorname{Ker}\varphi$. Namely, $F(\mathbf{X}) = 0$ in $\mathcal{A}$. By multiplying some monomial $m(\mathbf{X})$ to $F(\mathbf{X})$ and applying the mutation relations (the Plücker relations for $\mathbf{X}$), one can eliminate all non-initial cluster variables x_{24}, x_{25}, x_{35} and obtain a polynomial $G(\mathbf{x}_0)$ in the initial cluster variables $\mathbf{x}_0$. Namely, we have $G(\mathbf{x}_0) = m(\mathbf{X})F(\mathbf{X}) = 0$ in $\mathcal{A}$. Then, due to the algebraic independence of $\mathbf{x}_0$, $G(\mathbf{x}_0)$ is the zero polynomial. Applying the same operation to $F(\mathbf{P})$ in $\mathbb{C}[\mathbf{P}]$, we have $m(\mathbf{P})F(\mathbf{P}) \equiv 0 \mod I_R$. Since $m(\mathbf{P}) \not\equiv 0$, we have $F(\mathbf{P}) \equiv 0$ by the primeness of I_R. □

Remark 3.28. It was shown by Scott [Sco06] that the homogeneous coordinate ring of a general Grassmannian $\mathrm{Gr}(k, m)$ also has the cluster algebra structure.

4 Separation formulas

In this section we present the *separation formulas* given in CA4. The are particularly important to study the structure of seeds in a cluster pattern systematically.

4.1 Principal coefficients

We introduce the notion of cluster patterns with principal coefficients. They are a special class of cluster patterns of geometric type and play an important role in studying cluster patterns.

Definition 4.1. We say that a cluster pattern $\mathbf{\Sigma}$ is *with principal coefficients at* $t_0 \in \mathbb{T}_n$ if the following conditions are satisfied:

- The coefficient semifield of $\mathbf{\Sigma}$ is a tropical semifield $\mathrm{Trop}(\mathbf{y})$ with generators $\mathbf{y} = (y_1, \dots, y_n)$, where n is the rank of $\mathbf{\Sigma}$.
- The coefficient tuple $\mathbf{y}_{t_0}$ at t_0 coincides with $\mathbf{y}$.

The same remark in Remark 2.24 is applicable for the notation.

In other words, a cluster pattern with principal coefficients at t_0 is a cluster pattern of geometric type such that the extended exchange matrix in (3.30) at t_0 is given by

$$\tilde{B}_{t_0} = \begin{pmatrix} B_{t_0} \\ I \end{pmatrix}. \tag{4.1}$$

All properties of cluster patterns of geometric type hold to cluster patterns with principal coefficients.

Remark 4.2. In contrast to a cluster pattern of free coefficients in Definition 2.22, the notion of principal coefficients crucially depends on the choice of the base vertex t_0. To explain it in more detail, let $\mathbf{y}_t$ and $\mathbf{y}'_t$ ($t \in \mathbb{T}_n$) are principal coefficients at base vertices t_0 and t'_0, respectively. Then, the correspondence $\varphi : y_{i;t} \mapsto y'_{i;t}$ *cannot* be extended to a semifield homomorphism from $\mathrm{Trop}(\mathbf{y}_{t_0})$ to $\mathrm{Trop}(\mathbf{y}_{t'_0})$, in general.

Let $\mathbf{\Sigma}$ be a cluster pattern with principal coefficients at t_0. One may regard the coefficients of $\mathbf{\Sigma}$ as the *tropicalization of free coefficients* at t_0. Namely, let $\mathbf{\Sigma}'$ be a cluster pattern with free coefficients at t_0 such that it shares the common B-pattern with $\mathbf{\Sigma}$. Then, as a special case of Proposition 2.25, $\mathbf{y}'_t$ and $\mathbf{y}_t$ are related by the tropicalization homomorphism in (1.24) by

$$\begin{aligned} \pi_{\mathrm{trop}} : \quad \mathbb{Q}_{\mathrm{sf}}(\mathbf{y}) &\rightarrow \mathrm{Trop}(\mathbf{y}) \\ y'_{i;t} &\mapsto y_{i;t}. \end{aligned} \tag{4.2}$$

4.2 C- and G-matrices, and F-polynomials

For a given cluster pattern $\mathbf{\Sigma}$ with principal coefficients at t_0, one can define important quantities called C-matrices, G-matrices, and F-polynomials. They are the *building blocks* of seeds of a cluster pattern with coefficients in any semifield $\mathbb{P}$, and together with the separation formula in Theorem 4.16, they clarify the structure of seeds as well as the relation between cluster variables and coefficients.

(a). C-matrices. Let us start with C-matrices. They are the matrices defined in (3.26) specialized for a cluster pattern $\mathbf{\Sigma}$ with principal coefficients at t_0.

Definition 4.3 (C-matrix/c-vector)**.** For a given cluster pattern $\mathbf{\Sigma}$ with principal coefficients at t_0, the *C-matrix* $C_t = (c_{ij;t})$ of $\mathbf{y}_t$ is the $n \times n$ integer matrix defined by

$$y_{i;t} = \prod_{j=1}^{n} y_j^{c_{ji;t}}. \tag{4.3}$$

Equivalently, they are defined as the lower half of the $2n \times n$ extended exchange matrix $\tilde{B}_t$ of (3.30) with the initial condition (4.1). The ith column vector $\mathbf{c}_{i;t} = (c_{ji})_{j=1}^{n}$ of C_t is called the *c-vector* of $y_{i;t}$. We call the collection of the C-matrices $\mathbf{C}^{t_0} = \{C_t\}_{t\in\mathbb{T}_n}$ the *C-pattern* of $\mathbf{\Sigma}$.

The C-matrices are uniquely determined by the underlying B-pattern $\mathbf{B}$ of $\mathbf{\Sigma}$ and t_0, thus, eventually by t_0 and B_{t_0} only.

Proposition 4.4. *The C-pattern $\mathbf{C}^{t_0}$ of $\mathbf{\Sigma}$ is uniquely determined by the following initial condition and the mutation formula:*

$$C_{t_0} = I, \tag{4.4}$$

$$c_{ij;t'} = \begin{cases} -c_{ik;t} & j = k, \\ c_{ij;t} + c_{ik;t}[b_{kj;t}]_+ + [-c_{ik;t}]_+ b_{kj;t} & j \neq k, \end{cases} \tag{4.5}$$

where t and t' are k-adjacent.

Proof. The initial condition (4.4) holds by (4.1). The mutation (4.5) was established in (3.27). □

For the second case of (4.5), we have an alternative expression, which is parallel to the ε-expressions in Proposition 2.7.

Proposition 4.5. *The following expression does not depend on the choice of $\varepsilon \in \{1, -1\}$:*

$$c_{ij;t} + c_{ik;t}[\varepsilon b_{kj;t}]_+ + [-\varepsilon c_{ik;t}]_+ b_{kj;t}. \tag{4.6}$$

Proof. This follows from (1.36). Alternatively, one can derive each expression from (2.13) in the same way as (3.27). □

(b). *G*-matrices. Next, we define G-matrices. By Theorem 3.23, we have

$$x_{i;t} \in \mathbb{Z}[\mathbf{x}^{\pm 1}, \mathbf{y}], \tag{4.7}$$

where we set $\mathbf{x}_{t_0} = \mathbf{x}$. We introduce a certain degree vector in $\mathbb{Z}^n$ for each cluster variables $x_{i;t}$.

Definition 4.6 (Principal $\mathbb{Z}^n$-grading)**.** For each monomial in $\mathbb{Z}[\mathbf{x}^{\pm 1}, \mathbf{y}]$, we define its $\mathbb{Z}^n$-grading by

$$\deg(x_i) = \mathbf{e}_i, \quad \deg(y_i) = -\mathbf{b}_{i;t_0}, \tag{4.8}$$

where $\mathbf{e}_i$ is the ith unit vector and $\mathbf{b}_{i;t_0}$ is the ith column vector of B_{t_0}. We call it the *principal $\mathbb{Z}^n$-grading.*

A seemingly artificial degree of y_i is designed to ensures the following property:

$$\deg(\hat{y}_i) = \deg\left(y_i \prod_{j=1}^{n} x_j^{b_{ji;t_0}} \right) = -\mathbf{b}_{i;t_0} + \mathbf{b}_{i;t_0} = \mathbf{0}. \tag{4.9}$$

Lemma 4.7. *Every $x_{i;t} \in \mathbb{Z}[\mathbf{x}^{\pm 1}, \mathbf{y}]$ is homogenous with respect to the principal $\mathbb{Z}^n$-grading.*

Proof. We first note that, thanks to Proposition 2.6, any $\hat{y}$-variable $\hat{y}_{i;t}$ is written as a rational function of the initial $\hat{y}$-variables $\hat{\mathbf{y}}$. Then, we can effectively treat it as $\deg(\hat{y}_{i;t}) = \mathbf{0}$ in the following calculation, even though $\hat{y}_{i;t}$ does not belong to $\mathbb{Z}[\mathbf{x}^{\pm 1}, \mathbf{y}]$. We prove the claim by the induction on $d = d(t_0, t)$. The claim is trivial for the initial cluster variables x_i. Suppose that claim is true up to $d = d(t_0, t)$. Let $t' \in \mathbb{T}_n$ be k-adjacent to t. We look at the mutation formula in the form (2.2),

$$x_{k;t'} = x_{k;t}^{-1} \left(\prod_{j=1}^{n} x_{j;t}^{[-b_{jk;t}]_+} \right) \frac{1 + \hat{y}_{k;t}}{1 \oplus y_{k;t}}. \tag{4.10}$$

Then, the binomial $1 + \hat{y}_{k;t}$ is homogeneous with degree $\mathbf{0}$. Also, $1 \oplus y_{k;t}$ is actually a monomial in $\mathbf{y}$ because it belongs to $\mathrm{Trop}(\mathbf{y})$. Thus, the right hand side of (4.10) is homogeneous thanks to the induction assumption. □

Based on Lemma 4.7, we define G-matrices as follows.

Definition 4.8 (G-matrix/g-vector)**.** For a given cluster pattern $\mathbf{\Sigma}$ with principal coefficients at t_0, we define a $\mathbb{Z}^n$-vector

$$\mathbf{g}_{i;t} = \deg(x_{i;t}) \in \mathbb{Z}^n, \tag{4.11}$$

where deg is the principal $\mathbb{Z}^n$-grading in Definition 4.6. We call it the *g-vector* of $x_{i;t}$. For each $t \in \mathbb{T}_n$, we define the *G-matrix* $G_t = (g_{ij;t})$ of $\mathbf{x}_t$ such that its ith column vector is $\mathbf{g}_{i;t} = (g_{ji;t})_{j=1}^n$. We call the collection of the G-matrices $\mathbf{G}^{t_0} = \{G_t\}_{t\in\mathbb{T}_n}$ the *G-pattern* of $\mathbf{\Sigma}$.

Again, the G-pattern $\mathbf{G}^{t_0}$ is uniquely determined by the underlying B-pattern $\mathbf{B}$ and t_0 (together with the C-pattern $\mathbf{C}^{t_0}$ determined from them).

Proposition 4.9. *The G-pattern $\mathbf{G}^{t_0}$ of $\mathbf{\Sigma}$ is uniquely determined by the following initial condition and the mutation formula:*

$$G_{t_0} = I, \tag{4.12}$$

$$g_{ij;t'} = \begin{cases} -g_{ik;t} + \sum_{\ell=1}^n g_{i\ell;t}[-b_{\ell k;t}]_+ - \sum_{\ell=1}^n b_{i\ell;t_0}[-c_{\ell k;t}]_+ & j = k, \\ g_{ij;t} & j \neq k, \end{cases} \tag{4.13}$$

where t and t' are k-adjacent.

Proof. The initial condition (4.12) follows from the fact $\deg x_i = \mathbf{e}_i$. The mutation (4.13) follows from (4.10) and the formula

$$\deg\left(\frac{1}{1 \oplus y_{k;t}}\right) = \deg\left(\prod_{j=1}^n y_j^{[-c_{jk;t}]_+}\right) = -\sum_{j=1}^n [-c_{jk;t}]_+ \mathbf{b}_{j;t_0}, \tag{4.14}$$

where we used (3.32) in the first equality. □

We have the following duality relation between C- and G-matrices.

Proposition 4.10. *The following equality holds:*

$$G_t B_t = B_{t_0} C_t. \tag{4.15}$$

Proof. This follows from the formula

$$\begin{aligned} \deg(\hat{y}_{i;t}) &= \deg\left(y_{i;t} \prod_{j=1}^n x_{j;t}^{b_{ji;t}}\right) = \deg\left(\prod_{j=1}^n y_j^{c_{ji;t}} \prod_{j=1}^n x_{j;t}^{b_{ji;t}}\right) \\ &= -B_{t_0}\mathbf{c}_{i;t} + G_t \mathbf{b}_{i;t}. \end{aligned} \tag{4.16}$$

and the fact $\deg(\hat{y}_{i;t}) = \mathbf{0}$. □

For the first case of (4.13), we have an alternative expression, which is parallel to the ε-expressions in Proposition 2.7.

Proposition 4.11. *The following expression does not depend on the choice of* $\varepsilon \in \{1, -1\}$*:*

$$-g_{ik;t} + \sum_{\ell=1}^{n} g_{i\ell;t}[-\varepsilon b_{\ell k;t}]_+ - \sum_{\ell=1}^{n} b_{i\ell;t_0}[-\varepsilon c_{\ell k;t}]_+. \tag{4.17}$$

Proof. We take the difference of two expressions. Then, it vanishes thanks to (1.35) and (4.15). □

(c). F-polynomials.

The following definition relies on the fact (4.7).

Definition 4.12 (F-polynomial)**.** For a given cluster pattern $\mathbf{\Sigma}$ with principal coefficients at t_0, we define polynomials $F_{i;t}(\mathbf{y}) \in \mathbb{Z}[\mathbf{y}]$ in formal variables $\mathbf{y} = (y_1, \dots, y_n)$ by specializing Laurent polynomials $x_{i;t}(\mathbf{x}, \mathbf{y}) \in \mathbb{Z}[\mathbf{x}^{\pm 1}, \mathbf{y}]$ with $x_1 = \cdots = x_n = 1$. We call $F_{i;t}(\mathbf{y})$ the *F-polynomial* of $x_{i;t}$. Let $\mathbf{F}_t = (F_{1;t}(\mathbf{y}), \dots, F_{n;t}(\mathbf{y}))$. We call the collection of the F-polynomials $\mathbf{F}^{t_0} = \{\mathbf{F}_t\}_{t \in \mathbb{T}_n}$ the *F-pattern* of $\mathbf{\Sigma}$.

Again, the F-pattern $\mathbf{F}^{t_0}$ is uniquely determined by the underlying B-pattern $\mathbf{B}$ and t_0 (together with the C-pattern $\mathbf{C}^{t_0}$ determined from them).

Proposition 4.13. *The F-pattern $\mathbf{F}^{t_0}$ of $\mathbf{\Sigma}$ is uniquely determined by the following initial condition and the mutation formula:*

$$F_{i;t_0}(\mathbf{y}) = 1, \tag{4.18}$$

$$F_{i;t'}(\mathbf{y}) = \begin{cases} \dfrac{M_{k;t}(\mathbf{y})}{F_{k;t}(\mathbf{y})} & i = k, \\ F_{i;t}(\mathbf{y}) & i \neq k, \end{cases} \tag{4.19}$$

where t and t' are k-adjacent, and

$$M_{k;t}(\mathbf{y}) = \prod_{j=1}^{n} y_j^{[c_{jk;t}]_+} \prod_{j=1}^{n} F_{j;t}(\mathbf{y})^{[b_{jk;t}]_+} + \prod_{j=1}^{n} y_j^{[-c_{jk;t}]_+} \prod_{j=1}^{n} F_{j;t}(\mathbf{y})^{[-b_{jk;t}]_+} \tag{4.20}$$

$$= \left(\prod_{j=1}^{n} y_j^{[-c_{jk;t}]_+} \prod_{j=1}^{n} F_{j;t}(\mathbf{y})^{[-b_{jk;t}]_+} \right) \left(1 + \prod_{j=1}^{n} y_j^{c_{jk;t}} \prod_{j=1}^{n} F_{j;t}(\mathbf{y})^{b_{jk;t}} \right). \tag{4.21}$$

Proof. This is obtained by specializing the mutation of $\mathbf{x}_t$ in the form (3.33) with $x_1 = \cdots = x_n = 1$. □

We may regard F-polynomials also as elements in $\mathbb{Q}_{\mathrm{sf}}(\mathbf{y})$ because the mutation in (4.19) is a subtraction-free operation. Then, we may apply the tropicalization homomorphism $\pi_{\mathrm{trop}} : \mathbb{Q}_{\mathrm{sf}}(\mathbf{y}) \to \mathrm{Trop}(\mathbf{y})$ in (4.2).

Proposition 4.14. *The following fact holds:*

$$\pi_{\text{trop}}(F_{i;t}(\mathbf{y})) = 1. \tag{4.22}$$

Proof. We prove it by the induction on the distance $d = d(t_0, t)$. For $t = t_0$, the claim holds by (4.18). Assume that the claim holds for $d = d(t_0, t)$. Let t' be k-adjacent to t. Note that, for each j, either $[c_{jk;t}]_+$ or $[-c_{jk;t}]_+$ is zero. Then, $\pi_{\text{trop}}(M_{k;t}(\mathbf{y})) = 1$ for $M_{k;t}(\mathbf{y})$ in (4.20). So, the claim also holds for t' by (4.19) □

Remark 4.15. The fact (4.22) does not imply that $F_{i;t}(\mathbf{y})$ has a nonzero constant term as a polynomial in $\mathbf{y}$. For example, $\pi_{\text{trop}}(y_1 + y_2) = 1$.

In summary, even if the patterns $\mathbf{C}^{t_0}$, $\mathbf{G}^{t_0}$, $\mathbf{F}^{t_0}$ are originally extracted from a cluster pattern $\mathbf{\Sigma}$ with principal coefficients at t_0, it turned out that they are uniquely and directly defined from the underlying B-pattern $\mathbf{B}$ and the base vertex t_0. Therefore, one can associate these patterns $\mathbf{C}^{t_0}$, $\mathbf{G}^{t_0}$, $\mathbf{F}^{t_0}$ to any cluster pattern $\mathbf{\Sigma}$ with coefficients in *any semifield* $\mathbb{P}$.

4.3 Separation formulas

Let $\mathbf{\Sigma}$ be any cluster pattern with coefficients in any semifield $\mathbb{P}$. Let t_0 be a given initial vertex t_0, and let $\mathbf{C}^{t_0}$, $\mathbf{G}^{t_0}$, $\mathbf{F}^{t_0}$ be the associated C-, G-, F-patterns, respectively. As already mentioned, each F-polynomial $F_{i;t}(\mathbf{y})$ belongs to $\mathbb{Q}_{\text{sf}}(\mathbf{y})$. Following Definition 1.9, let $F_{i;t}|_{\mathbb{P}}(\mathbf{y})$ be the specialization of $F_{i;t}(\mathbf{y})$ in $\mathbb{P}$ at the initial coefficient $\mathbf{y}_{t_0} = \mathbf{y}$ of $\mathbf{\Sigma}$, where the abuse of the symbol $\mathbf{y}$ does not cause a serious problem.

The cluster variables and coefficients of any cluster pattern are expressed by the initial cluster variables and coefficients together with C- and G-matrices, and F-polynomials. This is one of the most fundamental and useful properties of cluster patterns.

Theorem 4.16 (Separation Formulas [FZ07]). *Let $\mathbf{\Sigma}$ be any cluster pattern with coefficients in any semifield $\mathbb{P}$ and with a given initial vertex t_0. Let*

$$\mathbf{x}_{t_0} = \mathbf{x}, \quad \mathbf{y}_{t_0} = \mathbf{y}, \quad \hat{\mathbf{y}}_{t_0} = \hat{\mathbf{y}} \tag{4.23}$$

be the initial cluster variables, coefficients, and $\hat{y}$-variables. Then, the following formulas hold.

$$x_{i;t} = \left(\prod_{j=1}^{n} x_j^{g_{ji;t}}\right) \frac{F_{i;t}(\hat{\mathbf{y}})}{F_{i;t}|_{\mathbb{P}}(\mathbf{y})}, \tag{4.24}$$

$$y_{i;t} = \left(\prod_{j=1}^{n} y_j^{c_{ji;t}}\right) \prod_{j=1}^{n} F_{j;t}|_{\mathbb{P}}(\mathbf{y})^{b_{ji;t}}, \tag{4.25}$$

$$\hat{y}_{i;t} = \left(\prod_{j=1}^{n} \hat{y}_j^{c_{ji;t}} \right) \prod_{j=1}^{n} F_{j;t}(\hat{\mathbf{y}})^{b_{ji;t}}. \tag{4.26}$$

Proof. The formulas hold for $t = t_0$ due to the initial condition of C- and G-matrices and F-polynomials in Propositions 4.4, 4.9, and 4.13. Therefore, it is enough to show that their right hand sides mutate in the same way as $x_{i;t}$, $y_{i;t}$, $\hat{y}_{i;t}$ based on the mutations of C_t, G_t, and $F_{i;t}$.

We first treat the formula (4.25). Let $y_{i;t}$ temporarily denote the one in the right hand side of (4.25). Let $t' \in \mathbb{T}_n$ be k-adjacent to t. For $i = k$,

$$y_{k;t'} = \left(\prod_{j=1}^{n} y_j^{-c_{jk;t}} \right) \prod_{j=1}^{n} F_{j;t}|_{\mathbb{P}}(\mathbf{y})^{-b_{jk;t}} = y_{k;t}^{-1}, \tag{4.27}$$

where in the first equality we used the fact $b_{kk;t} = 0$. (Below we do not repeat this remark.) For $i \neq k$,

$$\begin{aligned} y_{i;t'} = {} & \left(\prod_{j=1}^{n} y_j^{c_{ji;t} + c_{jk;t}[b_{ki;t}]_+ + [-c_{jk;t}]_+ b_{ki;t}} \right) \\ & \times \left(\prod_{j \neq k} F_{j;t}|_{\mathbb{P}}(\mathbf{y})^{b_{ji;t} + b_{jk;t}[b_{ki;t}]_+ + [-b_{jk;t}]_+ b_{ki;t}} \right) \\ & \times \left\{ F_{k;t}|_{\mathbb{P}}(\mathbf{y})^{-1} \left(\prod_{j=1}^{n} y_j^{[-c_{jk;t}]_+} \prod_{j=1}^{n} F_{j;t}|_{\mathbb{P}}(\mathbf{y})^{[-b_{jk;t}]_+} \right) \right. \\ & \times \left. \left(1 \oplus \prod_{j=1}^{n} y_j^{c_{jk;t}} \prod_{j=1}^{n} F_{j;t}|_{\mathbb{P}}(\mathbf{y})^{b_{jk;t}} \right) \right\}^{-b_{ki;t}} \\ = {} & y_{i;t} y_{k;t}^{[b_{ki;t}]_+} (1 \oplus y_{k;t})^{-b_{ki;t}}. \end{aligned} \tag{4.28}$$

The formula (4.26) is proved in the exactly same manner as above.

Now we treat (4.24). Let $x_{i;t}$ temporarily denote the one in the right hand side of (4.24). For $i \neq k$, $x_{i;t'} = x_{i;t}$ holds. For $x_{k;t'}$, we have

$$\begin{aligned} x_{k;t'} = {} & \left(\prod_{j=1}^{n} x_j^{-g_{jk;t} + \sum_{\ell=1}^{n} g_{j\ell;t}[-b_{\ell k;t}]_+ - \sum_{\ell=1}^{n} b_{j\ell;t_0}[-c_{\ell k;t}]_+} \right) \\ & \times \frac{F_{k;t}(\hat{\mathbf{y}})^{-1} \prod_{j=1}^{n} \hat{y}_j^{[-c_{jk;t}]_+} \prod_{j=1}^{n} F_{j;t}(\hat{\mathbf{y}})^{[-b_{jk;t}]_+}}{F_{k;t}|_{\mathbb{P}}(\mathbf{y})^{-1} \prod_{j=1}^{n} y_j^{[-c_{jk;t}]_+} \prod_{j=1}^{n} F_{j;t}|_{\mathbb{P}}(\mathbf{y})^{[-b_{jk;t}]_+}} \end{aligned}$$

$$\times \frac{1 + \prod_{j=1}^{n} \hat{y}_j^{c_{jk;t}} \prod_{j=1}^{n} F_{j;t}(\hat{\mathbf{y}})^{b_{jk;t}}}{1 \oplus \prod_{j=1}^{n} y_j^{c_{jk;t}} \prod_{j=1}^{n} F_{j;t}|_{\mathbb{P}}(\mathbf{y})^{b_{jk;t}}}$$
$$= x_{k;t}^{-1} \left(\prod_{j=1}^{n} x_{j;t}^{[-b_{jk;t}]_+} \right) \frac{1 + \hat{y}_{k;t}}{1 \oplus y_{k;t}},$$

where we used (4.25) and (4.26) in the last equality. □

Remark 4.17. Originally in [FZ07], the formula (4.24) was referred to as the separation formula, where the additions $\oplus/+$ in $\mathbb{P}/\mathcal{F}$ are *separated* in the denominator and the numerator. Here we also included the formulas (4.25) and (4.26) in the family, because they together exhibit a duality relation between cluster variables and coefficients/$\hat{y}$-variables.

Example 4.18 (Type A_2)**.** Let us consider the cluster pattern of type A_2 in Example 2.18. By comparing the result therein with the separation formulas (4.24) and (4.25), one can easily read off C- and G-matrices and F-polynomials as follows:

$$C_0 = \begin{pmatrix} 1 & 0 \\ 0 & 1 \end{pmatrix}, \quad G_0 = \begin{pmatrix} 1 & 0 \\ 0 & 1 \end{pmatrix}, \quad \begin{cases} F_{1;0}(\mathbf{y}) = 1, \\ F_{2;0}(\mathbf{y}) = 1, \end{cases} \tag{4.29}$$

$$C_1 = \begin{pmatrix} -1 & 0 \\ 0 & 1 \end{pmatrix}, \quad G_1 = \begin{pmatrix} -1 & 0 \\ 0 & 1 \end{pmatrix}, \quad \begin{cases} F_{1;1}(\mathbf{y}) = 1 + y_1, \\ F_{2;1}(\mathbf{y}) = 1, \end{cases} \tag{4.30}$$

$$C_2 = \begin{pmatrix} -1 & 0 \\ 0 & -1 \end{pmatrix}, \quad G_2 = \begin{pmatrix} -1 & 0 \\ 0 & -1 \end{pmatrix}, \quad \begin{cases} F_{1;2}(\mathbf{y}) = 1 + y_1, \\ F_{2;2}(\mathbf{y}) = 1 + y_2 + y_1 y_2, \end{cases} \tag{4.31}$$

$$C_3 = \begin{pmatrix} 1 & -1 \\ 0 & -1 \end{pmatrix}, \quad G_3 = \begin{pmatrix} 1 & 0 \\ -1 & -1 \end{pmatrix}, \quad \begin{cases} F_{1;3}(\mathbf{y}) = 1 + y_2, \\ F_{2;3}(\mathbf{y}) = 1 + y_2 + y_1 y_2, \end{cases} \tag{4.32}$$

$$C_4 = \begin{pmatrix} 0 & 1 \\ -1 & 1 \end{pmatrix}, \quad G_4 = \begin{pmatrix} 1 & 1 \\ -1 & 0 \end{pmatrix}, \quad \begin{cases} F_{1;4}(\mathbf{y}) = 1 + y_2, \\ F_{2;4}(\mathbf{y}) = 1, \end{cases} \tag{4.33}$$

$$C_5 = \begin{pmatrix} 0 & 1 \\ 1 & 0 \end{pmatrix}, \quad G_5 = \begin{pmatrix} 0 & 1 \\ 1 & 0 \end{pmatrix}, \quad \begin{cases} F_{1;5}(\mathbf{y}) = 1, \\ F_{2;5}(\mathbf{y}) = 1. \end{cases} \tag{4.34}$$

We have the following observations.

(a). *Duality.* The following relation between C- and G-matrices holds:

$$G_t^T C_t = I, \tag{4.35}$$

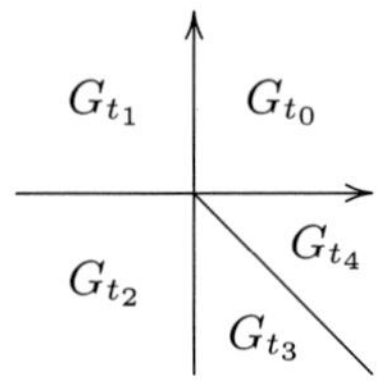

Figure 3: The G-fan of a cluster pattern of type A_2.

where G_t^T is the transpose of G_t.

(b). *G-fan.* For each G-matrix G_t, let

$$\sigma(G_t) := \mathbb{R}_{\geq 0}\mathbf{g}_{1;t} + \mathbb{R}_{\geq 0}\mathbf{g}_{2;t} \tag{4.36}$$

be the cone in $\mathbb{R}^2$ spanned by its g-vectors, which we call a *G-cone*. G-cones intersect only in their boundaries, thereby forming a *fan* (the *G-fan*, or the *g-vector fan* of $\mathbf{\Sigma}$). See Figure 3.

In view of the definitions of C- and G-matrices and F-polynomials, we define the action of a permutation $\sigma \in S_n$ on them as follows:

$$\sigma C_t = C', \quad c'_{ij} := c_{i\sigma^{-1}(j);t}, \quad \sigma G_t = G', \quad g'_{ij} := g_{i\sigma^{-1}(j);t}, \tag{4.37}$$

$$\sigma \mathbf{F}_t = \mathbf{F}', \quad F'_i(\mathbf{y}) := F_{\sigma^{-1}(i);t}(\mathbf{y}). \tag{4.38}$$

The following is a consequence of the separation formulas.

Proposition 4.19. *Let $\mathbf{\Sigma}$ be a cluster pattern with free coefficients at t_0. Let $\mathbf{\Sigma}'$ be a cluster pattern with principal coefficients at t_0 sharing the common B-pattern with $\mathbf{\Sigma}$. Then, for any $t, t' \in \mathbb{T}_n$ and a permutation $\sigma \in S_n$, the following fact holds:*

$$\Sigma_t = \sigma\Sigma_{t'} \iff \Sigma'_t = \sigma\Sigma'_{t'}. \tag{4.39}$$

In other words, the periodicities of $\mathbf{\Sigma}$ and $\mathbf{\Sigma}'$ coincide.

Proof. The implication $\Longrightarrow$ was given in Proposition 2.26. Let us show the implication $\Longleftarrow$. Assume the equality $\Sigma'_t = \sigma\Sigma'_{t'}$. Then, it implies the same periodicity for C- and G-matrices and F-polynomials,

$$C_t = \sigma C_{t'}, \quad G_t = \sigma G_{t'}, \quad \mathbf{F}_t = \sigma \mathbf{F}_{t'}. \tag{4.40}$$

Also, we have $B_t = \sigma B_{t'}$. Then, applying the separation formulas (4.24) and (4.25) to $\mathbf{\Sigma}$, we have

$$\mathbf{x}_t = \sigma \mathbf{x}_{t'}, \quad \mathbf{y}_t = \sigma \mathbf{y}_{t'}. \tag{4.41}$$

□

In summary, through the separation formulas, a cluster pattern of principal coefficients controls a cluster pattern of any coefficients (including free coefficients) sharing the common B-pattern. This is the reason why it is called *principal* in CA4.

4.4 Further results

In the rest of the section, we present without proof some advanced results on C- and G-matrices and F-polynomials, which are fundamentally important in cluster algebra theory.

Definition 4.20. We say that a vector $\mathbf{v} \in \mathbb{Z}^n$ is *positive (resp. negative)* if it is a nonzero vector, and all nonzero components are positive (resp. negative). We say that a square matrix is *column sign-coherent* if each column vector is either positive or negative.

The following theorems were conjectured by Fomin-Zelevinsky [FZ02a, FZ07] and proved by Gross-Hacking-Keel-Kontsevich [GHKK18] by the scattering diagram method.

Theorem 4.21 (Sign-coherence of C-matrices [GHKK18]). *Every c-vector $\mathbf{c}_{i;t}$ is either positive or negative. In other words, every C-matrix C_t is column sign-coherent.*

Theorem 4.22 (Unit constant property [GHKK18]). *Every F-polynomial $F_{i;t}(\mathbf{y})$ has constant term* 1.

Theorem 4.23 (Laurent positivity [GHKK18]). *Every F-polynomial $F_{i;t}(\mathbf{y})$ has no negative coefficients.*

It was shown in [FZ07] that Theorems 4.21 and 4.22 are equivalent to each other.

Proof of equivalence between Theorems 4.21 and 4.22. Let $t, t' \in \mathbb{T}_n$ be k-adjacent. Assume that the constant term of $F_{i;t}(\mathbf{y})$ is 1. Then, by (4.19), the following two conditions are equivalent:

(a). The constant term of $F_{k;t'}(\mathbf{y})$ is 1.

(b). The c-vector $\mathbf{c}_{k;t}$ is either positive or negative.

Therefore, Theorem 4.21 follows from Theorem 4.22. Conversely, Theorem 4.22 can be shown from Theorem 4.21 and the fact $F_{i;t_0}(\mathbf{y}) = 1$ by the induction on $d = d(t_0, t)$. □

There are several important consequences of Theorems 4.21, 4.22, 4.23 together with the separation formulas.

The duality between C- and G-matrices observed in Example 4.18 is a consequence of Theorem 4.21.

Theorem 4.24 (Duality [NZ12]). *Let D be a common skew-symmetrizer of B-pattern $\mathbf{B}$ of $\mathbf{\Sigma}$. Then, the following equality holds:*

$$D^{-1}G_t^T DC_t = I. \tag{4.42}$$

Proof. Thanks to Theorem 4.21, for each $t \in \mathbb{T}_n$ and $k \in \{1, \dots, n\}$, the sign $\varepsilon_{k;t} \in \{1, -1\}$ of the c-vector $\mathbf{c}_{k;t}$ is ambiguously determined. We set $\varepsilon = \varepsilon_{k;t}$ in (4.6) and (4.17). Note that $[-\varepsilon_{k;t} c_{\ell k}]_+ = 0$ for any $\ell = 1, \dots, n$ due to the sign-coherence. Then, the mutation formulas in (4.5) and (4.13) are simplified as follows:

$$c_{ij;t'} = \begin{cases} -c_{ik;t} & j = k, \\ c_{ij;t} + c_{ik;t}[\varepsilon_{k;t} b_{kj;t}]_+ & j \neq k, \end{cases} \tag{4.43}$$

$$g_{ij;t'} = \begin{cases} -g_{ik;t} + \displaystyle\sum_{\ell=1}^{n} g_{i\ell;t}[-\varepsilon_{k;t} b_{\ell k;t}]_+ & j = k, \\ g_{ij;t} & j \neq k. \end{cases} \tag{4.44}$$

They are written in the matrix form

$$C_{t'} = C_t P, \quad G_{t'} = G_t Q, \tag{4.45}$$

$$P = \begin{pmatrix} 1 & & & & & & \\ & \ddots & & & & & \\ & & 1 & & & & \\ [\varepsilon_{k;t} b_{k1;t}]_+ & \cdots & [\varepsilon_{k;t} b_{k,k-1;t}]_+ & -1 & [\varepsilon_{k;t} b_{k,k+1;t}]_+ & \cdots & [\varepsilon_{k;t} b_{kn;t}]_+ \\ & & & & 1 & & \\ & & & & & \ddots & \\ & & & & & & 1 \end{pmatrix}, \tag{4.46}$$

$$Q = \begin{pmatrix} 1 & & & [-\varepsilon_{k;t} b_{1k;t}]_+ & & & \\ & \ddots & & \vdots & & & \\ & & 1 & [-\varepsilon_{k;t} b_{k-1,k;t}]_+ & & & \\ & & & -1 & & & \\ & & & [-\varepsilon_{k;t} b_{k+1,k;t}]_+ & 1 & & \\ & & & \vdots & & \ddots & \\ & & & [-\varepsilon_{k;t} b_{nk;t}]_+ & & & 1 \end{pmatrix}. \tag{4.47}$$

By (1.27), we have

$$DP = Q^T D. \tag{4.48}$$

Also, it is easy to see that

$$P^2 = I. \tag{4.49}$$

Now, we prove (4.42) by the induction on the distance $d = d(t_0, t)$. For $t = t_0$, the claim holds by (4.4) and (4.12). Assume that the claim holds for $d = d(t_0, t)$. Let t' be k-adjacent to t. Then, we have

$$D^{-1}G_{t'}^T DC_{t'} = D^{-1}(Q^T G_t^T)D(C_t P) = P(D^{-1}G_t^T DC_t)P = I. \tag{4.50}$$

□

We have the following immediate consequences of Theorem 4.24.

Proposition 4.25 ([NZ12]). *(a). (Unimodularity)*

$$\det C_t = \det G_t \in \{1, -1\}. \tag{4.51}$$

(b). The matrix B_t is determined by B_{t_0} and C_t as follows:

$$DB_t = C_t^T (DB_{t_0})C_t. \tag{4.52}$$

Proof. (a). Since C_t and G_t are integer matrices, the equality (4.42) implies the claim (4.51). (This can be also shown more directly by (4.42) and the fact $\det P = \det Q = -1$.) (b). By (4.15) and (4.42), we have

$$C_t^T DB_{t_0} C_t = C_t^T DG_t B_t = DB_t. \tag{4.53}$$

□

The following result was shown by Cao-Huang-Li [CHL20] using Theorems 4.21, 4.22, 4.23, 4.24 and the separation formulas all together.

Theorem 4.26 (Detropicalization [CHL20]). *Let $\mathbf{\Sigma}$ be any cluster pattern with a given initial vertex t_0. For any $t, t' \in \mathbb{T}_n$, the following implications hold:*

$$G_t = G_{t'} \quad \Longrightarrow \quad \mathbf{x}_t = \mathbf{x}_{t'}, \tag{4.54}$$

$$C_t = C_{t'} \quad \Longrightarrow \quad \mathbf{y}_t = \mathbf{y}_{t'}. \tag{4.55}$$

In other words, the cluster variables and coefficients are uniquely determined by their tropical parts, respectively.

The above theorem is refined and completed as the following equivalences of periodicities.

Theorem 4.27 (Synchronicity [Nak21]). *Let $\mathbf{\Sigma}$ be any cluster pattern with a given initial vertex t_0. For any $t, t' \in \mathbb{T}_n$ and any permutation $\sigma \in S_n$, the following four conditions are equivalent:*

(a). $G_t = \sigma G_{t'}$.

(b). $C_t = \sigma C_{t'}$.

(c). $\mathbf{x}_t = \sigma \mathbf{x}_{t'}$.

(d). $\Sigma_t = \sigma\Sigma_{t'}$.

Moreover, if $\mathbf{\Sigma}$ *has free or principal coefficients at some* t'_0, *then the above conditions are also equivalent to the following condition:*

(e). $\mathbf{y}_t = \sigma\mathbf{y}_{t'}$.

As a corollary of Theorem 4.27, we obtain the following result, which answer the problem in CA4.

Corollary 4.28 ([Nak21]). *In Proposition 2.26, the opposite implication in* (2.107) *holds for any cluster pattern* $\mathbf{\Sigma}'$.

Proof. By Theorem 4.27, the both conditions in (2.107) is equivalent to the condition $G_t = \sigma G_{t'}$ for their common G-matrices.

□

5 Upper cluster algebras

So far, we mainly consider the properties of seeds and cluster patterns. In this section we study the structure of cluster algebras via the notion of *upper cluster algebras.* All results are taken from CA3.

5.1 Upper cluster algebras

Upper cluster algebras was introduced in CA3. The motivation was two-fold as follows:

- They are useful to study the structure of cluster algebras.
- The coordinate rings of some algebraic varieties (e.g., double Bruhat cells) are isomorphic to some upper cluster algebras, not cluster algebras. In other words, cluster algebras are too small in some situation.

Let $\mathbf{\Sigma}$ be any cluster pattern with coefficients in any semifield $\mathbb{P}$, and let $\mathcal{A} = \mathcal{A}(\mathbf{\Sigma})$ be the associated cluster algebra in the ambient filed $\mathcal{F}$.

Definition 5.1 (Upper cluster algebra). The *upper cluster algebra* $\overline{\mathcal{A}} = \overline{\mathcal{A}}(\mathbf{\Sigma})$ is a $\mathbb{ZP}$-subalgebra of $\mathcal{F}$ defined by

$$\overline{\mathcal{A}} = \bigcap_{t \in \mathbb{T}_n} \mathbb{ZP}[\mathbf{x}_t^{\pm 1}]. \tag{5.1}$$

Thanks to the Laurent phenomenon in Theorem 3.1, any cluster variable $x_{i;t}$ belong to $\overline{\mathcal{A}}$. Therefore,

$$\mathcal{A} \subset \overline{\mathcal{A}}. \tag{5.2}$$

In some cases, $\mathcal{A} = \overline{\mathcal{A}}$ occurs, but, in general, $\mathcal{A} \neq \overline{\mathcal{A}}$.

Let us introduce some related notions.

Definition 5.2. For each $t \in \mathbb{T}_n$, let t_i $(i = 1, \dots, n)$ be the vertex that is i-adjacent to t. We define $\mathbb{ZP}$-subalgebras of $\mathcal{F}$,

$$\mathcal{U}_t = \mathbb{ZP}[\mathbf{x}_t^{\pm 1}] \cap \bigcap_{i=1}^{n} \mathbb{ZP}[\mathbf{x}_{t_i}^{\pm 1}], \tag{5.3}$$

$$\mathcal{L}_t = \mathbb{ZP}[\mathbf{x}_t, \mathbf{x}_{t_1}, \dots, \mathbf{x}_{t_n}]. \tag{5.4}$$

We call them the *upper bound* and *lower bound* of $\mathcal{A}$ at t, respectively.

Clearly, we have

$$\mathcal{L}_t \subset \mathcal{A} \subset \overline{\mathcal{A}} \subset \mathcal{U}_t, \tag{5.5}$$

which explains their names.

We introduce the following technical condition.

Definition 5.3. For a given seed Σ_t and $k = 1, \dots, n$, let

$$P_{k;t} := \frac{1}{1 \oplus y_{k;t}} \left(y_{k;t} \prod_{j=1}^{n} x_{j;t}^{[b_{jk;t}]_+} + \prod_{j=1}^{n} x_{j;t}^{[-b_{jk;t}]_+} \right) \in \mathbb{ZP}[\mathbf{x}_t] \tag{5.6}$$

be the polynomial appearing in the mutation formula (2.5). We say that a seed Σ_t is *coprime* if $P_{1;t}, \dots, P_{n;t}$ are pairwise coprime in $\mathbb{ZP}[\mathbf{x}_t]$, namely, any common factor of $P_{i;t}$ and $P_{j;t}$ $(i \neq j)$ belongs to $\mathbb{ZP}^\times = \{\pm 1\}\mathbb{P}$, where $\mathbb{ZP}^\times$ is the unit group of the ring $\mathbb{ZP}$.

For our purpose, the following sufficient condition for coprimeness is useful.

Lemma 5.4. *Let $\boldsymbol{\Sigma}$ be any cluster pattern with free coefficients at t_0. Then, Σ_t is coprime for any $t \in \mathbb{T}_n$.*

Proof. We see in (5.6) that, up to a multiplicative factor $1 \oplus y_{k;t}$ in $\mathbb{ZP}[\mathbf{x}_t]$, $P_{k;t}$ is a binomial of degree 1 in $y_{k;t}$. Thus, for $\mathbb{P} = \mathbb{Q}_{\mathrm{sf}}(\mathbf{y}_{t_0}) \simeq \mathbb{Q}_{\mathrm{sf}}(\mathbf{y}_t)$, $P_{k;t}$ is clearly irreducible. Therefore, $P_{1;t}, \dots, P_{n;t}$ are pairwise coprime. □

Remark 5.5. In [BFZ05], instead of free coefficients, the following condition is used to guarantee the coprimeness of Σ_t:

- A cluster pattern Σ is of geometric type and its initial extended exchange matrix $\tilde{B}_{t_0}$ has *full* rank.

5.2 Rank 2 case

Let us concentrate on a cluster pattern $\boldsymbol{\Sigma}$ of rank 2 with a given initial vertex t_0. let t_i $(i = 1, 2)$ be the vertex that is i-adjacent to t_0. Let

$$\mathbf{x}_{t_0} = \mathbf{x} = (x_1, x_2), \quad x'_1 = x_{1;t_1}, \; x'_2 = x_{2;t_2}, \tag{5.7}$$

so that

$$\mathcal{U}_{t_0} = \mathbb{ZP}[x_1^{\pm 1}, x_2^{\pm 1}] \cap \mathbb{ZP}[x_1'^{\pm 1}, x_2^{\pm 1}] \cap \mathbb{ZP}[x_1^{\pm 1}, x_2'^{\pm 1}], \tag{5.8}$$
$$\mathcal{L}_{t_0} = \mathbb{ZP}[x_1, x_2, x'_1, x'_2]. \tag{5.9}$$

We take the initial exchange matrix B_{t_0} as in (2.56), where we include the case $a = b = 0$ therein. Then, the mutations at t_0 in directions 1 and 2 are written explicitly as

$$x_1 x'_1 = P_{1;t_0} = p_1^+ x_2^a + p_1^-, \quad x_2 x'_2 = P_{2;t_0} = p_2^+ + p_2^- x_1^b, \tag{5.10}$$
$$p_k^+ = \frac{y_k}{1 \oplus y_k}, \quad p_k^- = \frac{1}{1 \oplus y_k}, \tag{5.11}$$

where we set $\mathbf{y}_{t_0} = \mathbf{y}$. For $a, b > 0$, it is clear that Σ_{t_0} is coprime. For $a = b = 0$, there are cases that Σ_{t_0} is not coprime.

Example 5.6. Suppose that $a = b = 0$ and $y_2 = y_1^3$ in $\mathbb{P}$. Then,

$$P_{1;t_0} = \frac{1 + y_1}{1 \oplus y_1}, \quad P_{2;t_0} = \frac{1 + y_1^3}{1 \oplus y_1^3} \tag{5.12}$$

have a non-monomial common factor $1 + y_1$. Thus, Σ_{t_0} is not coprime.

Proposition 5.7 ([BFZ05]). *Let $\mathbf{\Sigma}$ be any cluster pattern of rank 2 such that the initial seed Σ_{t_0} is coprime. Then, the following equality holds:*

$$\mathcal{L}_{t_0} = \mathcal{U}_{t_0}. \tag{5.13}$$

This further implies that

$$\mathcal{L}_{t_0} = \mathcal{A} = \overline{\mathcal{A}} = \mathcal{U}_{t_0}. \tag{5.14}$$

It is clear that (5.13) implies (5.14) thanks to (5.5).

First we present a consequence of Proposition 5.7.

Theorem 5.8 ([BFZ05]). *Let $\mathbf{\Sigma}$ be any cluster pattern of rank 2 with coefficients in any semifield $\mathbb{P}$. Then, the cluster algebra $\mathcal{A}$ is generated by x_1, x_2, x'_1, x'_2. In particular, it is finitely generated.*

Proof. First we assume that $\mathbf{\Sigma}$ has free coefficients at t_0. Then, by Lemma 5.4, Σ_{t_0} is coprime. Thus, by Proposition 5.7, $\mathcal{A} = \mathcal{L}_{t_0} = \mathbb{ZP}[x_1, x_2, x'_1, x'_2]$. Therefore, the claim holds. This means that any cluster variable $x_{i;t}$ is expressed as a polynomial in x_1, x_2, x'_1, x'_2. This expression holds under any specialization of coefficients including the non-coprime case. (Alternatively, one can prove the claim for the case $a = b = 0$, where the non-coprime case happens, directly from the result in Example 2.16. See also Proposition 5.9.) □

Now we prove Proposition 5.7.

Proof of Proposition 5.7. The goal is to show that (5.8) reduces to (5.9).

First, we show that

$$\mathbb{ZP}[x_1^{\pm 1}, x_2^{\pm 1}] \cap \mathbb{ZP}[{x'_1}^{\pm 1}, x_2^{\pm 1}] = \mathbb{ZP}[x_1, x'_1, x_2^{\pm 1}]. \tag{5.15}$$

The inclusion $\supset$ is clear by (5.10). Let us show the inclusion $\subset$. Let $L \in \mathbb{ZP}[x_1^{\pm 1}, x_2^{\pm 1}]$. Then, we have

$$\begin{aligned} L &= \sum_{m \in \mathbb{Z}} x_1^m Q_m(x_2) \qquad (Q_m(x_2) \in \mathbb{ZP}[x_2^{\pm 1}]) \\ &= \sum_{m \in \mathbb{Z}} {x'_1}^{-m} (p_1^+ x_2^a + p_1^-)^m Q_m(x_2), \end{aligned} \tag{5.16}$$

where the sum is finite. Imposing that L also belongs to $\mathbb{ZP}[x_1'^{\pm1}, x_2^{\pm1}]$, we have, for any $m < 0$,

$$(p_1^+ x_2^a + p_1^-)^m Q_m(x_2) \in \mathbb{ZP}[x_2^{\pm1}]. \tag{5.17}$$

Then,

$$L = \sum_{m\geq 0} x_1^m Q_m(x_2) + \sum_{m<0} x_1'^{-m}(p_1^+ x_2^a + p_1^-)^m Q_m(x_2) \in \mathbb{ZP}[x_1, x_1', x_2^{\pm1}], \tag{5.18}$$

which proves (5.15). By (5.15) and the one obtained by interchanging x_1 and x_2, we obtain the first step reduction

$$\mathcal{U}_{t_0} = \mathbb{ZP}[x_1, x_1', x_2^{\pm1}] \cap \mathbb{ZP}[x_2, x_2', x_1^{\pm1}]. \tag{5.19}$$

Now our goal is to prove the following claim.

Claim. The following equality holds:

$$\mathbb{ZP}[x_1, x_1', x_2^{\pm1}] \cap \mathbb{ZP}[x_2, x_2', x_1^{\pm1}] = \mathbb{ZP}[x_1, x_2, x_1', x_2']. \tag{5.20}$$

The inclusion $\supset$ is clear. We separate the proof of the inclusion $\subset$ into two cases.

Case 1: $a, b > 0$. First we prove the following equality.

$$\mathbb{ZP}[x_1, x_1', x_2^{\pm1}] = \mathbb{ZP}[x_1, x_2, x_1', x_2'] + \mathbb{ZP}[x_1, x_2^{\pm1}]. \tag{5.21}$$

The inclusion $\supset$ is clear. Let us show the inclusion $\subset$. It is enough to prove that, for $k, \ell > 0$,

$$x_1'^k x_2^{-\ell} \in \mathbb{ZP}[x_1, x_1', x_2'] + \mathbb{ZP}[x_1, x_2^{\pm1}]. \tag{5.22}$$

Let $q := -p_2^-/p_2^+$. Then, by (5.10), we have

$$x_2^{-1} \equiv q x_1^b x_2^{-1} \mod \mathbb{ZP}[x_2']. \tag{5.23}$$

Applying it repeatedly, we have, for any $k > 0$,

$$x_2^{-1} \equiv q x_1^b x_2^{-1} \equiv q^2 x_1^{2b} x_2^{-1} \equiv \cdots \equiv q^k x_1^{kb} x_2^{-1} \mod \mathbb{ZP}[x_1, x_2']. \tag{5.24}$$

Therefore,

$$x_2^{-\ell} \in \mathbb{ZP}[x_1, x_2'] + x_1^k \mathbb{ZP}[x_1, x_2^{\pm1}], \tag{5.25}$$

where we used the assumption $b > 0$. Noticing that $x_1' x_1 \in \mathbb{ZP}[x_2]$, we obtain (5.22). Now, applying the equality (5.21) to the left hand side of (5.20), we have

$$\begin{aligned} &\mathbb{ZP}[x_1, x_1', x_2^{\pm1}] \cap \mathbb{ZP}[x_2, x_2', x_1^{\pm1}] \\ &= (\mathbb{ZP}[x_1, x_2, x_1', x_2'] + \mathbb{ZP}[x_1, x_2^{\pm1}]) \cap \mathbb{ZP}[x_2, x_2', x_1^{\pm1}] \\ &= \mathbb{ZP}[x_1, x_2, x_1', x_2'] + (\mathbb{ZP}[x_1, x_2^{\pm1}] \cap \mathbb{ZP}[x_2, x_2', x_1^{\pm1}]). \end{aligned} \tag{5.26}$$

Then, the claim (5.20) follows from the following equality:

$$(\mathbb{ZP}[x_1, x_2^{\pm 1}] \cap \mathbb{ZP}[x_2, x_2', x_1^{\pm 1}]) = \mathbb{ZP}[x_1, x_2, x_2']. \tag{5.27}$$

The inclusion $\supset$ is clear. We show the inclusion $\subset$. Let $L \in \mathbb{ZP}[x_2, x_2', x_1^{\pm 1}]$. We have

$$\begin{aligned} L = \sum_{m,k,l} c_{mk\ell} x_1^m x_2^k x_2'^{\ell} &= \sum_{m,k,l} c_{mk\ell} x_1^m x_2^k x_2^{-\ell} (p_2^+ + p_2^- x_1^b)^{\ell}, \\ &(m \in \mathbb{Z},\ k, \ell \geq 0,\ c_{mk\ell} \in \mathbb{ZP}). \end{aligned} \tag{5.28}$$

Imposing that L also belong to $\mathbb{ZP}[x_1, x_2^{\pm 1}]$, we see that the powers m in the above should be nonnegative. This is the desired result.

Case 2: $a = b = 0$. In this case, one can directly show (5.20). We only need to show the inclusion $\subset$. Let $L \in \mathbb{ZP}[x_2, x_2', x_1^{\pm 1}]$. Then, we have

$$\begin{aligned} L = \sum_{m,k,l} c_{mk\ell} x_1^m x_2^k x_2'^{\ell} &= \sum_{m,k,l} c_{mk\ell} x_1^m x_2^k x_2^{-\ell} (p_2^+ + p_2^-)^{\ell}, \\ &(c_{mk\ell} \in \mathbb{ZP},\ m \in \mathbb{Z},\ k, \ell \geq 0). \end{aligned} \tag{5.29}$$

A similar expression holds for $L' \in \mathbb{ZP}[x_1, x_1', x_2^{\pm 1}]$. It follows that

$$L'' = \sum_{m,k} c_{mk} x_1^m x_2^k \in \mathbb{ZP}[x_1^{\pm 1}, x_2^{\pm 1}], \quad (c_{mk} \in \mathbb{ZP}) \tag{5.30}$$

belongs to the left hand side of (5.20) if the following condition is satisfied:

(i). c_{mk} is divisible by $(p_1^+ + p_1^-)^{-m}$ if $m < 0$,
(ii). c_{mk} is divisible by $(p_2^+ + p_2^-)^{-k}$ if $k < 0$.

Here, we recall that $P_{1;t_0} = p_1^+ + p_1^-$ and $P_{2;t_0} = p_2^+ + p_2^-$ in (5.10) are coprime in $\mathbb{ZP}$ by assumption. (This is the only place where the coprimeness condition is concerned.) Then, the conditions (i) and (ii) ensure the following condition,

(iii). c_{mk} is divisible by $(p_1^+ + p_1^-)^{-m}(p_2^+ + p_2^-)^{-k}$ if $m, k < 0$.

The conditions (i)–(iii) guarantee that L'' belongs to $\mathbb{ZP}[x_1, x_2, x_1', x_2']$ after replacing negative powers of x_i with positive power of $x_i'/(p_i^+ + p_i^-)$.

This completes the proof of Proposition 5.7. □

In the non-coprime case, one can directly verify the following fact, using the result in Example 2.16,

Proposition 5.9. *Let* $\mathbf{\Sigma}$ *be any cluster pattern of rank 2 such that the initial seed* Σ_{t_0} *is not coprime. Then, we have*

$$\mathcal{L}_{t_0} = \mathcal{A} = \overline{\mathcal{A}} \subsetneqq \mathcal{U}_{t_0}. \tag{5.31}$$

Proof. Since Σ_{t_0} is not coprime, we have $a = b = 0$. Then, as shown in Example 2.16, there are four clusters (x_1, x_2), (x'_1, x_2), (x_1, x'_2), (x'_1, x'_2), where

$$x_1 x'_1 = P_{1;t_0} = p_1^+ + p_1^-, \quad x_2 x'_2 = P_{2;t_0} = p_2^+ + p_2^-. \tag{5.32}$$

Therefore, $\mathcal{L}_{t_0} = \mathbb{ZP}[x_1, x_2, x'_1, x'_2] = \mathcal{A}$ holds. By assumption, there is some common factor $R \in \mathbb{ZP}$ of $P_{1;t_0}$ and $P_{2;t_0}$ with $R \notin \mathbb{ZP}^\times$ such that

$$P_{1;t_0} = Q_1 R, \quad P_{2;t_0} = Q_2 R \quad (Q_1, Q_2 \in \mathbb{ZP}). \tag{5.33}$$

We have

$$x_1^{-1} x_2^{-1} Q_1 Q_2 R = x'_1 x_2^{-1} Q_2 = x_1^{-1} x'_2 Q_1 \in \mathcal{U}_{t_0}. \tag{5.34}$$

On the other hand, this element does not belong to $\mathcal{A}$, because one cannot eliminate the negative powers of x_1 and x_2 simultaneously. Therefore, $\mathcal{A} \neq \mathcal{U}_{t_0}$. On the other hand, any element L of $\overline{\mathcal{A}}$ is explicitly written as ($c_{mn} \in \mathbb{ZP}$)

$$\begin{aligned} L = {} & \sum_{m,n\geq 0} c_{mn} x_1^m x_2^n + \sum_{m\geq 0, n<0} c_{mn} P_{2;t_0}^{-n} x_1^m x_2^n \\ & + \sum_{m<0, n\geq 0} c_{mn} P_{1;t_0}^{-m} x_1^m x_2^n + \sum_{m,n<0} c_{mn} P_{1;t_0}^{-m} P_{2;t_0}^{-n} x_1^m x_2^n \\ = {} & \sum_{m,n\geq 0} c_{mn} x_1^m x_2^n + \sum_{m\geq 0, n<0} c_{mn} x_1^m {x'_2}^{-n} \\ & + \sum_{m<0, n\geq 0} c_{mn} {x'_1}^{-m} x_2^n + \sum_{m,n<0} c_{mn} {x'_1}^{-m} {x'_2}^{-n} \in \mathcal{A}. \end{aligned} \tag{5.35}$$

Therefore, $\mathcal{A} = \overline{\mathcal{A}}$. □

5.3 Alternative proof of Laurent phenomenon

As another application of Proposition 5.7 (and the results in the proof), we present alternative proof of the Laurent phenomenon in Section 3.1.

First, we prove the following fact.

Proposition 5.10. *Let $\mathbf{\Sigma}$ be any cluster pattern of rank 2 such that any seed Σ_t is coprime. Then, the following equality holds for any $t \in \mathbb{T}_2$:*

$$\mathcal{U}_t = \mathcal{L}_t = \mathcal{U}_{t_0} = \mathcal{L}_{t_0}. \tag{5.36}$$

In particular, the upper bound $\mathcal{U}_t$ is independent of t.

Proof. Since any seed Σ_t is coprime by assumption, Proposition 5.7 is applicable to any t, so that we have $\mathcal{L}_t = \mathcal{U}_t$. Therefore, it is enough to prove

$$\mathcal{L}_{t_0} = \mathcal{L}_{t_1}, \tag{5.37}$$

where

$$\mathcal{L}_{t_1} = \mathbb{ZP}[x_1, x_2, x'_1, x''_2], \tag{5.38}$$

and x''_2 is the mutation of $x_2 = x_{2;t_1}$ at t_1 in direction 2. Moreover, by the symmetry of x'_2 and x''_2, it is enough to prove that

$$x''_2 \in \mathcal{L}_{t_0} = \mathbb{ZP}[x_1, x_2, x'_1, x'_2]. \tag{5.39}$$

We apply (2.61) with $s = 1$ therein to obtain

$$x_2 x''_2 = q_2^+ {x'_1}^b + q_2^-, \tag{5.40}$$

$$q_2^+ = \frac{y'_2}{1 \oplus y'_2}, \quad q_2^- = \frac{1}{1 \oplus y'_2}, \quad y'_2 = y_2(1 \oplus y_1)^b. \tag{5.41}$$

The following relations hold:

$$q_2^+/q_2^- = y'_2 = (p_2^+/p_2^-)(p_1^-)^{-b}, \tag{5.42}$$

$$x_1 x'_1 = p_1^+ x_2^a + p_1^-, \tag{5.43}$$

$$x_2 x'_2 = p_2^+ + p_2^- x_1^b. \tag{5.44}$$

It follows that

$$\begin{aligned} x''_2 &= \frac{q_2^+ {x'_1}^b + q_2^-}{x_2} = \frac{q_2^+ {x'_1}^b (x_2 x'_2 - p_2^- x_1^b)}{p_2^+ x_2} + \frac{q_2^-}{x_2} \\ &= \frac{q_2^+ {x'_1}^b x'_2}{p_2^+} - \frac{q_2^+ p_2^-}{p_2^+ x_2} \left((x'_1 x_1)^b - \frac{q_2^- p_2^+}{q_2^+ p_2^-} \right) \\ &= \frac{q_2^+ {x'_1}^b x'_2}{p_2^+} - \frac{q_2^+ p_2^-}{p_2^+ x_2} \left((p_1^+ x_2^a + p_1^-)^b - (p_1^-)^b \right). \end{aligned} \tag{5.45}$$

The numerator of the second term in the last expression is divisible by x_2. Thus, we have the property (5.39). □

We once again prove the Laurent phenomenon in Theorem 3.1 based on Proposition 5.10.

Alternative proof of Theorem 3.1. Let $\mathbf{\Sigma}$ be any cluster pattern with rank n. As in the previous proof in Section 3.1, it is enough to prove Theorem 3.1 assuming that $\mathbf{\Sigma}$ has free coefficients at t_0. Then, by Lemma 5.4, any seed Σ is coprime. In particular, the equality (5.36) is applicable for any rank 2 restriction of $\mathbf{\Sigma}$, where the frozen variables are regarded as a part of coefficients. For example, when we consider a cluster cluster pattern restricted in direction 1 and 2, we replace $\mathbb{ZP}$ with $\mathbb{ZP}[x_3^{\pm 1}, \dots, x_n^{\pm 1}]$. Then, in the same notation in the proof of Proposition 5.10, we have the following equality corresponding to (5.37):

$$(\mathbb{ZP}[x_3^{\pm 1},\dots,x_n^{\pm 1}])[x_1,x_2,x_1',x_2'] = (\mathbb{ZP}[x_3^{\pm 1},\dots,x_n^{\pm 1}])[x_1,x_2,x_1',x_2''], \tag{5.46}$$

which is also written as

$$\mathbb{ZP}[x_1,x_2,x_1',x_2',x_3^{\pm 1},\dots,x_n^{\pm 1}] = \mathbb{ZP}[x_1,x_2,x_1',x_2'',x_3^{\pm 1},\dots,x_n^{\pm 1}]. \tag{5.47}$$

Let us prove that the upper bound $\mathcal{U}_t$ of $\mathbf{\Sigma}$ is independent of t. Let t_i ($i=1,\dots,n$) be the vertex that is i-adjacent to t_0. Let $x_i' = x_{i;t_i}$. It is enough to prove $\mathcal{U}_{t_0} = \mathcal{U}_{t_1}$. Indeed, we have

$$\begin{aligned}
\mathcal{U}_{t_0} &= \mathbb{ZP}[\mathbf{x}^{\pm 1}] \cap \bigcap_{i=1}^{n} \mathbb{ZP}[\mathbf{x}_{t_i}^{\pm 1}]\\
&= \bigcap_{i=1}^{n} (\mathbb{ZP}[\mathbf{x}^{\pm 1}] \cap \mathbb{ZP}[\mathbf{x}_{t_i}^{\pm 1}])\\
&= \bigcap_{i=1}^{n} \mathbb{ZP}[x_1^{\pm 1},\dots,x_{i-1}^{\pm 1},x_i,x_i',x_{i+1}^{\pm 1},\dots,x_n^{\pm 1}] \quad \text{(by (5.15))}\\
&= \bigcap_{i=2}^{n} (\mathbb{ZP}[x_1,x_1',x_2^{\pm 1},\dots,x_n^{\pm 1}] \cap \mathbb{ZP}[x_1^{\pm 1},\dots,x_{i-1}^{\pm 1},x_i,x_i',x_{i+1}^{\pm 1},\dots,x_n^{\pm 1}])\\
&= \bigcap_{i=2}^{n} \mathbb{ZP}[x_1,x_1',x_2^{\pm 1},\dots,x_{i-1}^{\pm 1},x_i,x_i',x_{i+1}^{\pm 1},\dots,x_n^{\pm 1}]\\
&= \bigcap_{i=2}^{n} \mathbb{ZP}[x_1,x_1',x_2^{\pm 1},\dots,x_{i-1}^{\pm 1},x_i,x_i'',x_{i+1}^{\pm 1},\dots,x_n^{\pm 1}] \quad \text{(by (5.47))}\\
&= \mathcal{U}_{t_1},
\end{aligned} \tag{5.48}$$

where the last equality is obtained by reversing the above procedure. Then, the upper cluster algebra $\overline{\mathcal{A}}$ is given by

$$\overline{\mathcal{A}} = \bigcap_{t\in\mathbb{T}_n} \mathbb{ZP}[\mathbf{x}_t^{\pm 1}] = \bigcap_{t\in\mathbb{T}_n} \mathcal{U}_t = \mathcal{U}_{t_0}. \tag{5.49}$$

On the other hand, for the cluster algebra $\mathcal{A}$, we have

$$\mathcal{A} = \mathbb{ZP}[\mathbf{x}_t \mid t \in \mathbb{T}_n] \subset \bigcup_{t\in\mathbb{T}_n} \mathcal{U}_t = \mathcal{U}_{t_0}. \tag{5.50}$$

Therefore, we obtain $\mathcal{A} \subset \overline{\mathcal{A}}$. □

5.4 Further results

Let us present generalizations of Proposition 5.7 and Theorem 5.8, which are part of the main result in CA3, without proof.

Definition 5.11. An $n \times n$ skew-symmetrizable matrix B is *acyclic* if the following condition holds:

- For any $3 \leq m \leq n$, there is no cyclic sequence $i_1, i_2, \dots, i_m, i_{m+1} = i_1$ in $\{1, \dots, n\}$ such that $b_{i_s i_{s+1}} > 0$ holds for any $s = 1, \dots, m$.

Example 5.12. (a). Any 2×2 skew-symmetrizable matrix B is acyclic.

(b). Let B be any skew-symmetrizable matrix such that the associated Cartan matrix $A(B)$ is of finite type given in Definition 3.9. Then, B is acyclic because the corresponding Dynkin diagram is a tree.

By a similar technique using upper and lower bounds, a parallel result to Proposition 5.7 and Theorem 5.8 hold for any cluster pattern with an acyclic initial exchange matrix.

Theorem 5.13 ([BFZ05]). *Let $\mathbf{\Sigma}$ be a cluster pattern with coefficients in any semifield $\mathbb{P}$ with a given initial vertex t_0. Suppose that the initial exchange matrix B_{t_0} is acyclic. Then, the following facts hold.*

(a). We have

$$\mathcal{L}_{t_0} = \mathcal{A}. \tag{5.51}$$

In particular, the cluster algebra $\mathcal{A}$ is finitely generated.

(b). If the initial seed Σ_{t_0} is coprime, we have

$$\mathcal{L}_{t_0} = \mathcal{A} = \overline{\mathcal{A}} = \mathcal{U}_{t_0}. \tag{5.52}$$

6 Generalized cluster algebras

To conclude this introductory guide, we present a generalization of cluster algebras introduced by Chekhov and Shapiro as an extra material beyond basics. Essentially all nice properties of cluster algebras are inherited to this generalization. Therefore, they extensively widen the perspective of cluster algebra theory.

6.1 Generalized cluster algebras

Chekhov and Shapiro [CS14] introduced a generalization of cluster algebras called *generalized cluster algebras* (GCA, for short), motivated by some examples naturally appeared in the study of Teichmüller space for the Riemann surface with orbifold points. Moreover, it turned out that they are indeed a natural generalization such that all essential properties of cluster algebras presented in this text are shown or conjectured to hold [CS14, Nak15].

Let us explain the idea shortly. The mutations of ordinary cluster algebras are defined by *binomials* $1 + \hat{y}_k$ and $1 \oplus y_k$ in (2.2) and (2.3). We may ask how/why this binomial condition is essential. It turns out that one can replace it with fairly general *polynomials*, and they are still as good as the ordinary one, *if* we adequately adjust the other parts of mutations.

Definition 6.1 (Mutation data)**.** Let $\mathbb{P}$ be a given semifield. We fix the following *mutation data* $(\mathbf{r}, \mathbf{z})$ in $\mathbb{P}$:

- an n-tuple of positive integers $\mathbf{r} = (r_1, \dots, r_n)$, which are called *mutation degrees.*
- a collection $\mathbf{z} = (z_{i,s})_{i=1,\dots,n; s=0,\dots,r_i}$ of elements in $\mathbb{P}$ satisfying the following conditions:

$$\text{(unit constant term/monic property)} \quad z_{i,0} = z_{i,r_i} = 1, \tag{6.1}$$

$$\text{(reciprocity)} \quad z_{i,s} = z_{i,r_i - s}. \tag{6.2}$$

They are equivalent to give an n-tuple of *monic* and *reciprocal* polynomials in a single formal variable y with coefficients in $\mathbb{P}$,

$$P_{i,\mathbf{z}}(y) = \sum_{s=0}^{r_i} z_{i,s} y^s, \quad (i = 1, \dots, n). \tag{6.3}$$

In the simplest case $\mathbf{r} = (1, \dots, 1)$, they reduce to $P_{i,\mathbf{z}}(y) = 1 + y$, which is the binomial for the ordinary mutations of seeds mentioned above.

Definition 6.2 (Seed mutation for GCA)**.** Let $(\mathbf{r}, \mathbf{z})$ be a mutation data in $\mathbb{P}$. Let $\Sigma = (\mathbf{x}, \mathbf{y}, B)$ be an (ordinary) seed with coefficients in $\mathbb{P}$ in Definition 2.1. For the polynomials in (6.3), let

$$P_{k,\mathbf{z}}(\hat{y}_k) = \sum_{s=0}^{r_k} z_{k,s} \hat{y}_k^s, \quad P_{k,\mathbf{z}}|_{\mathbb{P}}(y_k) = \bigoplus_{s=0}^{r_k} z_{k,s} y_k^s, \tag{6.4}$$

where $\hat{y}_k$ is the ordinary $\hat{y}$-variable for Σ. Then, the $(\mathbf{r},\mathbf{z})$*-mutation of* Σ *in direction* k, $\mu_k(\Sigma)=(\mathbf{x}',\mathbf{y}',B')$, is defined as follows:

$$x'_i = \begin{cases} x_k^{-1}\left(\prod_{j=1}^{n} x_j^{[-b_{jk}r_k]_+}\right)\dfrac{P_{k,\mathbf{z}}(\hat{y}_k)}{P_{k,\mathbf{z}}|_{\mathbb{P}}(y_k)} & i=k,\\ x_i & i\neq k, \end{cases} \tag{6.5}$$

$$y'_i = \begin{cases} y_k^{-1} & i=k,\\ y_i y_k^{[r_k b_{ki}]_+} P_{k,\mathbf{z}}|_{\mathbb{P}}(y_k)^{-b_{ki}} & i\neq k, \end{cases} \tag{6.6}$$

$$b'_{ij} = \begin{cases} -b_{ij} & i=k \text{ or } j=k,\\ b_{ij}+b_{ik}[r_k b_{kj}]_+ + [-b_{ik}r_k]_+ b_{kj} & i,j\neq k. \end{cases} \tag{6.7}$$

It is easy to check the following properties.

Proposition 6.3. *(a). Any skew-symmetrizer D of B is also a skew-symmetrizer of B'.*

(b). The mutation μ_k is an involution. In particular, $(\mathbf{x}',\mathbf{y}',B')$ is a seed.

(c). The $\hat{y}$-variables mutate as

$$\hat{y}'_i = \begin{cases} \hat{y}_k^{-1} & i=k,\\ \hat{y}_i \hat{y}_k^{[r_k b_{ki}]_+} P_{k,\mathbf{z}}(\hat{y}_k)^{-b_{ki}} & i\neq k. \end{cases} \tag{6.8}$$

(d). The ε-expressions, which are parallel to the ones in Proposition 2.7, hold, where we put ε in every $[\,\cdot\,]_+$ and also replace $P_{k,\mathbf{z}}(\hat{y}_k)$ and $P_{k,\mathbf{z}}|_{\mathbb{P}}(y_k)$ with $P_{k,\mathbf{z}}(\hat{y}_k^{\varepsilon})$ and $P_{k,\mathbf{z}}|_{\mathbb{P}}(y_k^{\varepsilon})$ in (6.5)–(6.7).

Proof. (a). This is proved in the same way as Proposition 1.18 (a).

(b). By the reciprocity condition (6.2), we have a parallel formula to (2.6),

$$\frac{P_{k,\mathbf{z}}(\hat{y}_k^{-1})}{P_{k,\mathbf{z}}|_{\mathbb{P}}(y_k^{-1})} = \frac{P_{k,\mathbf{z}}(\hat{y}_k)}{P_{k,\mathbf{z}}|_{\mathbb{P}}(y_k)}\prod_{j=1}^{n} x_j^{-b_{jk}r_k}. \tag{6.9}$$

Then, one can repeat the proofs of Propositions 1.18 (c) and 2.5 (a).

The properties (c) and (d) can be proved in a similar way as Propositions 2.6 and 2.7. □

Remark 6.4. One can lift the reciprocity condition (6.2) by putting the data $\mathbf{z}$ in a seed and introducing its mutation [NR16].

Definition 6.5. For a given mutation data $(\mathbf{r},\mathbf{z})$ in $\mathbb{P}$, a *generalized cluster pattern* $\mathbf{\Sigma}$ with coefficients in $\mathbb{P}$ are defined in the same way as the ordinary one by replacing mutations with $(\mathbf{r},\mathbf{z})$-mutations. The *generalized cluster algebra* $\mathcal{A}$ associated with $\mathbf{\Sigma}$ is defined in the same way as the ordinary one.

The first serious test for this generalization is the Laurent phenomenon, which is the *raison d'etre* of cluster algebras. This indeed holds literally in the same way as the ordinary ones.

Theorem 6.6 (Laurent phenomenon [CS14]). *Let $\mathbf{\Sigma}$ be any generalized cluster pattern with coefficients in any semifield $\mathbb{P}$. Let $t_0, t \in \mathbb{T}_n$ be any vertices. Then, any cluster variable $x_{i;t}$ is expressed as a Laurent polynomial in $\mathbf{x}_{t_0}$ with coefficients in $\mathbb{ZP}$.*

Proof. The proof is also the same as the proof of Theorem 3.1, just by replacing the binomial $1 + \hat{y}_k$ therein with the polynomial $P_{k,\mathbf{z}}(\hat{y}_k)$. □

Example 6.7. Let us consider a rank 2 example. As the simplest nontrivial mutation data $(\mathbf{r}, \mathbf{z})$, we take $\mathbf{r} = (2, 1)$, where the only nontrivial data in $\mathbf{z}$ is $z_{1,1}$. We choose $z_{1,1} = z$ to be arbitrary in $\mathbb{P}$. We also choose the simplest nontrivial initial exchange matrix

$$B_0 = \begin{pmatrix} 0 & -1 \\ 1 & 0 \end{pmatrix}. \tag{6.10}$$

Accordingly,

$$\hat{y}_1 = y_1 x_2, \quad \hat{y}_2 = y_2 x_1^{-1}. \tag{6.11}$$

We note that, for the diagonal matrix $R = \mathrm{diag}(r_1, r_2)$,

$$B_0 R = \begin{pmatrix} 0 & -1 \\ 2 & 0 \end{pmatrix}, \tag{6.12}$$

which is the exchange matrix for an ordinary cluster algebra of type B_2 we studied in Section 2.3. We repeat a similar calculation as in Section 2.3. The result is given below up to $t = 6$.

$$\begin{cases} x_{1;0} = x_1, \\ x_{2;0} = x_2, \end{cases} \qquad \begin{cases} y_{1;0} = y_1, \\ y_{2;0} = y_2, \end{cases} \tag{6.13}$$

$$\begin{cases} x_{1;1} = x_1^{-1} \dfrac{1 + z\hat{y}_1 + \hat{y}_1^2}{1 \oplus z y_1 \oplus y_1^2}, \\ x_{2;1} = x_2, \end{cases} \qquad \begin{cases} y_{1;1} = y_1^{-1}, \\ y_{2;1} = y_2 (1 \oplus z y_1 \oplus y_1^2), \end{cases} \tag{6.14}$$

$$\begin{cases} x_{1;2} = x_1^{-1} \dfrac{1 + z\hat{y}_1 + \hat{y}_1^2}{1 \oplus z y_1 \oplus y_1^2}, \\ x_{2;2} = x_2^{-1} \dfrac{1 + \hat{y}_2 + z\hat{y}_1\hat{y}_2 + \hat{y}_1^2\hat{y}_2}{1 \oplus y_2 \oplus z y_1 y_2 \oplus y_1^2 y_2}, \end{cases} \tag{6.15}$$

$$\begin{cases} y_{1;2} = y_1^{-1}(1 \oplus y_2 \oplus z y_1 y_2 \oplus y_1^2 y_2), \\ y_{2;2} = y_2^{-1}(1 \oplus z y_1 \oplus y_1^2)^{-1}, \end{cases}$$

$$\begin{cases} x_{1;3} = x_1 x_2^{-2} \dfrac{1 + 2\hat{y}_2 + \hat{y}_2^2 + z\hat{y}_1\hat{y}_2 + z\hat{y}_1\hat{y}_2^2 + \hat{y}_1^2\hat{y}_2^2}{1 \oplus 2y_2 \oplus y_2^2 \oplus zy_1y_2 \oplus zy_1y_2^2 \oplus y_1^2y_2^2}, \\ x_{2;3} = x_2^{-1} \dfrac{1 + \hat{y}_2 + z\hat{y}_1\hat{y}_2 + \hat{y}_1^2\hat{y}_2}{1 \oplus y_2 \oplus zy_1y_2 \oplus y_1^2y_2}, \end{cases} \tag{6.16}$$

$$\begin{cases} y_{1;3} = y_1(1 \oplus y_2 \oplus zy_1y_2 \oplus y_1^2y_2)^{-1}, \\ y_{2;3} = y_1^{-2}y_2^{-1}(1 \oplus 2y_2 \oplus y_2^2 \oplus zy_1y_2 \oplus zy_1y_2^2 \oplus y_1^2y_2^2), \end{cases}$$

$$\begin{cases} x_{1;4} = x_1 x_2^{-2} \dfrac{1 + 2\hat{y}_2 + \hat{y}_2^2 + z\hat{y}_1\hat{y}_2 + z\hat{y}_1\hat{y}_2^2 + \hat{y}_1^2\hat{y}_2^2}{1 \oplus 2y_2 \oplus y_2^2 \oplus zy_1y_2 \oplus zy_1y_2^2 \oplus y_1^2y_2^2}, \\ x_{2;4} = x_1 x_2^{-1} \dfrac{1 + \hat{y}_2}{1 \oplus y_2}, \end{cases} \tag{6.17}$$

$$\begin{cases} y_{1;4} = y_1^{-1}y_2^{-1}(1 \oplus y_2), \\ y_{2;4} = y_1^2y_2(1 \oplus 2y_2 \oplus y_2^2 \oplus zy_1y_2 \oplus zy_1y_2^2 \oplus y_1^2y_2^2)^{-1}, \end{cases}$$

$$\begin{cases} x_{1;5} = x_1, \\ x_{1;5} = x_1 x_2^{-1} \dfrac{1 + \hat{y}_2}{1 \oplus y_2}, \end{cases} \qquad \begin{cases} y_{1;5} = y_1y_2(1 \oplus y_2)^{-1}, \\ y_{1;5} = y_2^{-1}, \end{cases} \tag{6.18}$$

$$\begin{cases} x_{1;6} = x_1, \\ x_{2;6} = x_2, \end{cases} \qquad \begin{cases} y_{1;6} = y_1, \\ y_{2;6} = y_2. \end{cases} \tag{6.19}$$

We observe that all characteristic feature of ordinary cluster patterns are preserved. In particular, we see the same periodicity of the ordinary cluster pattern of type B_2 in Section 2.3. In fact, if we formally set $z = 0$ in the above (though such a specialization is prohibited in $\mathbb{P}$) we recover the result for the ordinary cluster pattern of type B_2 therein. This is an example of the general fact that any generalized cluster pattern reduces to some (ordinary) cluster pattern called the *right companion cluster pattern* under such a specialization [NR16].

6.2 Separation formulas for GCA

As one more confirmation of the rightness of GCA, we present the separation formulas for generalized cluster patterns, which are parallel to the ordinary ones.

Definition 6.8. We say that a generalized cluster pattern $\mathbf{\Sigma}$ with mutation data $(\mathbf{r}, \mathbf{z})$ is *with principal coefficients at* $t_0 \in \mathbb{T}_n$ if the following conditions are satisfied:

- The coefficient semifield of $\mathbf{\Sigma}$ is a tropical semifield $\mathrm{Trop}(\mathbf{y}, \mathbf{z})$ with generators $\mathbf{y} = (y_1, \dots, y_n)$, $\mathbf{z} = (z_{i,s})_{i=1,\dots,n;\, s=1,\dots,r_i-1}$ with $z_{i,s} = z_{i,r_i-s}$ as formal variables.
- The coefficient tuple $\mathbf{y}_{t_0}$ at t_0 coincides with $\mathbf{y}$.
- The mutation data $\mathbf{z}$ coincides with $\mathbf{z}$.

We need a stronger version of the Laurent phenomenon, which is parallel to Theorem 3.23.

Theorem 6.9 ([Nak15]). *For any generalized cluster pattern with principal coefficients at t_0, we have*

$$x_{i;t} \in \mathbb{Z}[\mathbf{x}_{t_0}^{\pm 1}, \mathbf{y}, \mathbf{z}]. \tag{6.20}$$

Proof. This is proved by showing a similar claim in the proof of Theorem 3.23 for both variables $\mathbf{y}$ and $\mathbf{z}$, where the property (6.1) of $\mathbf{z}$ is essential. □

For a generalized cluster pattern one can define its (generalized) C- and G-matrices and F-polynomials in two ways:

- Define them through a generalized cluster pattern with principal coefficients at t_0
- Define them by the underlying (generalized) B-pattern $\mathbf{B}$ and t_0.

Below we skip the first definition, which is parallel to the ordinary case, and only give the *second* one. One notable difference to the ordinary case is that the F-polynomials are now polynomials in *both* $\mathbf{y}$ *and* $\mathbf{z}$, where $\mathbf{z}$ are the formal variables $\mathbf{z}$ corresponding to the mutation data $\mathbf{z}$ (under our usual abuse of notations).

Definition 6.10. For a given generalized cluster pattern $\mathbf{\Sigma}$ with a given mutation data $(\mathbf{r}, \mathbf{z})$ and a given initial vertex t_0, the (generalized) C-, G, F-patterns $\mathbf{C}^{t_0}$, $\mathbf{G}^{t_0}$, $\mathbf{F}^{t_0}$ of $\mathbf{\Sigma}$ are uniquely determined by the following initial conditions and the mutation formulas, where t and t' are k-adjacent:

$$C_{t_0} = I, \tag{6.21}$$

$$c_{ij;t'} = \begin{cases} -c_{ik;t} & j = k, \\ c_{ij;t} + c_{ik;t}[r_k b_{kj;t}]_+ + [-c_{ik;t} r_k]_+ b_{kj;t} & j \neq k, \end{cases} \tag{6.22}$$

$$G_{t_0} = I, \tag{6.23}$$

$$g_{ij;t'} = \begin{cases} -g_{ik;t} + \sum_{\ell=1}^{n} g_{i\ell;t}[-b_{\ell k;t} r_k]_+ - \sum_{\ell=1}^{n} b_{i\ell;t_0}[-c_{\ell k;t} r_k]_+ & j = k, \\ g_{ij;t} & j \neq k, \end{cases} \tag{6.24}$$

$$F_{i;t_0}(\mathbf{y}, \mathbf{z}) = 1, \tag{6.25}$$

$$F_{i;t'}(\mathbf{y}, \mathbf{z}) = \begin{cases} \dfrac{M_{k;t}(\mathbf{y}, \mathbf{z})}{F_{k;t}(\mathbf{y}, \mathbf{z})} & i = k, \\ F_{i;t}(\mathbf{y}, \mathbf{z}) & i \neq k, \end{cases} \tag{6.26}$$

where

$$M_{k;t}(\mathbf{y}) = \left(\prod_{j=1}^{n} y_j^{[-c_{jk;t}r_k]_+} \prod_{j=1}^{n} F_{j;t}(\mathbf{y},\mathbf{z})^{[-b_{jk;t}r_k]_+} \right) \times \sum_{s=0}^{r_k} z_{k,s} \left(\prod_{j=1}^{n} y_j^{c_{jk;t}} \prod_{j=1}^{n} F_{j;t}(\mathbf{y},\mathbf{z})^{b_{jk;t}} \right)^s. \tag{6.27}$$

In the first definition of F-polynomials, $F_{i;t}(\mathbf{y},\mathbf{z})$ is defined from the cluster variable $x_{i;t}$ with principal coefficients in t_0 by the specialization $x_1 = \cdots = x_n = 1$. Therefore, thanks to Theorem 6.9, $F_{i;t}(\mathbf{y},\mathbf{z})$ is a polynomial in $\mathbf{y}$ and $\mathbf{z}$. .

As expected, the separation formulas for generalized cluster patterns are given exactly in the same form as the ordinary ones in Theorem 6.11 just by adding the variables $\mathbf{z}$ for F-polynomials.

Theorem 6.11 (Separation Formulas [Nak15]). *Let $\mathbf{\Sigma}$ be any generalized cluster pattern with coefficients in $\mathbb{P}$, mutation data $(\mathbf{r},\mathbf{z})$, and a given initial vertex t_0. Let*

$$\mathbf{x}_{t_0} = \mathbf{x}, \quad \mathbf{y}_{t_0} = \mathbf{y}, \quad \hat{\mathbf{y}}_{t_0} = \hat{\mathbf{y}} \tag{6.28}$$

be the initial cluster variables, coefficients, and $\hat{y}$-variables. Then, the following formulas hold.

$$x_{i;t} = \left(\prod_{j=1}^{n} x_j^{g_{ji;t}} \right) \frac{F_{i;t}(\hat{\mathbf{y}},\mathbf{z})}{F_{i;t}|_{\mathbb{P}}(\mathbf{y},\mathbf{z})}, \tag{6.29}$$

$$y_{i;t} = \left(\prod_{j=1}^{n} y_j^{c_{ji;t}} \right) \prod_{j=1}^{n} F_{j;t}|_{\mathbb{P}}(\mathbf{y},\mathbf{z})^{b_{ji;t}}, \tag{6.30}$$

$$\hat{y}_{i;t} = \left(\prod_{j=1}^{n} \hat{y}_j^{c_{ji;t}} \right) \prod_{j=1}^{n} F_{j;t}(\hat{\mathbf{y}},\mathbf{z})^{b_{ji;t}}. \tag{6.31}$$

Proof. One can repeat the proof of Theorem 4.16 taking care of the modification by the mutation degree $\mathbf{r}$. □

There is one notable feature of GCA compared with the ordinary one. For any generalized cluster pattern with mutation data $(\mathbf{r},\mathbf{z})$, let $\mathbf{B} = \{B_t\}_{t\in\mathbb{T}_n}$ be the (generalized) B-pattern of $\mathbf{\Sigma}$. Also, let $\mathbf{C}^{t_0} = \{C_t\}_{t\in\mathbb{T}_n}$ and $\mathbf{G}^{t_0} = \{G_t\}_{t\in\mathbb{T}_n}$ be the (generalized) C- and G-patterns of $\mathbf{\Sigma}$.

Let $R = (r_i\delta_{ij})_{i,j=1}^n$ be the diagonal matrix whose diagonal entries are given by the mutation degrees. Let $R\mathbf{B} = \{RB_t\}_{t\in\mathbb{T}_n}$ and $\mathbf{B}R = \{B_tR\}_{t\in\mathbb{T}_n}$. Then, one can verify from (6.7) by inspection that both $R\mathbf{B}$ and $\mathbf{B}R$ are B-pattern by the ordinary matrix mutations.

Let ${}^L\mathbf{C}^{t_0} = \{{}^LC_t\}_{t\in\mathbb{T}_n}$ and ${}^L\mathbf{G}^{t_0} = \{{}^LG_t\}_{t\in\mathbb{T}_n}$ be the (ordinary) C- and G-patterns associated with the (ordinary) B-pattern $R\mathbf{B}$. Similarly, let ${}^R\mathbf{C}^{t_0} = \{{}^RC_t\}_{t\in\mathbb{T}_n}$ and ${}^R\mathbf{G}^{t_0} = \{{}^RG_t\}_{t\in\mathbb{T}_n}$ be the (ordinary) C- and G-patterns associated with the (ordinary) B-pattern $\mathbf{B}R$.

Proposition 6.12 ([Nak15]). *The following relations holds:*

$$C_t = {}^LC_t = R({}^RC_t)R^{-1}, \tag{6.32}$$

$$G_t = {}^RG_t = R^{-1}({}^LG_t)R. \tag{6.33}$$

Proof. Let us prove (6.32). The first equality follows from the inspection of (6.22). To obtain the second equality, multiply $r_i^{-1}r_j$ to (6.22). It yields the equality $R^{-1}C_tR = {}^RC_t$. The proof of (6.33) is similar, where we also use (6.32). □

Finally, we present a parallel result to Theorem 4.24.

Theorem 6.13 (Duality [Nak15]). *(a). For a common skew-symmetrizer D of the (ordinary) B-pattern $R\mathbf{B}$, we have the equality*

$$D^{-1}R^{-1}G_t^TRDC_t = I. \tag{6.34}$$

(b). For a common skew-symmetrizer D of the (ordinary) B-pattern $\mathbf{B}R$, we have the equality

$$D^{-1}RG_t^TR^{-1}DC_t = I. \tag{6.35}$$

(c). For a common skew-symmetrizer D of the generalized B-pattern $\mathbf{B}$, we have the equality

$$D^{-1}G_t^TDC_t = I. \tag{6.36}$$

Proof. (a). By applying Theorem 4.24 to ${}^LC_t$ and ${}^LG_t$, we obtain the equality

$$D^{-1}({}^LG_t)^TD({}^LC_t) = I. \tag{6.37}$$

Then, by (6.32) and (6.33), it is written as (6.34).

(b). This is obtained in the same way as (a). Alternatively, note that, if D is a common skew-symmetrizer of the B-pattern $R\mathbf{B}$, DR^2 is a common skew-symmetrizer of the B-pattern $\mathbf{B}R$. Then, (6.35) follows from (6.34).

(c). If D is a common skew-symmetrizer of the B-pattern $R\mathbf{B}$, DR is a common skew-symmetrizer of the generalized B-pattern $\mathbf{B}$. Then, (6.36) follows from (6.34).

□

More results on generalized cluster patterns are found in [Nak21].

Part II
Cluster Patterns and Scattering Diagrams

0 Introduction to Part II

In this part we review some important results by Gross, Hacking, Keel, and Kontsevich [GHKK18] on cluster algebra theory, namely, the *column sign-coherence of C-matrices* and the *Laurent positivity*, both of which were conjectured by Fomin and Zelevinsky [FZ07, FZ02a].

Let us start by recalling the background. In the paper [FZ07] Fomin and Zelevinsky developed a new method to study *cluster patterns*, which are the underlying algebraic and combinatorial structure to define cluster algebras. The basic ingredients of the method are *C-matrices, G-matrices, and F-polynomials* associated with a given cluster pattern. Through the *separation formulas*, the cluster pattern is recovered from them. Moreover, one can regard C- and G-matrices as the *tropical part* of the cluster pattern, while F-polynomials as the *nontropical part*, so that the separation formulas naturally separate these two parts. In many applications of cluster algebras or cluster patterns, this separation plays a central role.

In [FZ07] several conjectures on the properties of C- and G-matrices and F-polynomials were also proposed. Among them, there are two prominent conjectures, namely,

- *The sign-coherence conjecture*: Every C-matrix is column sign-coherent.
- *The Laurent positivity conjecture*: Every F-polynomial has no negative coefficients. (It is equivalent to the earlier conjecture in [FZ02a] on the positivity of the coefficients in the Laurent polynomial expressions for cluster variables.)

Due to their importance, these conjectures, together with other conjectures in [FZ07], have been intensively studied, and proved in some special cases by various methods. Most notably, in the skew-symmetric case, the sign-coherence conjecture was proved by [DWZ10, Pla11, Nag13], and the Laurent positivity conjecture was proved by [LS15, Dav18]. However, the extending these proofs to the most general case, i.e., the skew-symmetrizable case, seemed difficult and demanding some new ideas.

In [GHKK18] Gross, Hacking, Keel, and Kontsevich finally proved both conjectures together in full generality by the *scattering diagram method.* The scattering diagrams were originally introduced in [KS06] to study the toric degenerations of Calabi-Yau manifolds in homological mirror symmetry in two dimensions, and they were generalized to higher dimensions by [GS11].

Roughly speaking, we have the correspondence between the notions in cluster patterns and scattering diagrams as in Table 2. Let us admit this correspondence. Then, by the definition/construction of scattering diagrams, any normal vector of a wall is either positive or negative. This proves the sign-coherence conjecture. On the other hand, any theta function has a manifestly

cluster patterns	scattering diagrams
G-matrices	chambers
C-matrices	normal vectors of walls
F-polynomials	wall-crossing automorphisms
cluster variables	theta functions

Table 2: Correspondence between the notions of cluster patterns and scattering diagrams

positive combinatorial description by *broken lines* as shown in [GHKK18]. This proves the Laurent positivity conjecture. *Therefore, establishing the above correspondence is the crucial step to prove the conjectures.*

In this part we digest and reconstruct the results and the proofs of the above theorems in [GHKK18] still based on their scattering diagram method, however, *without relying on toric/birational geometry therein.* To be more specific, we set the following guidelines while preparing the manuscript:

- We assume that readers are familiar with basic concepts in cluster algebras to some extent, for example, seeds and mutations, cluster variables, coefficients, the Laurent phenomenon, the finite type classification, etc.
- The goal of the part is to give reasonably self-contained proofs of the sign-coherence and the Laurent positivity theorems based on the formulation of cluster patterns by [FZ07] with the help of some key properties of scattering diagrams from [GHKK18].
- At the same time, we also give a detailed account of the correspondence between the notions of cluster patterns and scattering diagrams in Table 2.
- We employ the Fock-Goncharov decomposition of mutations *with tropical sign* to work on F-polynomials. This substitutes and avoids the toric/birational geometrical setting in [GHKK18], including cluster varieties, which plays a key role at several points in their original proofs of the conjectures.
- We rely crucially on some basic properties of scattering diagrams in [GHKK18] *that are independent of the geometrical setting therein.* The details and the proofs of these results will be given in Part III of this monograph.
- Our approach is entirely based on the formulation of cluster patterns in [FZ07]. However, after the paper [FZ07], considerable progress has been made in the subject. So, we also take it into account and reorganize and/or streamline some part of the formulation. Most notably, we do not rely on the notion of principal coefficients explicitly *except for* the polynomial property of F-polynomials.

Most of the results in this part are found in or translated from the known results in [GHKK18], and/or in other related literatures, especially, a series of papers by Reading [RS16, Rea14, Rea20b, Rea20a]. However, the approach, the construction of logic and proofs, and the overall presentation are new. For example, we prove the sign-coherence conjecture inductively along the n-regular tree graph *together with* establishing the correspondence in Table 2 for C- and G-matrices. This approach is completely different from the toric geometrical argument in [GHKK18]. On the other hand, it is close to the approach in [Rea14, Rea20b] using *mutation fans* therein. In fact, many of our results can be also obtained from the results on the mutations fans therein. A subtle but important difference is that we work on C- and G-matrices without referring to mutation fans at all. Therefore, our approach is more direct. To summarize, our presentation is complementary to the above references, and we believe it serves as an alternative guide for the scattering diagram formalism for cluster algebras.

Finally let us mention that there are already several excellent applications of the scattering diagram method to cluster algebra theory (e.g., [Mul16, CGM$^+$17, Bri17, CHL20, CL20, Zho20, Qin19, Mou19, DM21]).

1 Cluster patterns and separation formulas

In this section we recall several basic notions and facts on *cluster patterns* by following the formulation of Fomin and Zelevinsky especially in [FZ07]. Most of the materials are taken or adapted from [FZ07], and we assume that the readers are familiar with them to some extent. Thus, we skip the proofs of the results unless we think necessary.

1.1 Seeds and mutations

Below we fix a positive integer n, which is called the *rank* of the forthcoming seeds, cluster patterns, *etc.*

Definition 1.1 (Skew-symmetrizable matrix). An $n \times n$ integer matrix $B = (b_{ij})_{i,j=1}^n$ is said to be *skew-symmetrizable* if there is a positive rational diagonal matrix $D = (d_i \delta_{ij})_{i,j=1}^n$ such that DB is skew-symmetric, i.e,

$$d_i b_{ij} = -d_j b_{ji} \tag{1.1}$$

holds. The matrix D is called a *(left) skew-symmetrizer* of B.

Remark 1.2. Often a skew-symmetrizer D is assumed to be an *integer* matrix. However, it is natural to use a rational one in order to match with the convention in [GHKK18]. See (2.44). So, we employ the condition here.

For a skew-symmetrizable matrix B, its diagonal entries b_{ii} vanish due to (1.1). Also, the condition (1.1) is rephrased as

$$DBD^{-1} = -B^T, \tag{1.2}$$

where for any matrix M, M^T stands for the transpose of M.

For our purpose, we especially employ the following (nonstandard) definition of seeds.

Definition 1.3 (Seed). Let $\mathcal{F}_X$ and $\mathcal{F}_Y$ be a pair of fields both of which are isomorphic to the rational function field of n-variables with coefficients in $\mathbb{Q}$. A *(labeled) seed in* $(\mathcal{F}_X, \mathcal{F}_Y)$ is a triplet $\Sigma = (\mathbf{x}, \mathbf{y}, B)$, where $\mathbf{x} = (x_1, \dots, x_n)$ and $\mathbf{y} = (y_1, \dots, y_n)$ are n-tuples of free generating elements in $\mathcal{F}_X$ and $\mathcal{F}_Y$, respectively, and $B = (b_{ij})_{i,j=1}^n$ is an $n \times n$ skew-symmetrizable (integer) matrix. We call $\mathbf{x}$, $\mathbf{y}$, and B, respectively, the *x-variables*, the *y-variables*, and the *exchange matrix* of Σ. We also call the pair $(\mathcal{F}_X, \mathcal{F}_Y)$ the *ambient fields* of a seed Σ and the forthcoming cluster patterns, *etc.*

Remark 1.4. In the standard terminology of [FZ07], the x-variables above are *cluster variables without coefficients*, while the y-variables above are *coefficients in the universal semifield* $\mathbb{Q}_{\mathrm{sf}}(\mathbf{y}) \subset \mathcal{F}_Y$. Here, x- and y-variables are on an equal footing and independent of each other. In particular, y-variables do *not* serve as coefficients for x-variables.

For any seed $\Sigma = (\mathbf{x}, \mathbf{y}, B)$, we attach an n-tuple $\hat{\mathbf{y}} = (\hat{y}_1, \dots, \hat{y}_n)$ of elements in the ambient field $\mathcal{F}_X$ defined by

$$\hat{y}_i = \prod_{j=1}^{n} x_j^{b_{ji}}. \tag{1.3}$$

They play an important role in cluster algebra theory. We call them *$\hat{y}$-variables.* They are algebraically independent if and only if B is nonsingular.

For any integer a, we define

$$[a]_+ := \max(a, 0). \tag{1.4}$$

We have a useful identity:

$$a = [a]_+ - [-a]_+. \tag{1.5}$$

Definition 1.5 (Seed mutation). For any seed $\Sigma = (\mathbf{x}, \mathbf{y}, B)$ and $k \in \{1, \dots, n\}$, we define a new seed $\Sigma' = (\mathbf{x}', \mathbf{y}', B')$ by the following rule:

$$x'_i = \begin{cases} x_k^{-1} \left(\prod_{j=1}^{n} x_j^{[-b_{jk}]_+} \right) (1 + \hat{y}_k) & i = k, \\ x_i & i \neq k, \end{cases} \tag{1.6}$$

$$y'_i = \begin{cases} y_k^{-1} & i = k, \\ y_i y_k^{[b_{ki}]_+} (1 + y_k)^{-b_{ki}} & i \neq k, \end{cases} \tag{1.7}$$

$$b'_{ij} = \begin{cases} -b_{ij} & i = k \text{ or } j = k, \\ b_{ij} + b_{ik}[b_{kj}]_+ + [-b_{ik}]_+ b_{kj} & i, j \neq k, \end{cases} \tag{1.8}$$

where $\hat{y}_k$ in (1.6) is the $\hat{y}$-variable in (1.3). The seed Σ' is called the *mutation of Σ in direction k*, and denoted by $\mu_k(\Sigma) = \mu_k(\mathbf{x}, \mathbf{y}, B)$.

The following facts ensure that $(\mathbf{x}', \mathbf{y}', B')$ is indeed a seed.

(a). If D is a skew-symmetrizer of B, then it is also a skew-symmetrizer of B'. In particular, the matrix B' is skew-symmetrizable.

(b). Regard the mutation μ_k as a map from the set of seeds to itself. Then, μ_k is an involution. In particular, $\mathbf{x}'$ and $\mathbf{y}'$ are free generating elements in $\mathcal{F}_X$ and $\mathcal{F}_Y$, respectively.

Also, one can easily verify that $|B'| = |B|$, so that B' is nonsingular if and only if B is nonsingular.

The following fact can be easily verified.

Proposition 1.6 ([FZ07, Prop. 3.9]). *The $\hat{y}$-variables mutate as*

$$\hat{y}'_i = \begin{cases} \hat{y}_k^{-1} & i = k, \\ \hat{y}_i \hat{y}_k^{[b_{ki}]_+} (1 + \hat{y}_k)^{-b_{ki}} & i \neq k \end{cases} \tag{1.9}$$

in the ambient field $\mathcal{F}_X$ *by the same rule as* (1.7).

There is some flexibility to write the mutation formulas (1.6)–(1.8) as follows. (The choice $\varepsilon = 1$ reduces to the one in (1.6)–(1.8).)

Proposition 1.7 ([Nak12, Eqs. (2.9), (2.10)], [BFZ05, Eq. (3.1)]). *Let* $\varepsilon \in \{1, -1\}$. *Then, the right hand sides of the following equalities do not depend on the choice of* ε:

$$x'_i = \begin{cases} x_k^{-1}\left(\prod_{j=1}^{n} x_j^{[-\varepsilon b_{jk}]_+}\right)(1+\hat{y}_k^{\varepsilon}) & i = k, \\ x_i & i \neq k, \end{cases} \tag{1.10}$$

$$y'_i = \begin{cases} y_k^{-1} & i = k, \\ y_i y_k^{[\varepsilon b_{ki}]_+}(1+y_k^{\varepsilon})^{-b_{ki}} & i \neq k, \end{cases} \tag{1.11}$$

$$b'_{ij} = \begin{cases} -b_{ij} & i = k \text{ or } j = k, \\ b_{ij} + b_{ik}[\varepsilon b_{kj}]_+ + [-\varepsilon b_{ik}]_+ b_{kj} & i, j \neq k. \end{cases} \tag{1.12}$$

The claim can be easily checked with the identity (1.5). We call the above formulas the ε-*expressions* of mutations.

1.2 Cluster patterns

Let $\mathbb{T}_n$ be the *n-regular tree* graph, that is, a graph without cycles such that exactly n-edges are attached to each vertex. We assume that the edges are labeled by 1,…, n such that the edges attached to each vertex have different labels. By abusing the notation, the set of vertices of $\mathbb{T}_n$ is also denoted by $\mathbb{T}_n$. We say that a pair of vertices in t and t' in $\mathbb{T}_n$ are (k-)*adjacent*, or t' is (k-)*adjacent to* t, if they are connected with an edge labeled by k.

Definition 1.8 (Cluster pattern/B-pattern). A collection of seeds $\mathbf{\Sigma} = \{\Sigma_t = (\mathbf{x}_t, \mathbf{y}_t, B_t)\}_{t\in\mathbb{T}_n}$ in $(\mathcal{F}_X, \mathcal{F}_Y)$ indexed by $\mathbb{T}_n$ is called a *cluster pattern* if, for any vertices $t, t' \in \mathbb{T}_n$ that are k-adjacent, the equality $\Sigma_{t'} = \mu_k(\Sigma_t)$ holds. The collection of exchange matrices $\mathbf{B} = \{B_t\}_{t\in\mathbb{T}_n}$ extracted from a cluster pattern $\mathbf{\Sigma}$ is called the B-*pattern* of $\mathbf{\Sigma}$.

We arbitrary choose a distinguished vertex (the *initial vertex*) t_0 in $\mathbb{T}_n$. Any cluster pattern $\mathbf{\Sigma} = \{\Sigma_t = (\mathbf{x}_t, \mathbf{y}_t, B_t)\}_{t\in\mathbb{T}_n}$ is uniquely determined from the *initial seed* Σ_{t_0} at the initial vertex t_0 by repeating mutations.

Note that one can take a common skew-symmetrizer D for all B_t in $\mathbf{B}$. By (1.2), we have

$$DB_tD^{-1} = -B_t^T. \tag{1.13}$$

In particular, if B_t is skew-symmetric for some t, it is skew-symmetric for any t. Such a cluster pattern or a B-pattern is said to be *skew-symmetric*.

Similarly, if B_t is nonsingular for some t, it is nonsingular for any t. Such a cluster pattern or a B-pattern is said to be *nonsingular*.

For a seed $\Sigma_t = (\mathbf{x}_t, \mathbf{y}_t, B_t)$ $(t \in \mathbb{T}_n)$ in a cluster pattern $\mathbf{\Sigma}$, we use the notation

$$\mathbf{x}_t = (x_{1;t}, \dots, x_{n;t}), \quad \mathbf{y}_t = (y_{1;t}, \dots, y_{n;t}), \quad B_t = (b_{ij;t})_{i,j=1}^{n}. \tag{1.14}$$

For the initial seed $\Sigma_{t_0} = (\mathbf{x}_{t_0}, \mathbf{y}_{t_0}, B_{t_0})$, we often drop the indices t_0 as

$$\mathbf{x}_{t_0} = \mathbf{x} = (x_1, \dots, x_n), \quad \mathbf{y}_{t_0} = \mathbf{y} = (y_1, \dots, y_n), \quad B_{t_0} = B = (b_{ij})_{i,j=1}^{n}, \tag{1.15}$$

if there is no confusion. We also use similar notations for $\hat{y}$-variables $\hat{\mathbf{y}}_t = (\hat{y}_{1;t}, \dots, \hat{y}_{n;t})$ and the initial $\hat{y}$-variables $\hat{\mathbf{y}}_{t_0} = \hat{\mathbf{y}} = (\hat{y}_1, \dots, \hat{y}_n)$.

We recall the definition of cluster algebras, though the focus of this part is to study the properties of *cluster patterns* rather than cluster algebras.

Definition 1.9 (Cluster algebra)**.** For a cluster pattern $\mathbf{\Sigma}$, the associated *cluster algebra* $\mathcal{A}(\mathbf{\Sigma})$ is the $\mathbb{Z}$-subalgebra of the ambient field $\mathcal{F}_X$ generated by all x-variables (cluster variables) $x_{i,t}$ $(i = 1, \dots, n; t \in \mathbb{T}_n)$.

1.3 C-matrices, G-matrices, and F-polynomials

Following [FZ07], we introduce C-matrices, G-matrices, and F-polynomials. They play a crucial role in the study of cluster patterns.

1.3.1 Matrix notations

We start by introducing the following matrix notations (see, e.g., [NZ12]). For an $n \times n$ matrix A, let $[A]_+$ denote the matrix obtained from A by replacing the entry a_{ij} with $[a_{ij}]_+$. Let $A^{\bullet k}$ denote the matrix obtained from A by replacing the entry a_{ij} $(j \neq k)$ with 0. Similarly, let $A^{k\bullet}$ denote the matrix obtained from A by replacing the entry a_{ij} $(i \neq k)$ with 0.

For any $n \times n$ matrices, the following facts are easily checked, where for the last two equalities we use (1.5).

$$AB^{\bullet k} = (AB)^{\bullet k}, \quad A^{k\bullet}B = (AB)^{k\bullet}, \tag{1.16}$$

$$A^{\bullet k}B = A^{\bullet k}B^{k\bullet} = AB^{k\bullet}, \tag{1.17}$$

$$A[B]_+ + [-A]_+B = A[-B]_+ + [A]_+B, \tag{1.18}$$

$$\text{If } AB = CD, \text{ then } A[B]_+ - C[D]_+ = A[-B]_+ - C[-D]_+. \tag{1.19}$$

Let J_k be the $n \times n$ diagonal matrix obtained from the identity matrix by replacing the kth diagonal entry with -1. The following properties will be repeatedly used.

Lemma 1.10. *Let A be any $n \times n$ matrix whose diagonal entries are 0. Let $P = J_k + A^{k\bullet}$ or $J_k + A^{\bullet k}$. Then, we have*

$$A^{\bullet k} J_k = -A^{\bullet k}, \quad J_k A^{\bullet k} = A^{\bullet k}, \quad A^{k\bullet} J_k = A^{k\bullet}, \quad J_k A^{k\bullet} = -A^{k\bullet}, \tag{1.20}$$

$$P^2 = I, \tag{1.21}$$

$$|P| = -1. \tag{1.22}$$

Proof. Let us prove (1.21) for $P = J_k + A^{k\bullet}$. By $a_{kk} = 0$, we have

$$\begin{aligned} P^2 &= J_k^2 + J_k A^{k\bullet} + A^{k\bullet} J_k + A^{k\bullet} A^{k\bullet} \\ &= I - A^{k\bullet} + A^{k\bullet} = I. \end{aligned} \tag{1.23}$$

Also, we have $|P| = |J_k| = -1$. □

Example 1.11. The first case of the matrix mutation (1.8) is written as

$$(B')^{k\bullet} = -B^{k\bullet}, \quad (B')^{\bullet k} = -B^{\bullet k}. \tag{1.24}$$

Also, the entire mutation (1.8) is written as [BFZ05, Eq. (3.2)]

$$B' = (J_k + [-B]_+^{\bullet k}) B (J_k + [B]_+^{k\bullet}). \tag{1.25}$$

By (1.21), we have the relation

$$(J_k + [-B]_+^{\bullet k}) B' = B (J_k + [B]_+^{k\bullet}). \tag{1.26}$$

1.3.2 C-matrices

Let $\mathbf{\Sigma}$ be any cluster pattern, and let $\mathbf{B}$ be the B-pattern of $\mathbf{\Sigma}$. Let $t_0 \in \mathbb{T}_n$ be a given initial vertex.

We introduce a collection of matrices $\mathbf{C}^{t_0} = \mathbf{C} = \{C_t\}_{t \in \mathbb{T}_n}$ called a *C-pattern,* which is uniquely determined from $\mathbf{B}$ and t_0, therefore, eventually only from t_0 and B_{t_0}.

Definition 1.12 (C-matrices). *The C-matrices $C_t = (c_{ij;t})_{i,j=1}^n$ ($t \in \mathbb{T}_n$) of a cluster pattern $\mathbf{\Sigma}$ (or a B-pattern $\mathbf{B}$) with an initial vertex t_0* are $n \times n$ integer matrices that are uniquely determined by the following initial condition and the mutation rule:

$$C_{t_0} = I, \tag{1.27}$$

$$C_{t'} = C_t J_k + C_t [B_t]_+^{k\bullet} + [-C_t]_+^{\bullet k} B_t, \tag{1.28}$$

or more explicitly,

$$c_{ij;t'} = \begin{cases} -c_{ik;t} & j = k, \\ c_{ij;t} + c_{ik;t}[b_{kj;t}]_+ + [-c_{ik;t}]_+ b_{kj;t} & j \neq k, \end{cases} \tag{1.29}$$

where t and t' are k-adjacent. Each column vector $\mathbf{c}_{i;t} = (c_{ji;t})_{j=1}^n$ of a matrix C_t is called a *c-vector.*

Remark 1.13. The C-matrices appeared implicitly as the lower halves of the extended exchange matrices for the principal coefficients in [FZ07, Def. 3.1].

The mutation (1.28), in particular, implies

$$\mathbf{c}_{k;t'} = -\mathbf{c}_{k;t}, \tag{1.30}$$

or equivalently,

$$C_{t'}^{\bullet k} = -C_t^{\bullet k}. \tag{1.31}$$

One can easily check that the mutation (1.29) is involutive using (1.5), (1.8), and (1.30). Alternatively, in the matrix form, by multiplying J_k from the right to (1.28), we have

$$C_{t'} J_k = C_t + C_t^{\bullet k}[B_t]_+^{k\bullet} + [-C_t]_+^{\bullet k} B_t^{k\bullet}. \tag{1.32}$$

By (1.24) and (1.31), this can be written as

$$C_t = C_{t'} J_k + C_{t'}[-B_{t'}]_+^{k\bullet} + [C_{t'}]_+^{\bullet k} B_{t'}. \tag{1.33}$$

Then, by (1.18), we have

$$C_t = C_{t'} J_k + C_{t'}[B_{t'}]_+^{k\bullet} + [-C_{t'}]_+^{\bullet k} B_{t'}, \tag{1.34}$$

which is the desired result. Moreover, by (1.18) again, we have the following ε-expression of (1.28):

Proposition 1.14 ([NZ12, Eq. (2.4)]). *Let $\varepsilon \in \{1, -1\}$. Then, the right hand side of the following equality does not depend on the choice of ε:*

$$C_{t'} = C_t J_k + C_t[\varepsilon B_t]_+^{k\bullet} + [-\varepsilon C_t]_+^{\bullet k} B_t. \tag{1.35}$$

1.3.3 G-matrices

Under the same assumption of C-matrices, we introduce another collection of matrices $\mathbf{G}^{t_0} = \mathbf{G} = \{G_t\}_{t \in \mathbb{T}_n}$ called a *G-pattern*, which is uniquely determined from $\mathbf{B}$, t_0, and $\mathbf{C}^{t_0}$ defined above.

Definition 1.15 (G-matrices). The *G-matrices $G_t = (g_{ij;t})_{i,j=1}^n$ ($t \in \mathbb{T}_n$) of a cluster pattern $\mathbf{\Sigma}$ (or a B-pattern $\mathbf{B}$) with an initial vertex t_0* are $n \times n$ integer matrices that are uniquely determined by the following initial condition and the mutation rule:

$$G_{t_0} = I, \tag{1.36}$$
$$G_{t'} = G_t J_k + G_t[-B_t]_+^{\bullet k} - B_{t_0}[-C_t]_+^{\bullet k}, \tag{1.37}$$

or more explicitly,

$$g_{ij;t'} = \begin{cases} -g_{ik;t} + \sum_{\ell=1}^{n} g_{i\ell;t}[-b_{\ell k;t}]_+ - \sum_{\ell=1}^{n} b_{i\ell;t_0}[-c_{\ell k;t}]_+ & j = k, \\ g_{ij;t} & j \neq k, \end{cases} \tag{1.38}$$

where t and t' are k-adjacent. Each column vector $\mathbf{g}_{i;t} = (g_{ji;t})_{j=1}^{n}$ of a matrix G_t is called a *g-vector*.

Remark 1.16. The g-vectors were originally introduced as the degree vectors of cluster variables with principal coefficients in [FZ07].

The first nontrivial property so far is the following relation between C- and G-matrices.

Proposition 1.17 ([FZ07, Eq. (6.14)]). *The following equality holds for any* $t \in \mathbb{T}_n$.

(First duality)

$$G_t B_t = B_{t_0} C_t. \tag{1.39}$$

The proof in [FZ07, Eq. (6.14)] relied on the fact that g-vectors are the degree vectors of cluster variables with principal coefficients. We give an alternative proof based on the definitions of C- and G-matrices here. Let $d(t, t')$ denote the distance in $\mathbb{T}_n$ between two vertices $t, t' \in \mathbb{T}_n$.

Proof. We prove it by the induction on t along $\mathbb{T}_n$ from t_0. For $t = t_0$, the equality (1.39) holds because $C_{t_0} = G_{t_0} = I$. Suppose that the equality (1.39) holds for $t \in \mathbb{T}_n$ such that $d(t_0, t) = d$. Let t' be the vertex that is k-adjacent to t such that $d(t_0, t') = d + 1$. Then, we have

$$\begin{aligned} G_{t'} B_{t'} &= G_t(J_k + [-B_t]_+^{\bullet k})B_{t'} - B_{t_0}[-C_t]^{\bullet k} B_{t'}^{k\bullet} \\ &= G_t B_t(J_k + [B_t]_+^{k\bullet}) + B_{t_0}[-C_t]^{\bullet k} B_t^{k\bullet} \\ &= B_{t_0} C_t(J_k + [B_t]_+^{k\bullet}) + B_{t_0}[-C_t]^{\bullet k} B_t \\ &= B_{t_0} C_{t'}, \end{aligned} \tag{1.40}$$

where in the second equality we used (1.24) and (1.26). ☐

Remark 1.18. There has been no name on this equality so far in the literature; however, due to its importance, we name it as above, where the duality means a relation between C- and G-matrices. (Later we will have the second and the third dualities, which are more nontrivial.)

Thanks to (1.19) and (1.39), one can easily check that the mutation (1.37) is involutive as before. Moreover, by (1.19) and (1.39) again, we have the following ε-expression of (1.37).

Proposition 1.19 ([FZ07, Eqs. (6.12), (6.13)]). *Let $\varepsilon \in \{1, -1\}$. Then, the right hand side of the following equality does not depend on the choice of ε:*

$$G_{t'} = G_t J_k + G_t[-\varepsilon B_t]_+^{\bullet k} - B_{t_0}[-\varepsilon C_t]^{\bullet k}. \tag{1.41}$$

1.3.4 F-polynomials

Let $\mathbf{y} = (y_1, \dots, y_n)$ be an n-tuple of formal variables. Although it is not necessary at this moment, it is natural to identify them with the initial y-variables $\mathbf{y}_{t_0}$ of $\mathbf{\Sigma}$ as we will see soon. Under the same assumption of C-matrices, we introduce a collection of rational functions $\mathbf{F}^{t_0} = \mathbf{F} = \{F_{i;t}(\mathbf{y})\}_{i=1,\dots,n;\, t\in\mathbb{T}_n}$ in $\mathbf{y}$ called a *F-pattern,* which is uniquely determined from $\mathbf{B}$, t_0, and $\mathbf{C}^{t_0}$ as follows.

Definition 1.20 (F-polynomials). The *F-polynomials $F_{i;t}(\mathbf{y})$ ($i = 1, \dots, n$; $t \in \mathbb{T}_n$) of a cluster pattern $\mathbf{\Sigma}$ (or a B-pattern $\mathbf{B}$) with an initial vertex t_0* are rational functions in $\mathbf{y}$ with coefficients in $\mathbb{Q}$ that are uniquely determined by the following initial condition and the mutation rule:

$$F_{i;t_0}(\mathbf{y}) = 1, \tag{1.42}$$

$$F_{i;t'}(\mathbf{y}) = \begin{cases} \dfrac{M_{k;t}(\mathbf{y})}{F_{k;t}(\mathbf{y})} & i = k, \\ F_{i;t}(\mathbf{y}) & i \neq k, \end{cases} \tag{1.43}$$

where t and t' are k-adjacent, and $M_{k,t}(\mathbf{y})$ is a rational function in $\mathbf{y}$ defined as follows:

$$M_{k;t}(\mathbf{y}) = \prod_{j=1}^{n} y_j^{[c_{jk;t}]_+} \prod_{j=1}^{n} F_{j;t}(\mathbf{y})^{[b_{jk;t}]_+} + \prod_{j=1}^{n} y_j^{[-c_{jk;t}]_+} \prod_{j=1}^{n} F_{j;t}(\mathbf{y})^{[-b_{jk;t}]_+}. \tag{1.44}$$

In the situation in (1.43), we have

$$M_{k;t}(\mathbf{y}) = M_{k;t'}(\mathbf{y}). \tag{1.45}$$

Therefore, the mutation (1.43) is involutive. We do not discuss the ε-expression for (1.43), because we do not use it here.

The F-polynomials were originally introduced as the specialized x-variables with principal coefficients under the specialization $x_1 = \cdots = x_n = 1$ in [FZ07]. The Laurent phenomenon applied to them yields the following important fact, which justifies the name *F-polynomials.*

Theorem 1.21 ([FZ07, Prop. 3.6]). *For any $i = 1, \dots, n$ and $t \in \mathbb{T}_n$, the rational function $F_{i;t}(\mathbf{y})$ is a polynomial in $\mathbf{y}$ with coefficients in $\mathbb{Z}$.*

Remark 1.22. In case we need to clarify the choice of the initial vertex t_0, we write C-matrices and others as, $C_t^{t_0} = (c_{ij;t}^{t_0})_{i,j=1}^n$, $\mathbf{c}_{i;t}^{t_0}$, $G_t^{t_0} = (g_{ij;t}^{t_0})_{i,j=1}^n$, $\mathbf{g}_{i;t}^{t_0}$, $F_{i;t}^{t_0}(\mathbf{y})$. Conversely, if there is no confusion, we suppress the superscript t_0 as above for simplicity.

1.4 Separation formulas and tropicalization

Let us present one of the most fundamental results on cluster patterns in [FZ07].

Theorem 1.23 (Separation Formulas [FZ07, Prop. 3.13, Cor. 6.3]). *Let $\mathbf{\Sigma}$ be any cluster pattern, and let $t_0 \in \mathbb{T}_n$ be a given initial vertex. Let $\mathbf{C}$, $\mathbf{G}$, and $\mathbf{F}$ be the C-, G-, and F-patterns of $\mathbf{\Sigma}$ with the initial vertex t_0. Let $\mathbf{x}$, $\mathbf{y}$, and $\hat{\mathbf{y}}$ be the initial x-, y-variables, and $\hat{y}$-variables for the initial seed Σ_{t_0}. Then, the following formulas hold.*

$$x_{i;t} = \left(\prod_{j=1}^{n} x_j^{g_{ji;t}} \right) F_{i;t}(\hat{\mathbf{y}}), \tag{1.46}$$

$$y_{i;t} = \left(\prod_{j=1}^{n} y_j^{c_{ji;t}} \right) \prod_{j=1}^{n} F_{j;t}(\mathbf{y})^{b_{ji;t}}. \tag{1.47}$$

Proof. Here is an alternative proof to the one in [FZ07]. Using the mutations of C- and G-matrices, and F-polynomials, one can easily verify (1.46) and (1.47) by the induction on t along $\mathbb{T}_n$ from t_0 with the help of (1.39). We leave the details as an exercise to the readers. □

Since $\hat{y}$-variables mutate exactly in the same manner as y-variables, a parallel formula holds for them:

$$\hat{y}_{i;t} = \left(\prod_{j=1}^{n} \hat{y}_j^{c_{ji;t}} \right) \prod_{j=1}^{n} F_{j;t}(\hat{\mathbf{y}})^{b_{ji;t}}. \tag{1.48}$$

Alternatively, one can derive it from (1.46) with the help of the duality (1.39).

Thank to the separation formulas, the study of cluster patterns reduces to the study of C- and G-matrices, and F-polynomials. Moreover, the formulas play a central role in the interplay between the cluster algebraic structure and various subjects.

The original meaning of the separation formula in [FZ07, Cor. 6.3] is that a general version of the formula (1.46) for *x-variables with coefficients* naturally "separates" the additions of x-variables and coefficients. Unfortunately, we cannot observe it in (1.46), because our x-variables are *without coefficients.* Below we attach an additional meaning of "separation" to the formulas in view of the *tropicalization.*

First, we briefly recall the notion of tropicalization in the context of semifields [FZ07].

Definition 1.24. A *semifield* $\mathbb{P}$ is a multiplicative abelian group equipped with a binary operation $\oplus$ on $\mathbb{P}$ that is commutative, associative, and distributive, i.e., $(a \oplus b)c = ac \oplus bc$. The operation $\oplus$ is called the *addition* in $\mathbb{P}$.

Here we exclusively use the following two semifields.

Example 1.25. (a). *Universal semifield* $\mathbb{Q}_{\mathrm{sf}}(\mathbf{u})$. Let $\mathbf{u} = (u_1, \dots, u_n)$ be an n-tuple of commutative variables. We say that a rational function $f(\mathbf{u}) \in \mathbb{Q}(\mathbf{u})$ in u has a *subtraction-free expression* if it is expressed as $f(\mathbf{u}) = p(\mathbf{u})/q(\mathbf{u})$, where $p(\mathbf{u})$ and $q(\mathbf{u})$ are nonzero polynomials in $\mathbf{u}$ whose coefficients are nonnegative. Let $\mathbb{Q}_{\mathrm{sf}}(\mathbf{u})$ be the set of all rational functions in $\mathbf{u}$ having subtraction-free expressions. Then, $\mathbb{Q}_{\mathrm{sf}}(\mathbf{u})$ is a semifield by the usual multiplication and addition in $\mathbb{Q}(\mathbf{u})$.

(b). *Tropical semifield* $\mathrm{Trop}(\mathbf{u})$. Let $\mathbf{u} = (u_1, \dots, u_n)$ be an n-tuple of commutative variables. Let $\mathrm{Trop}(\mathbf{u})$ be the set of all Laurent monomials of $\mathbf{u}$ with coefficient 1, which is a multiplicative abelian group by the usual multiplication. We define the addition $\oplus$ by

$$\prod_{i=1}^{n} u_i^{a_i} \oplus \prod_{i=1}^{n} u_i^{b_i} := \prod_{i=1}^{n} u_i^{\min(a_i, b_i)}. \tag{1.49}$$

Then, $\mathrm{Trop}(\mathbf{u})$ becomes a semifield. The addition $\oplus$ is called the *tropical sum*.

Consider $\mathbb{Q}_{\mathrm{sf}}(\mathbf{u})$ and $\mathrm{Trop}(\mathbf{u})$ with common generating variables $\mathbf{u} = (u_1, \dots, u_n)$. Then, we have a unique semifield homomorphism

$$\pi_{\mathrm{trop}} : \mathbb{Q}_{\mathrm{sf}}(\mathbf{u}) \to \mathrm{Trop}(\mathbf{u}) \tag{1.50}$$

such that $\pi_{\mathrm{trop}}(u_i) = u_i$ for any $i = 1, \dots, n$. We call it the *tropicalization homomorphism*. Roughly speaking, it extracts the "leading monomial" of a rational function $f(\mathbf{u})$ in $\mathbb{Q}_{\mathrm{sf}}(\mathbf{u})$.

Example 1.26. For $\mathbf{u} = (u_1, u_2, u_3)$,

$$\pi_{\mathrm{trop}} \left(\frac{3u_1u_2^2u_3^2 + 2u_1^2u_2u_3}{3u_2^2 + u_1^2u_2^2 + u_1u_2^3u_3} \right) = \frac{u_1u_2u_3}{u_2^2} = u_1u_2^{-1}u_3. \tag{1.51}$$

Now let $\mathbf{\Sigma}$ be any cluster pattern, and let $t_0 \in \mathbb{T}_n$ be a given initial vertex. Let $\mathbf{y}_{t_0} = \mathbf{y}$ be the initial y-variables in the ambient field $\mathcal{F}_Y$. Then, the semifield generated by $\mathbf{y}$ in $\mathcal{F}_Y$ (by the usual multiplication and addition) is identified with the universal semifield $\mathbb{Q}_{\mathrm{sf}}(\mathbf{y})$. Moreover, any y-variable $y_{i;t}$ belongs to $\mathbb{Q}_{\mathrm{sf}}(\mathbf{y})$ because the mutation (1.7) is a subtraction-free operation. Let us apply the tropicalization homomorphism $\pi_{\mathrm{trop}}^{t_0} : \mathbb{Q}_{\mathrm{sf}}(\mathbf{y}) \to \mathrm{Trop}(\mathbf{y})$ to y-variables.

Definition 1.27 (Tropical y-variable). We call the image $\pi_{\mathrm{trop}}^{t_0}(y_{i;t})$ a *tropical y-variable* with respect to the initial vertex t_0.

The following fact can be easily deduced from (1.43) by the induction on t along $\mathbb{T}_n$ from t_0.

Proposition 1.28 ([FZ07, Eq. (5.5)]).

$$\pi_{\mathrm{trop}}^{t_0}(F_{i;t}(\mathbf{y})) = 1. \tag{1.52}$$

Remark 1.29. The equality (1.52) does not mean the polynomial $F_{i;t}(\mathbf{y})$ has a nonzero constant term. For example, $\pi_{\mathrm{trop}}(y_1 + y_2) = 1$.

Tropical y-variables are naturally identified with c-vectors as follows:

Proposition 1.30 ([FZ07, Eq. (3.14)]). *We have*

$$\pi_{\mathrm{trop}}^{t_0}(y_{i;t}) = y^{\mathbf{c}_{i;t}} := \prod_{j=1}^{n} y_j^{c_{ji;t}}. \tag{1.53}$$

Proof. This follows from the separation formula (1.47) and (1.52). □

Remark 1.31. In fact, (1.53) is the definition of c-vectors in [FZ07, Eq. (5.8)], where the tropical y-variables are identified with *principal coefficients*.

Now looking back the separation formula (1.47) for y-variables, we see that it naturally "separates" the *tropical* and the *nontropical* parts of y-variables.

On the other hand, the tropicalization of x-variables with respect to the initial x-variables $\mathbf{x}$ does not yield g-vectors; rather, it yields the so-called *d-vectors* (*denominator vectors* in [FZ07]), which are the leading exponents of the Laurent polynomial expression of x-variables. However, if we regard the operation of setting all F-polynomials $F_{i;t}(\mathbf{y})$ to 1 also as "a kind of tropicalization" for x-variables, then the separation formula (1.46) for x-variables can be viewed again as separating the "tropical" and the "nontropical" parts of x-variables, where a g-vector $\mathbf{g}_{i;t} = (g_{ji;t})_{j=1}^n$ is identified with a *tropical x-variable* defined by

$$x^{\mathbf{g}_{i;t}} := \prod_{j=1}^{n} x_j^{g_{ji;t}}. \tag{1.54}$$

In the same token, we call

$$\hat{y}^{\mathbf{c}_{i;t}} := \prod_{j=1}^{n} \hat{y}_j^{c_{ji;t}} \tag{1.55}$$

a *tropical $\hat{y}$-variable* with respect to the initial vertex t_0. Thanks to the duality (1.39), we see that the notion is compatible with a tropical x-variable as

$$\hat{y}^{\mathbf{c}_{i;t}} = \prod_{j=1}^{n} (x^{\mathbf{g}_{j;t}})^{b_{ji;t}}. \tag{1.56}$$

Alternatively, based on the duality (1.39), we introduce a *$\hat{c}$-vector* $\hat{\mathbf{c}}_{i;t}$ by

$$\hat{\mathbf{c}}_{i;t} = B_{t_0}\mathbf{c}_{i;t} = G_t\mathbf{b}_{i;t}, \tag{1.57}$$

where $\mathbf{b}_{i;t}$ is the ith column vector of B_t. In other words, it is the ith column vector of a *$\hat{C}$-matrix* defined by

$$\hat{C}_t := B_{t_0}C_t = G_tB_t. \tag{1.58}$$

Then, the tropical $\hat{y}$-variable in (1.56) is expressed also as

$$\hat{y}^{\mathbf{c}_{i;t}} = x^{\hat{\mathbf{c}}_{i;t}}. \tag{1.59}$$

These expressions and notions play a major role in the scattering diagram method later.

1.5 Sign-coherence of C-matrices and Laurent positivity

In the paper [FZ07] several conjectures on the properties of cluster patterns were proposed. Here we recall especially two (three, precisely speaking) conjectures among them, which are the theme of this part.

Let $\mathbf{C}$ and $\mathbf{F}$ be the C- and F-patterns of any cluster pattern $\mathbf{\Sigma}$ with a given initial point $t_0 \in \mathbb{T}_n$.

The first conjecture is the following:

Conjecture 1.32 (Unit constant property [FZ07, Conj. 5.4]). *Every F-polynomial $F_{i;t}(\mathbf{y})$ has constant term 1.*

This seemingly innocent conjecture turned out to be difficult to prove simply by the induction on t along $\mathbb{T}_n$. However, it was noticed by [FZ07] that Conjecture 1.32 is equivalent to another remarkable conjecture on C-matrices and c-vectors.

Definition 1.33 (Row/Column sign-coherence). We say that a vector $\mathbf{v} \in \mathbb{Z}^n$ is *positive* (resp. *negative*) if it is nonzero vector and all nonzero components are positive (resp. negative). We say that a matrix M is *column sign-coherent* (resp. *row sign-coherent*) if each column vector (resp. row vector) of M is either positive or negative.

Conjecture 1.34 (Column sign-coherence of C-matrices [FZ07, Prop. 5.6]). *Every C-matrix C_t is column sign-coherent. Equivalently, every c-vector $\mathbf{c}_{i;t}$ is either positive or negative.*

The equivalence between Conjectures 1.32 and 1.34 easily follows from the mutation of F-polynomials (1.43) [FZ07, Prop. 5.6].

The sign-coherence conjecture was proved in the skew-symmetric case by [DWZ10, Pla11, Nag13] with the representation/categorical methods, and in general by [GHKK18] with the scattering diagram method.

Theorem 1.35 ([GHKK18, Cor. 5.5]). *Conjectures 1.32 and 1.34 hold for any* $\mathbf{\Sigma}$ *and* t_0.

The second conjecture we consider is the following:

Conjecture 1.36 (Laurent positivity [FZ07, §3], [FZ02a, §3]). *Every F-polynomial* $F_{i;t}(\mathbf{y}) \in \mathbb{Z}[\mathbf{y}]$ *has no negative coefficients.*

Through the separation formula, this is equivalent to the earlier conjecture in [FZ02a] on the positivity of the coefficients in the Laurent polynomial expressions for x-variables. This conjecture was proved for surface type by [MSW11], for acyclic type by [KQ14], in the skew-symmetric case by [LS15, Dav18], with various methods, and in general by [GHKK18] with the scattering diagram method.

Theorem 1.37 ([GHKK18, Theorem 4.10]). *Conjecture 1.36 holds for any* $\mathbf{\Sigma}$ *and* t_0.

From these conjectures (now theorems) many other conjectures in [FZ07] and new results follow (e.g., [NZ12, CHL20, CL20, Nak21]). Therefore, they are considered to be in the heart of cluster algebra theory.

As mentioned in Introduction, the purpose of the part is to digest and reconstruct the proofs of these conjectures in [GHKK18]. Therefore, we temporally forget Theorems 1.35 and 1.37, and keep Conjectures 1.32, 1.34, and 1.36 as conjectures until we prove them here.

Remark 1.38. Theorem 1.35 was proved in [GHKK18] with some toric geometrical setting. In contrast, we prove Conjecture 1.34 without any toric geometrical setting.

2 More about C- and G-matrices

In this section we study more about C- and G-matrices, namely, the tropical part of a cluster pattern. Here we only consider a B-pattern $\mathbf{B}$, and we do not refer to a cluster pattern $\mathbf{\Sigma}$. Therefore, we regard that C- and G-matrices are directly associated with a B-pattern $\mathbf{B}$.

2.1 Mutations and second duality

Let $\mathbf{B}$ be any B-pattern, and let $\mathbf{C}^{t_0} = \mathbf{C}$ and $\mathbf{G}^{t_0} = \mathbf{G}$ are the C- and G-patterns of $\mathbf{B}$ with a given initial vertex t_0. In view of Proposition 1.30, we introduce the following notion.

Definition 2.1 (Tropical sign). Suppose that a c-vector $\mathbf{c}_{i;t}^{t_0} = \mathbf{c}_{i;t}$ is either positive or negative. Then, the common sign $\varepsilon_{i;t}^{t_0} = \varepsilon_{i;t} \in \{1, -1\}$ of all nonzero components of $\mathbf{c}_{i;t}$ is called the *tropical sign* of $\mathbf{c}_{i;t}$.

Example 2.2. (a). For the initial vertex t_0, we have $C_{t_0} = I$. Therefore, we have

$$\varepsilon_{i;t_0} = 1, \quad (i = 1, \dots, n). \tag{2.1}$$

(b). Let $t, t' \in \mathbb{T}_n$ be vertices that are k-adjacent. Then, by (1.30), we have

$$\varepsilon_{k;t'} = -\varepsilon_{k;t}. \tag{2.2}$$

Proposition 2.3. *Suppose that the sign-coherence conjecture holds. Then, the following facts hold.*

(a). For any $t \in \mathbb{T}_n$, we have
(Unimodularity [NZ12, Prop. 4.2])

$$|C_t| = |G_t| \in \{1, -1\}, \tag{2.3}$$

(Second duality [NZ12, Eq. (3.11)])

$$D^{-1} G_t^T D C_t = I, \tag{2.4}$$

or, equivalently,

$$D^{-1} C_t^T D G_t = I, \tag{2.5}$$

where D is a common skew-symmetrizer of $\mathbf{B}$.

(b). For any $t, t' \in \mathbb{T}_n$ that are k-adjacent, we have
(Mutations [NZ12, Prop. 1.3])

$$C_{t'} = C_t(J_k + [\varepsilon_{k;t} B_t]_+^{k\bullet}), \tag{2.6}$$

$$G_{t'} = G_t(J_k + [-\varepsilon_{k;t} B_t]_+^{\bullet k}). \tag{2.7}$$

Since we need some details of the proof later, we present it below.

Proof. We first prove (b). By the column sign-coherence of C_t, the tropical sign $\varepsilon_{k;t}$ is defined for any k. Then, we have

$$[-\varepsilon_{k;t}C_t]_+^{\bullet k} = O. \tag{2.8}$$

We set $\varepsilon = \varepsilon_{k;t}$ in the ε-expressions of mutations (1.35) and (1.41). The resulting relations are (2.6) and (2.7).

We prove (a) by the induction on t along $\mathbb{T}_n$ starting from t_0. Let us introduce the following statement:

$(a)_d$. The claim (a) holds for any $t \in \mathbb{T}_n$ such that $d(t_0, t) = d$.

First, $(a)_0$ holds because $C_{t_0} = G_{t_0} = I$. Next, suppose that $(a)_d$ holds for some d. Let $t, t' \in \mathbb{T}_n$ be vertices that are k-adjacent such that $d(t_0, t) = d$ and $d(t_0, t') = d+1$. Let us temporally set the matrices in (2.6) and (2.7) as

$$P = J_k + [\varepsilon_{k;t}B_t]_+^{k\bullet}, \quad Q = J_k + [-\varepsilon_{k;t}B_t]_+^{\bullet k}.$$

Then, by Lemma 1.10, we have

$$P^2 = I, \quad |P| = |Q| = -1. \tag{2.9}$$

Thus, by the induction assumption $|C_t| = |G_t| = \pm 1$, we have $|C_{t'}| = |G_{t'}| = \mp 1$. Also, by (1.13), we have

$$DPD^{-1} = Q^T. \tag{2.10}$$

Using the induction assumption $D^{-1}(G_t)^T DC_t = I$, we have

$$D^{-1}G_{t'}^T DC_{t'} = D^{-1}(Q^T G_t^T)D(C_t P) = P(D^{-1}G_t^T DC_t)P = P^2 = I. \tag{2.11}$$

Therefore, $(a)_{d+1}$ holds. □

Remark 2.4. For the later use, we record the following facts in the above proof.

- To prove (b), we only use the sign-coherence of C_t for t therein.
- To prove $(a)_d \Longrightarrow (a)_{d+1}$, we only use the sign-coherence of C_t such that $d(t_0, t) = d$.

The following expressions of formulas (2.6) and (2.7) by c- and g-vectors are also useful.

Proposition 2.5. *Suppose that the sign-coherence conjecture holds. Then, we have*

$$\mathbf{c}_{i;t'} = \begin{cases} -\mathbf{c}_{k;t} & i = k, \\ \mathbf{c}_{i;t} + [\varepsilon_{k;t}b_{ki;t}]_+\mathbf{c}_{k;t} & i \neq k, \end{cases} \tag{2.12}$$

$$\mathbf{g}_{i;t'} = \begin{cases} -\mathbf{g}_{k;t} + \sum_{j=1}^{n} [-\varepsilon_{k;t} b_{jk;t}]_+ \mathbf{g}_{j;t} & i = k, \\ \mathbf{g}_{i;t} & i \neq k. \end{cases} \tag{2.13}$$

We note that the matrices C_t and B_{t_0} in turn determine B_t.

Proposition 2.6 ([NZ12, Eq. (2.9)]). *Suppose that the sign-coherence conjecture holds. Then, the following equality holds:*

$$C_t^T (DB_{t_0}) C_t = DB_t. \tag{2.14}$$

Proof. By the dualities (1.39) and (2.5), we have

$$C_t^T DB_{t_0} C_t = C_t^T DG_t B_t = D(D^{-1} C_t^T DG_t) B_t = DB_t. \tag{2.15}$$

□

2.2 Dual mutations and third duality

Let $\mathbf{B} = \{B_t\}_{t \in \mathbb{T}_n}$ be any B-pattern, and let t_0 be a given initial vertex. Since we now vary the initial vertex below, we write the corresponding C- and G-matrices as $C_t^{t_0}$ and $G_t^{t_0}$ as in Remark 1.22.

We note that the mutation (1.12) is compatible with the matrix transpose. Namely, if $B' = \mu_k(B)$, then $B'^T = \mu_k(B^T)$. Therefore, $\mathbf{B}^T := \{B_t^T\}_{t \in \mathbb{T}_n}$ is also a B-pattern, which is called the *transpose of* $\mathbf{B}$. We temporarily write the associated C- and G-matrices as $\tilde{C}_t^{t_0}$, $\tilde{G}_t^{t_0}$.

We borrow the following notation from [FG19]: For any matrix M, $\varepsilon_{k\bullet}(M)$ (resp. $\varepsilon_{\bullet k}(M)$) stands for the common sign of the kth row (resp. kth column) assuming that it is either positive or negative. For example, the tropical sign $\varepsilon_{i;t}$ in Definition 2.1 is written as $\varepsilon_{\bullet i}(C_t^{t_0})$.

Proposition 2.7. *Suppose that the sign-coherence conjecture holds. Then, the following facts hold:*

(a). For any $t_0, t \in \mathbb{T}_n$, we have

(Third duality [NZ12, Eq. (1.13)])

$$C_t^{t_0} = (\tilde{G}_{t_0}^t)^T, \tag{2.16}$$

$$G_t^{t_0} = (\tilde{C}_{t_0}^t)^T. \tag{2.17}$$

In particular, each G-matrix $G_t^{t_0}$ is row sign-coherent, and the equality

$$\varepsilon_{k\bullet}(G_t^{t_0}) = \varepsilon_{\bullet k}(\tilde{C}_{t_0}^t) \tag{2.18}$$

holds.

(b). For any $t_0, t_1, t \in \mathbb{T}_n$ such that t_0 and t_1 are k-adjacent, we have

(Dual mutations [NZ12, Prop. 1.4, Eq. (4.2)], [FG19, Prop. 3.6])

$$C_t^{t_1} = (J_k + [-\varepsilon_{k\bullet}(G_t^{t_0})B_{t_0}]_+^{k\bullet})C_t^{t_0}, \tag{2.19}$$

$$G_t^{t_1} = (J_k + [\varepsilon_{k\bullet}(G_t^{t_0})B_{t_0}]_+^{\bullet k})G_t^{t_0}. \tag{2.20}$$

Here, $\varepsilon_{k\bullet}(G_t^{t_0})$ *is well-defined thanks to (a).*

Again, we need some details of the proof later. Furthermore, the statement and the proof in [NZ12] are a little indirect in the present context. Thus, we give a more transparent proof, still in the spirit of [NZ12].

Proof. We prove the claims by the induction on the distance $d(t_0, t)$ in $\mathbb{T}_n$, where we vary t_0 for (a) and t for (b).

Let us introduce the following claims:

$(a)_d$. The claim (a) holds for any $t_0, t \in \mathbb{T}_n$ such that $d(t_0, t) = d$.

$(b)_d$. The claim (b) holds for any $t_0, t_1, t \in \mathbb{T}_n$ such that $d(t_0, t) = d$ and t_0 and t_1 are k-adjacent.

We prove the claims in the following order,

$$(a)_0 \Longrightarrow (b)_0 \Longrightarrow (a)_1 \Longrightarrow (b)_1 \Longrightarrow (a)_2 \Longrightarrow \cdots, \tag{2.21}$$

assuming all preceding claims.

First, we show that $(a)_0$ and $(b)_0$ hold. The claim $(a)_0$, where $t_0 = t$, trivially holds, because all relevant matrices are the identity matrix I. Let us prove $(b)_0$, where $t_0 = t$. By assumption, t_0 and t_1 are k-adjacent. Then, by Proposition 2.3 (b) and (1.24), we have

$$C_{t_0}^{t_1} = C_{t_1}^{t_1}(J_k + [B_{t_1}]_+^{k\bullet}) = (J_k + [-B_{t_0}]_+^{k\bullet})C_{t_0}^{t_0}, \tag{2.22}$$

$$G_{t_0}^{t_1} = G_{t_1}^{t_1}(J_k + [-B_{t_1}]_+^{\bullet k}) = (J_k + [B_{t_0}]_+^{\bullet k})G_{t_0}^{t_0}. \tag{2.23}$$

Thus, $(b)_0$ holds.

Next, assuming the claims in (2.21) up to $(b)_d$, we show $(a)_{d+1}$. Assume $d(t_0, t) = d$ and $d(t_1, t) = d+1$ in (2.19). Take the transpose of (2.19). Then, we apply (2.16), which is valid thanks to $(a)_d$, to its right hand side. We have

$$(C_t^{t_1})^T = \tilde{G}_{t_0}^t (J_k + [-\varepsilon_{\bullet k}(\tilde{C}_{t_0}^t)B_{t_0}^T]_+^{\bullet k}). \tag{2.24}$$

The right hand side is $\tilde{G}_{t_1}^t$ by (2.7). Thus, we have (2.19). The other case is similar.

Finally, assuming the claims in (2.21) up to $(a)_{d+1}$, we show $(b)_{d+1}$. This is the nontrivial part. It is enough to concentrate on the following situation. Let t_0, t_1, t, $t' \in \mathbb{T}_n$ be vertices such that $d(t_0, t) = d$, $d(t_0, t') = d+1$, where

t_1 and t_0 are k-adjacent, while t and t' are ℓ-adjacent. They are depicted as follows, where t_1 could be between t_0 and t.

By (2.6) and (2.19) with $(b)_d$, we obtain

$$\begin{aligned} C_{t'}^{t_1} &= C_t^{t_1}(J_\ell + [\varepsilon_{\bullet\ell}(C_t^{t_1})B_t]_+^{\ell\bullet}) \\ &= (J_k + [-\varepsilon_{k\bullet}(G_t^{t_0})B_{t_0}]_+^{k\bullet})C_t^{t_0}(J_\ell + [\varepsilon_{\bullet\ell}(C_t^{t_1})B_t]_+^{\ell\bullet}). \end{aligned} \tag{2.25}$$

By $(a)_{d+1}$, the sign $\varepsilon_{k\bullet}(G_{t'}^{t_0})$ is well-defined. Let us temporally assume the identity:

$$\begin{aligned} &(J_k + [-\varepsilon_{k\bullet}(G_t^{t_0})B_{t_0}]_+^{k\bullet})C_t^{t_0}(J_\ell + [\varepsilon_{\bullet\ell}(C_t^{t_1})B_t]_+^{\ell\bullet}) \\ &= (J_k + [-\varepsilon_{k\bullet}(G_{t'}^{t_0})B_{t_0}]_+^{k\bullet})C_t^{t_0}(J_\ell + [\varepsilon_{\bullet\ell}(C_t^{t_0})B_t]_+^{\ell\bullet}). \end{aligned} \tag{2.26}$$

Then, the right hand side of (2.25) reduces to $(J_k + [-\varepsilon_{k\bullet}(G_{t'}^{t_0})B_{t_0}]_+^{k\bullet})C_{t'}^{t_0}$, which is the desired result for (2.19) of $(b)_{d+1}$. Similarly, from the identity

$$\begin{aligned} &(J_k + [\varepsilon_{k\bullet}(G_t^{t_0})B_{t_0}]_+^{\bullet k})G_t^{t_0}(J_\ell + [-\varepsilon_{\bullet\ell}(C_t^{t_1})B_t]_+^{\bullet\ell}) \\ &= (J_k + [\varepsilon_{k\bullet}(G_{t'}^{t_0})B_{t_0}]_+^{\bullet k})G_t^{t_0}(J_\ell + [-\varepsilon_{\bullet\ell}(C_t^{t_0})B_t]_+^{\bullet\ell}), \end{aligned} \tag{2.27}$$

we obtain (2.20) of $(b)_{d+1}$. It remains to prove (2.26) and (2.27). To do it, we need to know a precise relation between signs $\varepsilon_{k\bullet}(G_t^{t_0})$ and $\varepsilon_{k\bullet}(G_{t'}^{t_0})$, and also $\varepsilon_{\bullet\ell}(C_t^{t_1})$ and $\varepsilon_{\bullet\ell}(C_t^{t_0})$.

The following facts hold:

(i). By (2.7), $G_t^{t_0}$ and $G_{t'}^{t_0}$ differ only in the ℓth column.

(ii). By $(b)_d$, $C_t^{t_0}$ and $C_t^{t_1}$ differ only in the kth row.

(iii). By the duality (2.4), the kth row of $G_t^{t_0}$ has only a unique component ε at ℓth column if and only the ℓth column of $C_t^{t_0}$ has a unique nonzero component ε' at kth row. Furthermore, if it happens, $\varepsilon = \varepsilon' \in \{1, -1\}$, thanks to the unimodularity (2.3).

Now we consider two cases.

Case 1. Suppose that the kth row of $G_t^{t_0}$ has a nonzero component at some column other than the ℓth column. Then, from Facts (i)–(iii) above and the row and column sign-coherence of G- and C-matrices in $(a)_{d+1}$, we conclude that

$$\varepsilon_{k\bullet}(G_{t'}^{t_0}) = \varepsilon_{k\bullet}(G_t^{t_0}), \quad \varepsilon_{\bullet\ell}(C_t^{t_1}) = \varepsilon_{\bullet\ell}(C_t^{t_0}). \tag{2.28}$$

Therefore, (2.26) and (2.27) hold trivially.

Case 2. Suppose that the kth row of $G_t^{t_0}$ has only one nonzero component that is at the ℓth column. Let us write this component as $\varepsilon \in \{1, -1\}$. As mentioned in Fact (iii), ε coincides with the only one nonzero component in the ℓth column of $C_t^{t_0}$ that is at the kth row. Then, by (2.7) and $(b)_d$, we have

$$\varepsilon_{k\bullet}(G_{t'}^{t_0}) = -\varepsilon, \quad \varepsilon_{k\bullet}(G_t^{t_0}) = \varepsilon, \quad \varepsilon_{\bullet\ell}(C_t^{t_1}) = -\varepsilon, \quad \varepsilon_{\bullet\ell}(C_t^{t_0}) = \varepsilon. \tag{2.29}$$

Let us prove (2.26), which is now written as

$$\begin{aligned} &(J_k + [-\varepsilon B_{t_0}]_+^{k\bullet}) C_t^{t_0} (J_\ell + [-\varepsilon B_t]_+^{\ell\bullet}) \\ &= (J_k + [\varepsilon B_{t_0}]_+^{k\bullet}) C_t^{t_0} (J_\ell + [\varepsilon B_t]_+^{\ell\bullet}). \end{aligned} \tag{2.30}$$

By (1.21), it is rewritten as

$$\begin{aligned} &(J_k + [\varepsilon B_{t_0}]_+^{k\bullet})(J_k + [-\varepsilon B_{t_0}]_+^{k\bullet}) C_t^{t_0} \\ &= C_t^{t_0} (J_\ell + [\varepsilon B_t]_+^{\ell\bullet})(J_\ell + [-\varepsilon B_t]_+^{\ell\bullet}). \end{aligned} \tag{2.31}$$

After some manipulation, it reduces to

$$(B_{t_0} C_t^{t_0})^{k\bullet} = C_t^{t_0} (B_t)^{\ell\bullet}. \tag{2.32}$$

By the assumption, we have $(C_t^{t_0})^{\bullet\ell} = (G_t^{t_0})^{k\bullet} = \varepsilon E_{k\ell}$, where $E_{k\ell}$ is the matrix whose entries are zero except for the (k, ℓ)-entry that is 1. Then, the right hand side of (2.32) can be rewritten as

$$(C_t^{t_0})^{\bullet\ell} B_t = \varepsilon E_{k\ell} B_t = (G_t^{t_0})^{k\bullet} B_t = (G_t^{t_0} B_t)^{k\bullet}. \tag{2.33}$$

Thus, (2.32) reduces to the duality (1.39). Therefore, (2.26) is proved. Similarly, skipping details a little, (2.27) reduces to

$$(B_{t_0})^{\bullet k} G_t^{t_0} = (G_t^{t_0} B_t)^{\bullet\ell}. \tag{2.34}$$

Then, by repeating a similar argument, again (2.34) reduces to the duality (1.39). Therefore, (2.27) is proved. □

Remark 2.8. For the later use we record the following facts in the above proof.

- To prove $(a)_d$, $(b)_d \Longrightarrow (a)_{d+1}$, we only use the sign-coherence of $\tilde{C}_{t_0}^t$ such that $d(t_0, t) = d$.
- To prove $(b)_d$, $(a)_{d+1} \Longrightarrow (b)_{d+1}$, we only use the sign-coherence of $C_t^{t_0}$, $C_t^{t_1}$, $\tilde{C}_{t_0}^t$, $\tilde{C}_{t_0}^{t'}$ such that $d(t_0, t) = d$, $d(t_0, t') = d + 1$, $d(t_1, t) = d \pm 1$.

Remark 2.9. Combining three dualities (1.39), (2.4), (2.16), we have various relations among C- and G-matrices therein. For example, by (1.39) and (2.16), we have

$$B_{t_0} C_t^{t_0} = (B_t^T \tilde{C}_{t_0}^t)^T, \quad G_t^{t_0} B_{t_0} = (\tilde{G}_{t_0}^t B_t^T)^T. \tag{2.35}$$

The $\hat{C}$-matrices in (1.58) mutate as a "hybrid" of C- and G-matrices in the following sense.

Proposition 2.10. *Suppose that the sign-coherence conjecture holds. Let* $\hat{C}_t^{t_0} = B_{t_0} C_t^{t_0}$ *and* $\hat{C}_t^{t_1} = B_{t_1} C_t^{t_1}$. *Then, for any* $t_0, t_1, t, t' \in \mathbb{T}_n$ *such that* t_0 *and* t_1, *and also,* t *and* t' *are* k*-adjacent, respectively, we have*

$$\hat{C}_{t'}^{t_0} = \hat{C}_t^{t_0}(J_k + [\varepsilon_{\bullet k}(C_t^{t_0}) B_t]_+^{k\bullet}), \tag{2.36}$$

$$\hat{C}_t^{t_1} = (J_k + [\varepsilon_{k\bullet}(G_t^{t_0}) B_{t_0}]_+^{\bullet k}) \hat{C}_t^{t_0}. \tag{2.37}$$

Proof. The first formula is obtained from (1.58) by multiplying B_{t_0} from the left. The second one is obtained from (2.20) by multiplying B_t from the right. □

2.3 Principal extension of B-patterns

For any B-pattern $\mathbf{B}$ of rank n and a given initial vertex $t_0 \in \mathbb{T}_n$, we introduce a $2n \times 2n$ skew-symmetrizable matrix

$$\overline{B}_{t_0} = \begin{pmatrix} B_{t_0} & -I \\ I & O \end{pmatrix}, \tag{2.38}$$

which is called the *principal extension of* B_{t_0}. It is important that $\overline{B}_{t_0}$ is nonsingular. Choose any point $\overline{t}_0 \in \mathbb{T}_{2n}$. We naturally identify $\mathbb{T}_n$ as a subtree of $\mathbb{T}_{2n}$ by identifying t_0 with $\overline{t}_0$, and also the edges with labels $k = 1, \dots, n$ in $\mathbb{T}_n$ and $\mathbb{T}_{2n}$. Consider a B-pattern $\overline{\mathbf{B}} = \{\overline{B}_t\}_{t \in \mathbb{T}_{2n}}$ of rank $2n$ generated by $\overline{B}_{t_0}$, which is called *the principal extension of* $\mathbf{B}$ *with the initial vertex* t_0.

The following stability properties have repeatedly appeared in the literatures in various forms. Here, we quote from [FG19].

Proposition 2.11 (e.g., [FG19, Theorem 4.2, Remark 4.4]). *Let* C_t, G_t, $F_{i;t}(\mathbf{y})$ *be* C*- and* G*-matrices, and* F*-polynomials of* $\mathbf{B}$ *with the initial point* $t_0 \in \mathbb{T}_n$. *Let* $\overline{C}_t$, $\overline{G}_t$, $\overline{F}_{i;t}(\overline{\mathbf{y}})$ *be* C*- and* G*-matrices, and* F*-polynomials of* $\overline{\mathbf{B}}$ *with the same initial point* $t_0 \in \mathbb{T}_n \subset \mathbb{T}_{2n}$, *where* $\overline{\mathbf{y}} = (y_1, \dots, y_{2n})$. *Then, the following relations hold for any* $t \in \mathbb{T}_n \subset \mathbb{T}_{2n}$*:*

$$\overline{C}_t = \begin{pmatrix} C_t & Z_t \\ O & I \end{pmatrix}, \tag{2.39}$$

$$\overline{G}_t = \begin{pmatrix} G_t & O \\ O & I \end{pmatrix}, \tag{2.40}$$

$$\overline{F}_{i;t}(\overline{\mathbf{y}}) = \begin{cases} F_{i;t}(\mathbf{y}) & i = 1, \dots, n, \\ 1 & i = n+1, \dots, 2n, \end{cases} \tag{2.41}$$

where Z_t *is a certain matrix. Moreover,* $Z_t = O$ *if the sign-coherence of* C*-matrices for* $\mathbf{B}$ *holds.*

These equalities can be proved by the induction on t along $\mathbb{T}_n$ from t_0 [FG19, Theorem 4.2].

By Proposition 2.11, any information on C_t, G_t, $F_{i;t}(\mathbf{y})$ can be extracted from the one on $\overline{C}_t$, $\overline{G}_t$, $\overline{F}_{i;t}(\overline{\mathbf{y}})$.

2.4 G-cones and G-fans

From now on, we study the geometric aspect of G-matrices.

We recall some notions from convex geometry.

Definition 2.12 (Cone). Let $M \simeq \mathbb{Z}^n$ be a lattice of rank n. Let $M_{\mathbb{R}} = M \otimes_{\mathbb{Z}} \mathbb{R} \simeq \mathbb{R}^n$ be the extension to an n-dimensional vector space over $\mathbb{R}$. Let $N = \mathrm{Hom}_{\mathbb{Z}}(M, \mathbb{Z})$ and $N_{\mathbb{R}} = \mathrm{Hom}_{\mathbb{R}}(M_{\mathbb{R}}, \mathbb{R})$ be their duals. Let $\langle\ ,\ \rangle$ be the canonical paring $N_{\mathbb{R}} \times M_{\mathbb{R}} \to \mathbb{R}$.

- For given elements $a_1, \dots, a_r \in M \subset M_{\mathbb{R}}$, the subset of $M_{\mathbb{R}}$ defined by

$$\sigma(a_1, \dots, a_r) := \mathbb{R}_{\geq 0} a_1 + \cdots \mathbb{R}_{\geq 0} a_r \tag{2.42}$$

  is called a *convex rational polyhedral cone*, or simply, a *cone* generated by $a_1, \dots, a_r$. The set $\sigma(\emptyset) := \{0\}$ is also a cone.
- A cone σ is said to be *strongly convex* if $\sigma \cap (-\sigma) = \{0\}$.
- A cone is *simplicial* if it is generated by $\mathbb{Z}$-linearly independent elements in M. Thus, a simplicial cone is strongly convex.
- A cone is *nonsingular* if it is generated by a subset of a $\mathbb{Z}$-basis in M. Thus, a nonsingular cone is simplicial.
- The *dimension of a cone* σ is the dimension of the subspace of $M_{\mathbb{R}}$ spanned by σ.
- The *dual cone* $\sigma^{\vee}$ of a cone σ is defined by

$$\sigma^{\vee} = \{u \in N_{\mathbb{R}} \mid \langle u, a\rangle \geq 0 \text{ for any } a \in \sigma)\},$$

  which is a cone in $N_{\mathbb{R}}$.
- A subset τ of a cone σ is called a *face* of σ if there is some $u \in \sigma^{\vee}$ such that $\tau = \sigma \cap u^{\perp}$, where $u^{\perp} = \{a \in M_{\mathbb{R}} \mid \langle u, a\rangle = 0\}$. In particular, for a simplicial cone σ with a $\mathbb{Z}$-linearly independent generating set S, any face of σ is a cone generated by a subset of S.

Example 2.13. For $M = \mathbb{Z}^2$ and $M_{\mathbb{R}} = \mathbb{R}^2$, $\sigma(\mathbf{e}_1)$ and $\sigma(\mathbf{e}_1, \mathbf{e}_2)$ are strongly convex cones, while $\sigma(\mathbf{e}_1, -\mathbf{e}_1) = \mathbb{R}\mathbf{e}_1$ and $\sigma(\mathbf{e}_1, -\mathbf{e}_1, \mathbf{e}_2, -\mathbf{e}_2) = \mathbb{R}^2$ are cones that are not strongly convex. The cone $\sigma = \sigma(\mathbf{e}_1, \mathbf{e}_2)$ has four faces σ, $\sigma(\mathbf{e}_1)$, $\sigma(\mathbf{e}_2)$, $\{0\}$, while the cone $\sigma = \sigma(\mathbf{e}_1, -\mathbf{e}_1)$ has the unique face, which is σ.

Definition 2.14 (Fan).
- A *fan* Δ in $M_{\mathbb{R}} \simeq \mathbb{R}^n$ is a nonempty (possibly infinite) set of strongly convex cones in $M_{\mathbb{R}}$ satisfying the following conditions:

(i). If τ is a face of $\sigma \in \Delta$, then $\tau \in \Delta$.
(ii). If $\sigma, \tau \in \Delta$, then $\sigma \cap \tau \in \Delta$.

- A subset $|\Delta| = \bigcup_{\sigma \in \Delta} \sigma$ of $M_{\mathbb{R}}$ is called the *support* of a fan Δ.
- A fan Δ is *simplicial* if all cones of Δ are simplicial.
- A fan Δ is *nonsingular* if all cones of Δ are nonsingular. Thus, a nonsingular fan is simplicial.
- A fan Δ is *complete* if $|\Delta| = M_{\mathbb{R}}$.

We apply these geometric notions to G-matrices following [RS16, Rea14]. In the rest of this section *we temporarily assume the sign-coherence of C-matrices* and study some basic properties of G-cones.

Let $\mathbf{B}$ be any B-pattern, and let $\mathbf{G}^{t_0} = \mathbf{G}$ be the G-pattern of $\mathbf{B}$ with a given initial point t_0.

Definition 2.15 (G-cone)**.** For each G-matrix G_t, the cone spanned by its g-vectors in $\mathbb{R}^n$

$$\sigma(G_t) := \sigma(\mathbf{g}_{1;t}, \dots, \mathbf{g}_{n;t}) \tag{2.43}$$

is called a *G-cone.* The interior of $\sigma(G_t)$ is denoted by $\sigma^\circ(G_t)$. Each face of $\sigma(G_t)$ spanned by $\mathbf{g}_{j;t}$'s excluding the ith g-vector $\mathbf{g}_{i;t}$ is denoted by $\sigma_i(G_t)$ for $i = 1, \dots, n$.

By the unimodularity (2.3), every G-cone $\sigma(G_t)$ is a nonsingular cone, and its faces $\sigma_i(G_t)$ $(i = 1, \dots, n)$ are cones of codimension one.

From now on, let us especially take a common skew-symmetrizer D of the B-pattern $\mathbf{B}$ in the following form:

$$D = \mathrm{diag}(\delta_1^{-1} \dots, \delta_n^{-1}), \tag{2.44}$$

where $\delta_1, \dots, \delta_n$ are positive integers. (This seemingly awkward choice is suitable to the formulation of scattering diagrams in [GHKK18] we use later. See Example 5.4.) Accordingly, we introduce the following nonstandard inner product $(\ ,\ )_D$ in $\mathbb{R}^n$:

$$(\mathbf{u}, \mathbf{v})_D = \mathbf{u}^T D \mathbf{v} \quad (\mathbf{u}, \mathbf{v} \in \mathbb{R}^n). \tag{2.45}$$

Below, for a given B-pattern $\mathbf{B}$, we always assume this inner product $(\cdot,\cdot)_D$ in $\mathbb{R}^n$. For $\mathbf{n} \neq \mathbf{0} \in \mathbb{R}^n$, $\mathbf{n}^\perp$ is a hyperplane defined by

$$\mathbf{n}^\perp := \{\mathbf{v} \in \mathbb{R}^n \mid (\mathbf{n}, \mathbf{v})_D = 0\}. \tag{2.46}$$

We say that a normal vector $\mathbf{n}$ of a face $\sigma_i(G_t)$ of a cone $\sigma(G_t)$ is *inward for* $\sigma(G_t)$ if $(\mathbf{n}, \mathbf{g}_{i;t})_D > 0$.

In this geometrical setting, the duality (2.4)/(2.5) can be rephrased in the following manner.

Proposition 2.16. *Suppose that the sign-coherence conjecture holds. Then, the following dualities of vectors in* $\mathbb{R}^n$ *hold:*

$$(\delta_i \mathbf{c}_{i;t}, \mathbf{g}_{j;t})_D = \delta_{ij}, \tag{2.47}$$

$$(\mathbf{c}_{i;t}, \delta_j \mathbf{g}_{j;t})_D = \delta_{ij}. \tag{2.48}$$

In particular, each c-vector $\mathbf{c}_{i;t}$ *is a normal vector of the face* $\sigma_i(G_t)$ *of the cone* $\sigma(G_t)$ *with respect to the inner product* $(\cdot,\cdot)_D$. *Moreover, it is inward for* $\sigma(G_t)$.

Let $\Delta(\mathbf{G}^{t_0}) = \Delta(\mathbf{G})$ be the set of all faces of all G-cones $\sigma(G_t)$ ($t \in \mathbb{T}_n$) in the G-pattern $\mathbf{G}^{t_0}$. Let $t, t' \in \mathbb{T}_n$ be vertices that are k-adjacent. Then, by (2.13), two cones $\sigma(G_t)$ and $\sigma(G_{t'})$ intersect each other in their common face $\sigma_k(G_t) = \sigma_k(G_{t'})$ with codimension one. However, for a cone $\sigma(G_{t''})$ obtained from $\sigma(G_t)$ after several mutations, it is not clear at all that $\sigma(G_{t''})$ intersects $\sigma(G_t)$ only in their common face so that the set $\Delta(\mathbf{G})$ forms a fan. It was conjectured by [RS16, Conj. 3.14] that this always happens. Then, it was shown to be true by [Rea14] under the sign-coherence conjecture.

Theorem 2.17 ([Rea14, Theorem 8.7])**.** *Suppose that the sign-coherence conjecture holds. Then, for any B-pattern* $\mathbf{B}$ *and a given initial vertex* t_0, *the set of cones* $\Delta(\mathbf{G}^{t_0})$ *is a fan in* $\mathbb{R}^n$.

In [Rea14] this was proved in a more general setting of *mutation fans* therein. We provide a self-contained proof of Theorem 2.17 in Section 2.7 without referring to mutation fans, because the tools and techniques we use therein are also relevant to our main purpose.

The resulting fan $\Delta(\mathbf{G}^{t_0})$, which is nonsingular by (2.3), is called the *g-vector fan* in [RS16]. To simplify a little, we call it the *G-fan* of $\mathbf{B}$ (or $\mathbf{\Sigma}$) with the initial vertex t_0.

Let $\#\mathbf{G}^{t_0}$ be the number of *distinct* G-matrices in $\mathbf{G}^{t_0}$, which can be finite or infinite.

Proposition 2.18. *Suppose that the sign-coherence conjecture holds. Then, the G-fan* $\Delta(\mathbf{G}^{t_0})$ *is complete if and only if* $\#\mathbf{G}^{t_0}$ *is finite.*

Proof. This is true by the following topological reason. For each subset T of $\mathbb{T}_n$, we attach a subset Δ_T of $\Delta(\mathbf{G}^{t_0})$ consisting of the faces of all $\sigma(G_t)$ with $t \in T$. Let $|\Delta_T|^c$ be the complement of the support $|\Delta_T|$ in $\mathbb{R}^n$. Let S^{n-1} be the unit sphere in $\mathbb{R}^n$. Let $D_T = |\Delta_T|^c \cap S^{n-1}$. For $T = \{t_0\}$, D_T is topologically a $(n-1)$-dimensional open disk. We consider a strongly increasing sequence of subsets of $\mathbb{T}_n$, $T_1 = \{t_0\} \subsetneq T_2 \subsetneq T_3 \subsetneq \cdots$ with $|T_s| = s$. Correspondingly, we have a weakly decreasing sequence of subsets of S^{n-1}, $D_{T_1} \supset D_{T_2} \supset D_{T_3} \supset \cdots$. Then, D_{T_s} is a disjoint union of a finite number of open disks, such that at each process of increasing T_s to T_{s+1}, one of the followings occurs (by adding a G-cone):

- $D_{T_{s+1}} = D_{T_s}$.
- An open disk shrinks.
- An open disk splits into two open disks.
- An open disk disappears.

If $\#\mathbf{G}^{t_0}$ is finite, the decreasing process of D_{T_s} should terminate at some s. This occurs only when $D_{T_s} = \emptyset$. Thus, $\Delta(\mathbf{G}^{t_0})$ is complete. If $\#\mathbf{G}^{t_0}$ is infinite, the process never terminates. Then, when T_s converges to $\mathbb{T}_n$, D_{T_s} converges to the (possibly infinite) union of points and (not open, but not necessarily closed) disks. Therefore, $|\Delta(\mathbf{G}^{t_0})|^c$ is nonempty. □

Remark 2.19. We say that a cluster pattern $\mathbf{\Sigma}$ is *of finite type* if there are only finitely many distinct seeds in $\mathbf{\Sigma}$. It is known, for example, by [Nak21, Theorem 5.2], that a cluster pattern $\mathbf{\Sigma}$ is of finite type if and only if $\#\mathbf{G}^{t_0}$ is finite for $\mathbf{G}^{t_0}$ of $\mathbf{\Sigma}$. Then, combining it with Proposition 2.18, we have that a cluster pattern $\mathbf{\Sigma}$ is of finite type if and only if the G-fan $\Delta(\mathbf{G}^{t_0})$ is complete for $\mathbf{G}^{t_0}$ of $\mathbf{\Sigma}$. Also, we note that the proof of [Nak21, Theorem 5.2] is based on the sign-coherence of C-matrices and the Laurent positivity.

2.5 Rank 2 examples of G-fans

In the rank 2 case, one can confirm Theorem 2.17 and Proposition 2.18 by explicitly calculating G-matrices.

Let us arrange a B-pattern of rank 2 in the following way:

$$\cdots \overset{\mu_2}{\leftrightarrow} B_{t_{-2}} \overset{\mu_1}{\leftrightarrow} B_{t_{-1}} \overset{\mu_2}{\leftrightarrow} B_{t_0} \overset{\mu_1}{\leftrightarrow} B_{t_1} \overset{\mu_2}{\leftrightarrow} B_{t_2} \overset{\mu_1}{\leftrightarrow} \cdots . \tag{2.49}$$

We take B_{t_0} as the initial exchange matrix. Let us present the associated G-fan $\Delta(\mathbf{G})$ in $\mathbb{R}^2$ explicitly. Below the type of a B-pattern refers to the type of the Cartan matrix $A(B) = (a_{ij})$ associated with $B = B_{t_0}$ in the convention of [Kac90], where

$$a_{ij} = \begin{cases} 2 & i = j, \\ -|b_{ij}| & i \neq j. \end{cases} \tag{2.50}$$

(I). Finite type. The G-fan $\Delta(\mathbf{G})$ is complete as stated in Proposition 2.18.

(a). Type A_2: Let

$$B_{t_0} = \begin{pmatrix} 0 & -1 \\ 1 & 0 \end{pmatrix}. \tag{2.51}$$

Along the mutation sequence $\mu_1, \mu_2, \cdots$, the G-matrices are explicitly calculated as

$$\begin{aligned} G_{t_0} &= \begin{pmatrix} 1 & 0 \\ 0 & 1 \end{pmatrix}, \quad G_{t_1} = \begin{pmatrix} -1 & 0 \\ 0 & 1 \end{pmatrix}, \quad G_{t_2} = \begin{pmatrix} -1 & 0 \\ 0 & -1 \end{pmatrix}, \\ G_{t_3} &= \begin{pmatrix} 1 & 0 \\ -1 & -1 \end{pmatrix}, \quad G_{t_4} = \begin{pmatrix} 1 & 1 \\ -1 & 0 \end{pmatrix}, \quad G_{t_5} = \begin{pmatrix} 0 & 1 \\ 1 & 0 \end{pmatrix}. \end{aligned} \tag{2.52}$$

Both G_{t_5} and G_{t_0} define the same cone, corresponding to the celebrated *pentagon periodicity*. The G-fan $\Delta(\mathbf{G})$ is depicted in Figure 4 (a).

(b). Type B_2: Let

$$B_{t_0} = \begin{pmatrix} 0 & -1 \\ 2 & 0 \end{pmatrix}. \tag{2.53}$$

We start from the initial G-matrix $G_{t_0} = I$. Along the mutation sequence $\mu_1, \mu_2, \cdots$, the following g-vectors appear in this order, showing periodicity:

$$\begin{pmatrix} -1 \\ 0 \end{pmatrix}, \begin{pmatrix} 0 \\ -1 \end{pmatrix}, \begin{pmatrix} 1 \\ -2 \end{pmatrix}, \begin{pmatrix} 1 \\ -1 \end{pmatrix}, \begin{pmatrix} 1 \\ 0 \end{pmatrix}, \begin{pmatrix} 0 \\ 1 \end{pmatrix}. \tag{2.54}$$

The G-fan $\Delta(\mathbf{G})$ is depicted in Figure 4 (b).

(c). Type G_2: Let

$$B_{t_0} = \begin{pmatrix} 0 & -1 \\ 3 & 0 \end{pmatrix}. \tag{2.55}$$

We start from the initial G-matrix $G_{t_0} = I$. Along the mutation sequence $\mu_1, \mu_2, \cdots$, the following g-vectors appear in this order, showing periodicity:

$$\begin{pmatrix} -1 \\ 0 \end{pmatrix}, \begin{pmatrix} 0 \\ -1 \end{pmatrix}, \begin{pmatrix} 1 \\ -3 \end{pmatrix}, \begin{pmatrix} 1 \\ -2 \end{pmatrix}, \begin{pmatrix} 2 \\ -3 \end{pmatrix}, \begin{pmatrix} 1 \\ -1 \end{pmatrix}, \begin{pmatrix} 1 \\ 0 \end{pmatrix}, \begin{pmatrix} 0 \\ 1 \end{pmatrix}. \tag{2.56}$$

The G-fan $\Delta(\mathbf{G})$ is depicted in Figure 4 (c).

(II). Infinite type. The G-fan $\Delta(\mathbf{G})$ is not complete.

(d). Type $A_1^{(1)}$: Let

$$B_{t_0} = \begin{pmatrix} 0 & -2 \\ 2 & 0 \end{pmatrix}. \tag{2.57}$$

We start from the initial G-matrix $G_{t_0} = I$. Along the mutation sequence $\mu_1, \mu_2, \cdots$, the following g-vectors appear in this order:

$$\begin{pmatrix} -1 \\ 0 \end{pmatrix}, \begin{pmatrix} 0 \\ -1 \end{pmatrix}, \begin{pmatrix} 1 \\ -2 \end{pmatrix}, \begin{pmatrix} 2 \\ -3 \end{pmatrix}, \begin{pmatrix} 3 \\ -4 \end{pmatrix}, \cdots. \tag{2.58}$$

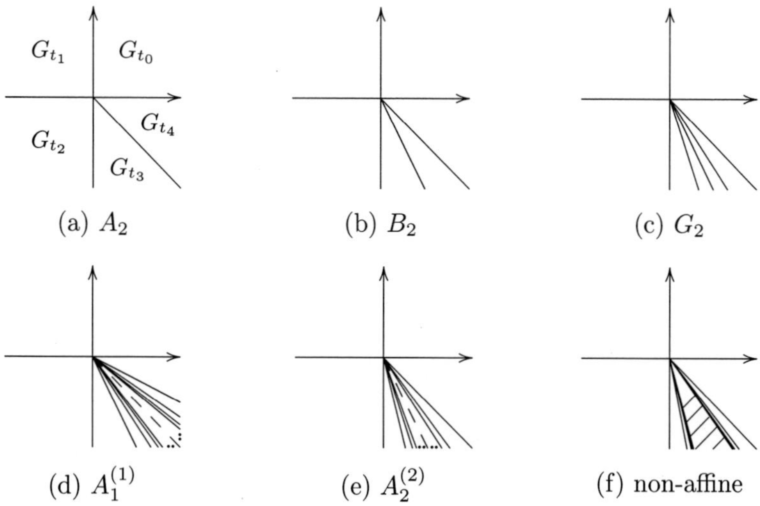

Figure 4: G-fans of rank 2.

On the other hand, along the mutation sequence μ_2, μ_1, $\cdots$, the following g-vectors appear in this order:

$$\begin{pmatrix} 2 \\ -1 \end{pmatrix}, \begin{pmatrix} 3 \\ -2 \end{pmatrix}, \begin{pmatrix} 4 \\ -3 \end{pmatrix}, \begin{pmatrix} 5 \\ -4 \end{pmatrix}, \cdots . \tag{2.59}$$

The G-fan $\Delta(\mathbf{G})$ is depicted in Figure 4 (d). It covers the region excluding the half line $\mathbb{R}_+(1,-1)$.

(e). Type $A_2^{(2)}$: Let

$$B_{t_0} = \begin{pmatrix} 0 & -1 \\ 4 & 0 \end{pmatrix}. \tag{2.60}$$

We start from the initial G-matrix $G_{t_0} = I$. Along the mutation sequence μ_1, μ_2, $\cdots$, the following g-vectors appear in this order:

$$\begin{pmatrix} -1 \\ 0 \end{pmatrix}, \begin{pmatrix} 0 \\ -1 \end{pmatrix}, \begin{pmatrix} 1 \\ -4 \end{pmatrix}, \begin{pmatrix} 1 \\ -3 \end{pmatrix}, \begin{pmatrix} 3 \\ -8 \end{pmatrix}, \begin{pmatrix} 2 \\ -5 \end{pmatrix}, \begin{pmatrix} 5 \\ -12 \end{pmatrix}, \begin{pmatrix} 3 \\ -7 \end{pmatrix}, \cdots . \tag{2.61}$$

On the other hand, along the mutation sequence μ_2, μ_1, $\cdots$, the following g-vectors appear in this order:

$$\begin{pmatrix} 1 \\ -1 \end{pmatrix}, \begin{pmatrix} 3 \\ -4 \end{pmatrix}, \begin{pmatrix} 2 \\ -3 \end{pmatrix}, \begin{pmatrix} 5 \\ -8 \end{pmatrix}, \begin{pmatrix} 3 \\ -5 \end{pmatrix}, \begin{pmatrix} 7 \\ -12 \end{pmatrix}, \cdots . \tag{2.62}$$

The G-fan $\Delta(\mathbf{G})$ is depicted in Figure 4 (e). It covers the region excluding the half line $\mathbb{R}_+(1,-2)$.

(f). Non-affine type: Let

$$B_{t_0} = \begin{pmatrix} 0 & -c \\ b & 0 \end{pmatrix}, \quad (b, c > 0,\ bc \geq 5). \tag{2.63}$$

We start from the initial G-matrix $G_{t_0} = I$. Along the mutation sequence $\mu_1, \mu_2, \cdots$, the following g-vectors appear in this order:

$$\mathbf{g}_1 = \begin{pmatrix} -1 \\ 0 \end{pmatrix}, \ \mathbf{g}_2 = \begin{pmatrix} 0 \\ -1 \end{pmatrix}, \ \mathbf{g}_3, \ \mathbf{g}_4, \cdots, \tag{2.64}$$

where $\mathbf{g}_i$'s obey the recursion,

$$\mathbf{g}_{i+2} = \begin{cases} -\mathbf{g}_i + b\mathbf{g}_{i+1} & i \text{ odd}, \\ -\mathbf{g}_i + c\mathbf{g}_{i+1} & i \text{ even}. \end{cases} \tag{2.65}$$

On the other hand, along the mutation sequence $\mu_2, \mu_1, \cdots$ from the initial G-matrix, the following g-vectors appear in this order:

$$\mathbf{g}'_1 = \begin{pmatrix} c \\ -1 \end{pmatrix}, \ \mathbf{g}'_2 = \begin{pmatrix} bc-1 \\ -b \end{pmatrix}, \ \mathbf{g}'_3, \ \mathbf{g}'_4, \cdots, \tag{2.66}$$

where $\mathbf{g}'_i$'s obey the recursion,

$$\mathbf{g}'_{i+2} = \begin{cases} -\mathbf{g}'_i + c\mathbf{g}'_{i+1} & i \text{ odd}, \\ -\mathbf{g}'_i + b\mathbf{g}'_{i+1} & i \text{ even}. \end{cases} \tag{2.67}$$

It is known [Rea14, Eqs. (9.11), (9.12)] that, in the limit $i \to \infty$, the rays $\sigma(\mathbf{g}_i)$ and $\sigma(\mathbf{g}'_i)$ monotonically converge to $\sigma(\mathbf{v})$ and $\sigma(\mathbf{v}')$, respectively, where

$$\mathbf{v} = \begin{pmatrix} bc - \sqrt{bc(bc-4)} \\ -2b \end{pmatrix}, \quad \mathbf{v}' = \begin{pmatrix} bc + \sqrt{bc(bc-4)} \\ -2b \end{pmatrix}. \tag{2.68}$$

The G-fan $\Delta(\mathbf{G})$ is depicted in Figure 4 (f). It covers the region excluding $\sigma(\mathbf{v}, \mathbf{v}') \setminus \{0\}$ (the hatched region in the figure).

2.6 Piecewise linear isomorphisms between G-fans

In this subsection we study a canonical bijection between the sets $\Delta(\mathbf{G}^{t_0})$ and $\Delta(\mathbf{G}^{t_1})$ for any $t_0, t_1 \in \mathbb{T}_n$. The idea comes from [GHKK18], but this is independent of their results. Also, the result of this and the next subsections has considerable overlap with the results in [Rea14, Rea20b], in which G-fans were studied as a part of mutation fans.

Let $\mathbf{B}$ be any B pattern. Let us start with the following consequence of the sign-coherence.

Proposition 2.20. *Suppose that the sign-coherence conjecture holds. Then, for any $t_0, t \in \mathbb{T}_n$, the cone $\sigma(G_t^{t_0})$ intersects the hyperplanes $\mathbf{e}_i^\perp$ ($i = 1, \dots, n$) only in the boundary of $\sigma(G_t^{t_0})$.*

Proof. Suppose that a cone $\sigma(G_t^{t_0})$ intersects with $\mathbf{e}_i^\perp$ for some i in its interior $\sigma^\circ(G_t^{t_0})$. This means that the ith row of the G-matrix $G_t^{t_0}$ has both positive and negative components. This contradicts the row sign-coherence of $G_t^{t_0}$ in Proposition 2.7. □

Let $t_0, t_1 \in \mathbb{T}_n$ be vertices that are k-adjacent. Recall the formula (2.20), which is valid under the assumption of the sign-coherence of C-matrices. It is written in the vector form as follows.

$$\mathbf{g}_{i;t}^{t_1} = (J_k + [\varepsilon_{k\bullet}(G_t^{t_0})B_{t_0}]_+^{\bullet k})\mathbf{g}_{i;t}^{t_0}, \quad (i = 1, \dots, n). \tag{2.69}$$

Let us introduce the half spaces of $\mathbb{R}^n$ as

$$\mathbb{R}_{k,+}^n = \{\mathbf{v} \in \mathbb{R}^n \mid v_k \geq 0\}, \quad \mathbb{R}_{k,-}^n = \{\mathbf{v} \in \mathbb{R}^n \mid v_k \leq 0\}. \tag{2.70}$$

The matrix $(J_k + [\varepsilon_{k\bullet}(G_t^{t_0})B_{t_0}]_+^{\bullet k})$ in (2.69) is regarded as a piecewise linear map

$$\begin{aligned} \varphi_{t_0}^{t_1} : \mathbb{R}^n &\to \mathbb{R}^n \\ \mathbf{v} &\mapsto \begin{cases} (J_k + [B_{t_0}]_+^{\bullet k})\mathbf{v} & \mathbf{v} \in \mathbb{R}_{k,+}^n, \\ (J_k + [-B_{t_0}]_+^{\bullet k})\mathbf{v} & \mathbf{v} \in \mathbb{R}_{k,-}^n, \end{cases} \end{aligned} \tag{2.71}$$

where we note that $\varphi_{t_0}^{t_1}(\mathbf{v}) = \mathbf{v}$ if $v_k = 0$. We have, by (2.69) and (2.71), for any $t \in \mathbb{T}_n$ and $i = 1, \dots, k$,

$$\varphi_{t_0}^{t_1}(\mathbf{g}_{i;t}^{t_0}) = \mathbf{g}_{i;t}^{t_1}. \tag{2.72}$$

The map $\varphi_{t_1}^{t_0}$ is also defined in the same way by replacing B_{t_0} with B_{t_1}.

Proposition 2.21. *The following relation holds:*

$$\varphi_{t_1}^{t_0} \circ \varphi_{t_0}^{t_1} = \varphi_{t_0}^{t_1} \circ \varphi_{t_1}^{t_0} = \mathrm{id}. \tag{2.73}$$

Proof. Note that, for $\mathbf{v}' = \varphi_{t_0}^{t_1}(\mathbf{v})$, we have $v'_k = -v_k$ due to $b_{kk}^{t_0} = 0$. We also recall that $(B_{t_1})^{\bullet k} = -(B_{t_0})^{\bullet k}$. Then, for $\mathbf{v} \in \mathbb{R}_{k,+}^n$,

$$\begin{aligned} \varphi_{t_1}^{t_0} \circ \varphi_{t_0}^{t_1}(\mathbf{v}) &= (J_k + [-B_{t_1}]_+^{\bullet k})(J_k + [B_{t_0}]_+^{\bullet k})\mathbf{v} \\ &= (J_k + [B_{t_0}]_+^{\bullet k})^2\mathbf{v} = \mathbf{v} \end{aligned} \tag{2.74}$$

thanks to Lemma 1.10. The other case is similar.

□

Proposition 2.22 (cf. [Rea14, Prop. 8.13])**.** *Suppose that the sign-coherence conjecture holds. Then, we have*

$$\varphi_{t_0}^{t_1}(\sigma(G_t^{t_0})) = \sigma(G_t^{t_1}), \quad (t \in \mathbb{T}_n). \tag{2.75}$$

Therefore, the correspondence $\varphi_{t_0}^{t_1} : \mathbf{g}_{i;t}^{t_0} \mapsto \mathbf{g}_{i;t}^{t_1}$ gives a bijection between $\Delta(\mathbf{G}^{t_0})$ and $\Delta(\mathbf{G}^{t_1})$ preserving the intersection and the inclusion of cones.

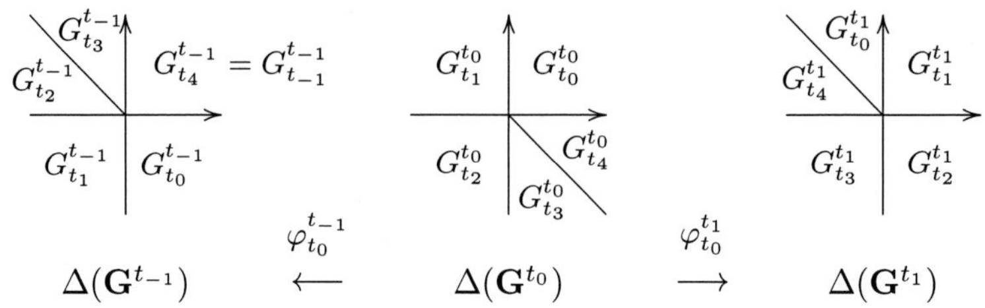

Figure 5: Bijections between G-fans for type A_2.

Proof. We only need to prove (2.75). By Proposition 2.20, each cone $\sigma(G_t^{t_0})$ belongs to either $\mathbb{R}^n_{k,+}$ or $\mathbb{R}^n_{k,-}$, so that it is *linearly* transformed under $\varphi_{t_0}^{t_1}$. Therefore, by (2.72), its image is $\sigma(G_t^{t_1})$. □

Remark 2.23. For the later use, we record the following fact in the above proof.

- To prove (2.75), we only use the formula (2.69) for t_0, t_1 and t therein, and the row sign-coherence of $G_t^{t_0}$ therein.

Example 2.24. Let us clarify Proposition 2.22 explicitly for type A_2 based on the convention in Section 2.5. The fans $\Delta(\mathbf{G}^{t_{-1}})$, $\Delta(\mathbf{G}^{t_0})$, $\Delta(\mathbf{G}^{t_1})$ are depicted in Figure 5. Also, the piecewise linear maps $\varphi_{t_0}^{t_1}$ and $\varphi_{t_0}^{t_{-1}}$ are given by combining the following linear maps:

$$J_1 + [B_{t_0}]_+^{\bullet 1} = \begin{pmatrix} -1 & 0 \\ 1 & 1 \end{pmatrix}, \quad J_1 + [-B_{t_0}]_+^{\bullet 1} = \begin{pmatrix} -1 & 0 \\ 0 & 1 \end{pmatrix}, \tag{2.76}$$

$$J_2 + [B_{t_0}]_+^{\bullet 2} = \begin{pmatrix} 1 & 0 \\ 0 & -1 \end{pmatrix}, \quad J_2 + [-B_{t_0}]_+^{\bullet 2} = \begin{pmatrix} 1 & 1 \\ 0 & -1 \end{pmatrix}. \tag{2.77}$$

Then, the correspondence in Proposition 2.22 can be easily confirmed in Figure 5.

By applying the above isomorphism repeatedly along the tree $\mathbb{T}_n$, we obtain the following important fact.

Proposition 2.25. *Suppose that the sign-coherence conjecture holds. Then, for any vertices $t_0, t_1 \in \mathbb{T}_n$, which are not necessarily adjacent, the correspondence $\mathbf{g}_{i;t}^{t_0} \mapsto \mathbf{g}_{i;t}^{t_1}$ gives a bijection between $\Delta(\mathbf{G}^{t_0})$ and $\Delta(\mathbf{G}^{t_1})$ preserving the intersection and the inclusion of cones.*

For the later use, we also describe the relation between $\varphi_{t_0}^{t_1}$ here and a certain piecewise linear isomorphism in [GHKK18].

Again, let $t_0, t_1 \in \mathbb{T}_n$ be vertices that are k-adjacent. First, for B_{t_0} and

k, we introduce a basis $\mathbf{a}_1, \dots, \mathbf{a}_n$ of $\mathbb{R}^n$ as

$$\mathbf{a}_i = \begin{cases} -\mathbf{e}_i + \sum_{j=1}^{n} [-b_{jk;t_0}]_+ \mathbf{e}_j & i = k, \\ \mathbf{e}_i & i \neq k. \end{cases} \tag{2.78}$$

In fact, by specializing the formula in (2.13) with $t = t_0$, and applying the fact $\varepsilon_{k;t_0} = 1$ therein, we see that

$$\mathbf{a}_i = \mathbf{g}_{i;t_1}^{t_0}. \tag{2.79}$$

The change of a basis of $\mathbb{R}^n$ from $\mathbf{e}_1, \dots, \mathbf{e}_n$ to $\mathbf{a}_1, \dots, \mathbf{a}_n$ induces the following linear isomorphism

$$\begin{array}{rccc} \eta_{t_0}^{t_1}: & \mathbb{R}^n & \rightarrow & \mathbb{R}^n \\ & \mathbf{v} & \mapsto & (J_k + [-B_{t_0}]_+^{\bullet k})\mathbf{v} \end{array} \tag{2.80}$$

such that $\eta_{t_0}^{t_1}(\mathbf{a}_i) = \mathbf{e}_i$.

Next, for the same B_{t_0} and k, we define a piecewise linear map $T_{k;t_0}$ on $\mathbb{R}^n$ following [GHKK18] as

$$\begin{array}{rccc} T_{k;t_0}: & \mathbb{R}^n & \rightarrow & \mathbb{R}^n \\ & \mathbf{v} & \mapsto & \begin{cases} (I + (B_{t_0})^{\bullet k})\mathbf{v} & \mathbf{v} \in \mathbb{R}^n_{k,+}, \\ \mathbf{v} & \mathbf{v} \in \mathbb{R}^n_{k,-}. \end{cases} \end{array} \tag{2.81}$$

The following fact appeared in [Mul16, Rea20b].

Proposition 2.26 ([Mul16, Lemma 5.2.1], [Rea20b, Cor. 4.4]). *The above maps are related as follows:*

$$\varphi_{t_0}^{t_1} = \eta_{t_0}^{t_1} \circ T_{k;t_0}. \tag{2.82}$$

Proof. By the definitions of these maps,

$$\eta_{t_0}^{t_1} \circ T_{k;t_0}(\mathbf{v}) = \begin{cases} (J_k + [-B_{t_0}]_+^{\bullet k})(I + (B_{t_0})^{\bullet k})\mathbf{v} & \mathbf{v} \in \mathbb{R}^n_{k,+}, \\ (J_k + [-B_{t_0}]_+^{\bullet k})\mathbf{v} & \mathbf{v} \in \mathbb{R}^n_{k,-}. \end{cases} \tag{2.83}$$

Thanks to (1.5), we have

$$(J_k + [-B_{t_0}]_+^{\bullet k})(I + (B_{t_0})^{\bullet k}) = J_k + (B_{t_0})^{\bullet k} + [-B_{t_0}]_+^{\bullet k} = J_k + [B_{t_0}]_+^{\bullet k}.$$

□

2.7 Proof of Theorem 2.17

Let us prove Theorem 2.17 under the assumption of the sign-coherence of C-matrices.

We recall that the condition for $\Delta(\mathbf{G}^{t_0})$ to form a fan is stated as follows:

Condition 2.27. For any $t, t' \in \mathbb{T}_n$, the cones $\sigma(G_t^{t_0})$ and $\sigma(G_{t'}^{t_0})$ intersect in their common face.

Thanks to Proposition 2.22, Condition 2.27 is rephrased as follows, where we used the bijection between $\Delta(\mathbf{G}^{t_0})$ and $\Delta(\mathbf{G}^{t})$ therein.

Condition 2.28. For any $t, t' \in \mathbb{T}_n$, the cones $\sigma(G_{t'}^{t})$ and $\sigma(I)$ intersect in their common face, where $\sigma(I) = \sigma(\mathbf{e}_1, \dots, \mathbf{e}_n)$.

By changing the notation, we may prove Condition 2.28 for $\sigma(G_t^{t_0})$ instead of $\sigma(G_{t'}^{t})$ therein. Then, the condition can be further rephrased as follows:

Claim. For any $t_0, t \in \mathbb{T}_n$, if a g-vector $\mathbf{g}_{i;t}^{t_0}$ is contained in $\sigma(I)$, then $\mathbf{g}_{i;t}^{t_0} = \mathbf{e}_\ell$ for some ℓ.

Let us prove the claim. For $J \subset \{1, \dots, n\}$, let $J^c = \{1, \dots, n\} \setminus J$ be its compliment, and let $\sigma_J(G_t^{t_0})$ be the face of $\sigma(G_t^{t_0})$ generated by $\mathbf{g}_{j;t}^{t_0}$'s ($j \in J^c$). For example, we have $\sigma_{\{1,\dots,n\}}(G_t^{t_0}) = \{0\}$, $\sigma_{\{i\}^c}(G_t^{t_0}) = \sigma(\mathbf{g}_{i;t}^{t_0})$.

Thanks to Proposition 2.16, we have

$$\sigma_J(G_t^{t_0}) = \sigma(G_t^{t_0}) \cap \left(\bigcap_{j \in J} (\mathbf{c}_{j;t}^{t_0})^\perp \right). \tag{2.84}$$

On the other hand, since $\mathbf{c}_{j;t}^{t_0}$ is either positive or negative, we have

$$\sigma(I) \cap (\mathbf{c}_{j;t}^{t_0})^\perp = \sigma(I) \cap \bigcap_{k \in K[j]} (\mathbf{e}_k)^\perp = \sigma_{K[j]}(I), \tag{2.85}$$

where $K[j]$ is the set of k such that the kth component of $\mathbf{c}_{j;t}^{t_0}$ is nonzero. Therefore, we have

$$\sigma_J(G_t^{t_0}) \cap \sigma(I) = \sigma(G_t^{t_0}) \cap \left(\bigcap_{j \in J} \sigma_{K[j]}(I) \right) = \sigma(G_t^{t_0}) \cap \sigma_{K[J]}(I), \tag{2.86}$$

where $K[J]$ is the set of k such that the kth component of $\mathbf{c}_{j;t}^{t_0}$ is nonzero for some $j \in J$. Now we set $J = \{i\}^c$. Then, we have $\sigma_{\{i\}^c}(G_t^{t_0}) = \sigma(\mathbf{g}_{i;t}^{t_0})$ and $K[J] = \{1, \dots, n\}$ or $\{\ell\}^c$ for some ℓ. In the former case, we see from (2.86) that $\sigma(\mathbf{g}_{i;t}^{t_0})$ intersects $\sigma(I)$ trivially. In the latter case, we have

$$\sigma(\mathbf{g}_{i;t}^{t_0}) \cap \sigma(I) = \sigma(G_t^{t_0}) \cap \sigma(\mathbf{e}_\ell). \tag{2.87}$$

Therefore, if $\sigma(\mathbf{g}_{i;t}^{t_0}) \cap \sigma(I) \neq \{0\}$, we have $\sigma(\mathbf{g}_{i;t}^{t_0}) = \sigma(\mathbf{e}_\ell)$. Then, by the unimodularity (2.3), we have $\mathbf{g}_{i;t}^{t_0} = \mathbf{e}_\ell$ as desired. This completes the proof of Theorem 2.17.

2.8 Properties of $\hat{c}$-vectors

In this subsection we observe some properties of $\hat{c}$-vectors that are relevant to the scattering diagram method.

The following properties do not depend on the sign-coherence conjecture.

Proposition 2.29. *The following facts hold:*

(a). Any $\hat{c}$-vector $\hat{\mathbf{c}}_{i;t}$ is orthogonal to the c-vector $\mathbf{c}_{i;t}$.

(b). Any $\hat{c}$-vector $\hat{\mathbf{c}}_{i;t}$ is in the subspace in $\mathbb{R}^n$ spanned by $\sigma_i(G_t)$.

Proof. (a). Since DB_{t_0} is skew-symmetric, we have

$$(\mathbf{c}_{i;t}, \hat{\mathbf{c}}_{i;t})_D = (\mathbf{c}_{i;t})^T DB_{t_0}\mathbf{c}_{i;t} = 0. \tag{2.88}$$

(b). By the second expression of $\hat{c}$-vectors in (1.57), we have

$$\hat{\mathbf{c}}_{i;t} = \sum_{j=1}^{n} b_{ji;t}\mathbf{g}_{j;t}, \tag{2.89}$$

where we recall that $b_{ii;t} = 0$. □

Definition 2.30. Suppose that the sign-coherence conjecture holds. Let $\mathbf{c}_{i;t}$ be a c-vector, and let $\varepsilon_{i;t}$ be its tropical sign in Definition 2.1. We call the vectors $\mathbf{c}^+_{i;t} = \varepsilon_{i;t}\mathbf{c}_{i;t}$ and $\hat{\mathbf{c}}^+_{i;t} = \varepsilon_{i;t}\hat{\mathbf{c}}_{i;t}$ a *c^+-vector* and a *$\hat{c}^+$-vector*, respectively.

Remark 2.31. Any c^+-vector $\mathbf{c}^+_{i;t}$ is *positive* due to the sign-coherence assumption, while a $\hat{c}^+$-vector $\hat{\mathbf{c}}^+_{i;t}$ is not so at all in general.

The following fact is related to the *incoming/outgoing property* of walls appearing later in Definition 5.14.

Proposition 2.32. *Suppose that the sign-coherence conjecture holds. Then, if*

$$\hat{\mathbf{c}}^+_{i;t} \in \sigma_i(G_t), \tag{2.90}$$

then $\mathbf{c}^+_{i;t} = \mathbf{e}_\ell$ for some ℓ.

Proof. In view of (2.89), the condition (2.90) is equivalent to the following condition:

$$\text{The vector } \varepsilon_{i;t}\mathbf{b}_{i;t} \text{ is either positive or zero.} \tag{2.91}$$

Suppose that (2.91) happens, where we replace i with k therein for the sake of convenience. Let t' be the vertex that is k-adjacent to t. Then, the mutation (2.13) is simplified by (2.91) as

$$\mathbf{g}_{i;t'} = \begin{cases} -\mathbf{g}_{k;t} & i = k, \\ \mathbf{g}_{i;t} & i \neq k. \end{cases} \tag{2.92}$$

Now we recall the row sign-coherence of G-matrices in Proposition 2.7. Then, (2.92) happens only when the following condition is satisfied:

- For each $\ell = 1, \dots, n$, if the ℓth element of $\mathbf{g}_{k;t}$ is nonzero, then the ℓth element of other $\mathbf{g}_{i;t}$ $(i \neq k)$ vanishes.

Thanks to the unimodularity (2.3) of G_t, one can sharpen the condition as follows:

- There is some ℓ such that $\mathbf{g}_{k;t} = \pm\mathbf{e}_\ell$, and the ℓth element of other $\mathbf{g}_{i;t}$ $(i \neq k)$ vanishes.

Then, by the duality (2.4), $\mathbf{c}_{k;t} = \pm\mathbf{e}_\ell$. Thus, $\mathbf{c}^+_{k;t} = \mathbf{e}_\ell$. □

3 Scattering diagrams and C- and G-matrices

In this section we present a proof of the sign-coherence conjecture based on the results on scattering diagrams in [GHKK18]. At the same time we also establish the correspondence between a G-fan and the support of a scattering diagram.

3.1 Supports of scattering diagrams

To write down a precise definition of scattering diagrams in [GHKK18] requires considerable pages, and it will be postponed in Section 5. Here we summarize the minimal information and results [GHKK18] that are needed to prove the sign-coherence conjecture.

Informally speaking, for a certain vector space $M_{\mathbb{R}} \simeq \mathbb{R}^n$, a *scattering diagram* $\mathfrak{D} = \{\mathbf{w}_\lambda = (\mathfrak{d}_\lambda, f_\lambda)\}_{\lambda\in\Lambda}$ in $M_{\mathbb{R}}$ is a finite or countably infinite collection of *walls*, where Λ is an index set, and each wall $\mathbf{w}_\lambda = (\mathfrak{d}_\lambda, f_\lambda)$ is a pair such that $\mathfrak{d}_\lambda$ is a (not necessarily strongly convex) cone of codimension one in $M_{\mathbb{R}}$ called the *support* of $\mathbf{w}_\lambda$, and f_λ is a certain formal power series called the *wall function* of $\mathbf{w}_\lambda$. The union of the supports of walls

$$\mathrm{Supp}(\mathfrak{D}) = \bigcup_{\lambda\in\Lambda} \mathfrak{d}_\lambda \subset M_{\mathbb{R}} \tag{3.1}$$

is called the *support of a scattering diagram* $\mathfrak{D}$.

For any *nonsingular* skew-symmetrizable matrix B, one can construct a scattering diagram $\mathfrak{D}_{\mathfrak{s}}$ called a *cluster scattering diagram*, where $\mathfrak{s}$ is a *seed data* in [GHKK18] corresponding to B. There is a canonical isomorphism $\phi_{\mathfrak{s}} : M_{\mathbb{R}} \xrightarrow{\sim} \mathbb{R}^n$ such that the image of $\mathrm{Supp}(\mathfrak{D}_{\mathfrak{s}})$ in $\mathbb{R}^n$ only depends on B. We write this image as $\mathfrak{S}(B)$, and we call it the *support diagram* for B, for simplicity. A connected component of $\mathbb{R}^n \setminus \mathfrak{S}(B)$ is called a *chamber* of $\mathfrak{S}(B)$.

From the viewpoint of cluster patterns, the supports and the wall functions of walls contain the information of the tropical and the nontropical part of a cluster pattern, respectively. Since at this moment we are interested only in the tropical part, i.e., C- and G-matrices, we may safely forget wall functions, and concentrate on the support diagram $\mathfrak{S}(B)$.

Let D be a skew-symmetrizer of B in the form (2.44), and let $(\ ,\ )_D$ be the inner product in $\mathbb{R}^n$ defined in (2.45). The following facts are direct consequences of the definition/theorem of the scattering diagram $\mathfrak{D}_{\mathfrak{s}}$ in [GHKK18, Theorems 1.12 &1.28], which will appear later as Theorem 5.27.

Proposition 3.1 ([GHKK18, Theorems 1.12 &1.28]). *For any nonsingular skew-symmetrizable matrix B, the support diagram $\mathfrak{S}(B)$ has the following properties.*

(a). The support $\mathfrak{d}_\lambda$ of each wall of $\mathfrak{S}(B)$ has a positive normal vector $\mathbf{n}_\lambda$ with respect to the inner product $(\, , \,)_D$.

(b). Each hyperplane $\mathbf{e}_i^\perp$ $(i = 1, \dots, n)$ with respect to the inner product $(\, , \,)_D$ is a subset of $\mathfrak{S}(B)$.

(c). The interior $\sigma^\circ(I)$ of the cone $\sigma(I) = \sigma(\mathbf{e}_1, \dots, \mathbf{e}_n)$ is a chamber *of $\mathfrak{S}(B)$. In other words, $\sigma^\circ(I)$ does not intersect $\mathfrak{S}(B)$, while its boundary belongs to $\mathfrak{S}(B)$. (This is a consequence of (a) and (b).)*

Also, the following property of the support diagram $\mathfrak{S}(B)$ is obtained from [GHKK18, Theorem 1.24], which will appear later as Theorem 6.7, together with Proposition 2.26. See the explanation after Theorem 6.7 for details.

Proposition 3.2 ([GHKK18, Theorem 1.24], [Mul16, Lemma 5.2.1],[Rea20b, Cor. 4.4]). *Let $\mathbf{B}$ be any nonsingular B-pattern. Let $t_0, t_1 \in \mathbb{T}_n$ be vertices that are k-adjacent. Let $\varphi_{t_0}^{t_1}$ be the map defined in (2.71). Then, the support diagrams $\mathfrak{S}(B_{t_0})$ and $\mathfrak{S}(B_{t_1})$ are related by*

$$\varphi_{t_0}^{t_1}(\mathfrak{S}(B_{t_0})) = \mathfrak{S}(B_{t_1}). \tag{3.2}$$

This is clearly a parallel result for G-fans in Proposition 2.22, which was proved under the assumption of the sign-coherence of C-matrices.

3.2 Proof of sign-coherence conjecture

Let us present a proof of Conjecture 1.34 relying on Propositions 3.1 and 3.2. Our strategy is to establish both

- the identification of the interior $\sigma^\circ(G_t^{t_0})$ of each G-cone as a chamber in $\mathfrak{S}(B_{t_0})$,
- the column sign-coherence of C-matrices $C_t^{t_0}$,

simultaneously by the induction on the distance $d(t_0, t)$ in $\mathbb{T}_n$, where both t_0 and t vary.

Now we prove the following theorem. See [Rea20b, Theorem 4.10] for a closely related result for mutation fans.

Theorem 3.3. *Let $\mathbf{B} = \{B_t\}_{t\in\mathbb{T}_n}$ be any nonsingular B-pattern. Then, for any $t_0, t \in \mathbb{T}_n$, the following facts hold:*

(a). The set $\sigma^\circ(G_t^{t_0})$ is a chamber of $\mathfrak{S}(B_{t_0})$.

(b). Each c-vector $\mathbf{c}_{i;t}^{t_0}$ is a normal vector of the support of some wall of $\mathfrak{S}(B_{t_0})$.

(c). The column sign-coherence holds for the C-matrix $C_t^{t_0}$.

Proof. As in Section 2.2, we introduce the transpose $\mathbf{B}^T = \{(B_t)^T\}_{t\in\mathbb{T}_n}$ of $\mathbf{B}$. Let $\tilde{C}_t^{t_0}$, $\tilde{G}_t^{t_0}$ be C- and G-matrices of $\mathbf{B}^T$.

Consider the following claims $(a)_d$, $(b)_d$, $(c)_d$ for $d = 0, 1, 2, \dots$.

$(a)_d$. For any $t_0, t \in \mathbb{T}_n$ such that $d(t_0, t) = d$, the following facts hold:

$$|C_t^{t_0}| = |G_t^{t_0}| \in \{1, -1\}, \tag{3.3}$$
$$D^{-1}(G_t^{t_0})^T D C_t^{t_0} = I, \tag{3.4}$$
$$C_t^{t_0} = (\tilde{G}_{t_0}^{t})^T, \tag{3.5}$$
$$G_t^{t_0} = (\tilde{C}_{t_0}^{t})^T. \tag{3.6}$$

$(b)_d$. For any $t_0, t \in \mathbb{T}_n$ such that $d(t_0, t) = d$, the following facts hold:

(i). The set $\sigma^\circ(G_t^{t_0})$ is a chamber of $\mathfrak{S}(B_{t_0})$.

(ii). Each c-vector $\mathbf{c}_{i;t}^{t_0}$ is a normal vector of the support of a wall of $\mathfrak{S}(B_{t_0})$.

(iii). The column sign-coherence holds for $C_t^{t_0}$.

(iv). The row sign-coherence holds for $G_t^{t_0}$.

$(c)_d$. For any $t_0, t_1, t, t' \in \mathbb{T}_n$ such that

- $d(t_0, t) = d$,
- t_0 and t_1 are k-adjacent,
- t and t' are k-adjacent,

the following facts hold:

$$C_{t'}^{t_0} = C_t^{t_0}(J_k + [\varepsilon_{\bullet k}(C_t^{t_0})B_t]_+^{k\bullet}), \tag{3.7}$$
$$G_{t'}^{t_0} = G_t^{t_0}(J_k + [-\varepsilon_{\bullet k}(C_t^{t_0})B_t]_+^{\bullet k}), \tag{3.8}$$
$$C_t^{t_1} = (J_k + [-\varepsilon_{k\bullet}(G_t^{t_0})B_{t_0}]_+^{k\bullet})C_t^{t_0}, \tag{3.9}$$
$$G_t^{t_1} = (J_k + [\varepsilon_{k\bullet}(G_t^{t_0})B_{t_0}]_+^{\bullet k})G_t^{t_0}. \tag{3.10}$$

We prove the claims in the following order,

$$(a)_0 \Longrightarrow (b)_0 \Longrightarrow (c)_0 \Longrightarrow (a)_1 \Longrightarrow (b)_1 \Longrightarrow (c)_1 \Longrightarrow (a)_2 \Longrightarrow \cdots, \tag{3.11}$$

assuming all preceding claims. Moreover, we run the same induction procedure for C- and G-matrices $\tilde{C}_t^{t_0}$, $\tilde{G}_t^{t_0}$ of $\mathbf{B}^T$ in the background.

We first prove $(a)_0$, $(b)_0$, and $(c)_0$. The claims $(a)_0$ and $(c)_0$ were proved in Propositions 2.3 and 2.7. The claim $(b)_0$ holds because $C_{t_0}^{t_0} = G_{t_0}^{t_0} = I$, and Proposition 3.1 (c), where we have already received an important input from scattering diagrams.

Now, assuming the claims in (3.11) up to $(c)_d$, we prove $(a)_{d+1}$. This was already proved in Propositions 2.3 and 2.7 by taking account of Remarks 2.4 and 2.8.

Next, assuming the claims in (3.11) up to $(a)_{d+1}$, we prove $(b)_{d+1}$. (This is the highlights of the proof of the sign-coherence conjecture.) Let $t_0, t_1, t \in \mathbb{T}_n$ be vertices such that $d(t_0, t) = d$, $d(t_1, t) = d+1$, and t_0 and t_1 are k-adjacent.

By $(b)_d$, $\sigma^\circ(G_t^{t_0})$ is a chamber in $\mathfrak{S}(B_{t_0})$. By $(a)_d$, $(c)_d$, Proposition 2.22, and Remark 2.23, we have $\sigma(G_t^{t_1}) = \varphi_{t_0}^{t_1}(\sigma(G_t^{t_0}))$. Meanwhile, by Proposition 3.2, the piecewise linear isomorphism $\varphi_{t_0}^{t_1}$, which is homeomorphism, maps a chamber in $\mathfrak{S}(B_{t_0})$ to a chamber in $\mathfrak{S}(B_{t_1})$. Therefore, $\sigma^\circ(G_t^{t_1})$ is a chamber in $\mathfrak{S}(B_{t_1})$. This proves (i). By (i), each face $\sigma_i(G_t^{t_1})$ $(i = 1, \dots, n)$, which is a cone of codimension one by $(a)_{d+1}$, is a subset of the union of the supports of some walls of $\mathfrak{S}(B_{t_1})$. Meanwhile, by $(a)_{d+1}$, each c-vector $\mathbf{c}_{i;t}^{t_1}$ is a normal vector of the face $\sigma_i(G_t^{t_1})$ with respect to the inner product $(\ ,\ )_D$. Therefore, each c-vector $\mathbf{c}_{i;t}^{t_1}$ is a normal vector of the support of some wall of $\mathfrak{S}(B_{t_1})$. See Proposition 2.16. This proves (ii). Then, by Proposition 3.1 (a), the c-vector $\mathbf{c}_{i;t}^{t_1}$ is either positive or negative. This proves (iii), i.e., the column sign-coherence of the C-matrix $C_t^{t_1}$. The same result is proved in the background for the C-matrix $\tilde{C}_{t_1}^{t}$ of the B-pattern $\mathbf{B}^T$. Thus, by $(a)_{d+1}$, the G-matrix $G_t^{t_1}$ is row sign-coherent. This proves (iv).

Finally, assuming the claims in (3.11) up to $(b)_{d+1}$, we prove $(c)_{d+1}$. This was already proved in Propositions 2.3 and 2.7 by taking account of Remarks 2.4 and 2.8. □

We may roughly describe the results in Theorem 3.3 (a) that *the G-fan $\Delta(\mathbf{G}^{t_0})$ is embedded in the corresponding cluster scattering diagram $\mathfrak{D}_{\mathfrak{s}}$.*

We immediately extend Theorem 3.3 (c) for any (possibly singular) B-pattern by Proposition 2.11,

Theorem 3.4. *Let $\mathbf{B} = \{B_t\}_{t \in \mathbb{T}_n}$ be any (possibly singular) B-pattern. Then, for any $t_0, t \in \mathbb{T}_n$, the column sign-coherence for the C-matrix $C_t^{t_0}$ holds.*

Proof. Let $\overline{C}_t$ and $\overline{G}_t$ be C- and G-matrices of the principal extension $\overline{\mathbf{B}}$ of $\mathbf{B}$ with the initial vertex t_0 in Section 2.3. The matrix $\overline{C}_t$ is column sign-coherent by Theorem 3.3 (c). Then, the claim follows from (2.39). □

This completes the proof of Conjecture 1.34.

From now on, we use the column sign-coherence of C-matrices not as an assumption but as a theorem.

Remark 3.5. In Part III we show that Propositions 3.1 and 3.2 actually holds for any (possibly singular) B-pattern. Thus, Theorem 3.3 also holds for any (possibly singular) B-pattern by the same proof. This gives a direct proof of the sign-coherence without using Proposition 2.11.

4 More about F-polynomials

Let us turn to study F-polynomials, namely, the nontropical part of a cluster pattern.

4.1 Fock-Goncharov decomposition

Let $\mathbf{\Sigma}$ be any cluster pattern. We come back to the mutations of x- and y-variables, especially in the ε-expression (1.10) and (1.11). Let $t, t' \in \mathbb{T}_n$ be vertices that are k-adjacent. One can regard mutations (1.10) and (1.11) as isomorphisms between rational function fields as follows, where $\mathbf{x}_t$, $\mathbf{x}_{t'}$, $\mathbf{y}_t$, $\mathbf{y}_{t'}$ are n-tuple of formal variables:

$$\mu_{k;t} : \mathbb{Q}(\mathbf{x}_{t'}) \to \mathbb{Q}(\mathbf{x}_t),\ \mathbb{Q}(\mathbf{y}_{t'}) \to \mathbb{Q}(\mathbf{y}_t), \tag{4.1}$$

$$\mu_{k;t}(x_{i;t'}) = \begin{cases} x_{k;t}^{-1} \left(\prod_{j=1}^{n} x_{j;t}^{[-\varepsilon b_{jk;t}]_+} \right) (1 + \hat{y}_{k;t}^{\varepsilon}) & i = k, \\ x_{i;t} & i \neq k, \end{cases} \tag{4.2}$$

$$\mu_{k;t}(y_{i;t'}) = \begin{cases} y_{k;t}^{-1} & i = k, \\ y_{i;t} y_{k;t}^{[\varepsilon b_{ki;t}]_+} (1 + y_{k;t}^{\varepsilon})^{-b_{ki;t}} & i \neq k. \end{cases} \tag{4.3}$$

Note that the map is in the opposite direction, namely, from t' to t. Here and below, we use common symbols for maps for x- and y-variables as above in view of the parallelism between them.

To make use of the flexibility of the choice of sign ε, we first choose a given initial vertex t_0. Since we have already established the sign-coherence of C-matrices $C_t^{t_0} = C_t$, the tropical sign $\varepsilon_{i;t}^{t_0} = \varepsilon_{i;t}$ in Definition 2.1 is defined for any t and i. Now we set $\varepsilon = \varepsilon_{k;t}$ in (4.2) and (4.3).

Then, modifying the idea of [FG09a, §2.1] with tropical signs, we consider the following decompositions of the maps $\mu_{k;t}$:

$$\mu_{k;t} = \rho_{k;t} \circ \tau_{k;t}. \tag{4.4}$$

Here, $\tau_{k;t}^{t_0} = \tau_{k;t}$ are the following isomorphisms,

$$\tau_{k;t} : \mathbb{Q}(\mathbf{x}_{t'}) \to \mathbb{Q}(\mathbf{x}_t),\ \mathbb{Q}(\mathbf{y}_{t'}) \to \mathbb{Q}(\mathbf{y}_t), \tag{4.5}$$

$$\tau_{k;t}(x_{i;t'}) = \begin{cases} x_{k;t}^{-1} \prod_{j=1}^{n} x_{j;t}^{[-\varepsilon_{k;t} b_{jk;t}]_+} & i = k, \\ x_{i;t} & i \neq k, \end{cases} \tag{4.6}$$

$$\tau_{k;t}(y_{i;t'}) = \begin{cases} y_{k;t}^{-1} & i = k, \\ y_{i;t} y_{k;t}^{[\varepsilon_{k;t} b_{ki;t}]_+} & i \neq k, \end{cases} \tag{4.7}$$

while $\rho^{t_0}_{k;t} = \rho_{k;t}$ are the following automorphisms,

$$\rho_{k;t} : \mathbb{Q}(\mathbf{x}_t) \to \mathbb{Q}(\mathbf{x}_t),\ \mathbb{Q}(\mathbf{y}_t) \to \mathbb{Q}(\mathbf{y}_t), \tag{4.8}$$

$$\rho_{k;t}(x_{i;t}) = x_{i;t}(1 + \hat{y}_{k;t}^{\varepsilon_{k;t}})^{-\delta_{ik}}, \tag{4.9}$$

$$\rho_{k;t}(y_{i;t}) = y_{i;t}(1 + y_{k;t}^{\varepsilon_{k;t}})^{-b_{ki;t}}. \tag{4.10}$$

Note that the maps $\tau_{k;t}$ and $\rho_{k;t}$ depend on the choice of the initial vertex t_0 through tropical signs. Also, they are defined only after establishing the sign-coherence of C-matrices. We call the decomposition (4.4) the *Fock-Goncharov decomposition* of a mutation $\mu_{k;t}$ (*with tropical sign*) with respect to the initial vertex t_0. We also call $\tau_{k;t}$ and $\rho_{k;t}$ the *tropical part* and the *nontropical part* of $\mu_{k;t}$, respectively. Indeed, it is clear that (4.6) and (4.7) are the exponential form of the mutations of g-vectors (2.13) and c-vectors (2.12), respectively.

Remark 4.1. The decomposition (4.4) was introduced in [FG09a] for $\varepsilon = 1$. The signed version here were used, for example, in the application to the Stokes automorphisms [IN14] and dilogarithm identities [GNR17].

Proposition 4.2. *Let $t, t' \in \mathbb{T}_n$ be vertices that are k-adjacent. Then, the following relations hold.*

$$\mu_{k;t'} \circ \mu_{k;t} = \mathrm{id}, \tag{4.11}$$

$$\tau_{k;t'} \circ \tau_{k;t} = \mathrm{id}. \tag{4.12}$$

Proof. The equality (4.11) is nothing but the involution of the mutations of x- and y-variables. The equality (4.12) for both x- and y-variables can be verified by the fact,

$$\varepsilon_{k;t'} = -\varepsilon_{k;t}, \quad b_{ki;t'} = -b_{ki;t}, \quad b_{jk;t'} = -b_{jk;t}. \tag{4.13}$$

□

For any $t \in \mathbb{T}_n$, consider a sequence of vertices $t_0, t_1, \dots, t_{r+1} = t \in \mathbb{T}_n$ such that they are sequentially adjacent with edges labeled by $k_0, \dots, k_r$. Then, we define isomorphisms

$$\mu^{t_0}_t := \mu_{k_0;t_0} \circ \mu_{k_1;t_1} \circ \cdots \circ \mu_{k_r;t_r} : \mathbb{Q}(\mathbf{x}_t) \to \mathbb{Q}(\mathbf{x}),\ \mathbb{Q}(\mathbf{y}_t) \to \mathbb{Q}(\mathbf{y}), \tag{4.14}$$

$$\tau^{t_0}_t := \tau_{k_0;t_0} \circ \tau_{k_1;t_1} \circ \cdots \circ \tau_{k_r;t_r} : \mathbb{Q}(\mathbf{x}_t) \to \mathbb{Q}(\mathbf{x}),\ \mathbb{Q}(\mathbf{y}_t) \to \mathbb{Q}(\mathbf{y}), \tag{4.15}$$

where we set $\mathbf{x}_{t_0} = \mathbf{x}$, $\mathbf{y}_{t_0} = \mathbf{y}$. Thanks to Proposition 4.2, $\mu^{t_0}_t$ and $\tau^{t_0}_t$ depend only on t_0 and t. Namely, we do not have to care about the redundancy $k_{s+1} = k_s$ in the sequence $k_0, \dots, k_r$.

Recall the notions of tropical x- and y-variables in (1.54) and (1.53). The following proposition tells that the tropical parts $\tau^{t_0}_t$ are nothing but the mutations of tropical x- and y-variables.

Proposition 4.3. *The following formulas hold:*

$$\mu_t^{t_0}(x_{i;t}) = x^{\mathbf{g}_{i;t}} F_{i;t}(\hat{\mathbf{y}}), \tag{4.16}$$

$$\mu_t^{t_0}(y_{i;t}) = y^{\mathbf{c}_{i;t}} \prod_{j=1}^{n} F_{j;t}(\mathbf{y})^{b_{ji;t}}, \tag{4.17}$$

$$\tau_t^{t_0}(x_{i;t}) = x^{\mathbf{g}_{i;t}}, \tag{4.18}$$

$$\tau_t^{t_0}(y_{i;t}) = y^{\mathbf{c}_{i;t}}. \tag{4.19}$$

Proof. The first two equalities are the separation formulas in Theorem 1.23. Let us prove the equality (4.18) for x-variables by the induction on t along $\mathbb{T}_n$ starting from t_0. For $t = t_0$, $x^{\mathbf{g}_{i;t_0}} = x^{\mathbf{e}_i} = x_i$. Therefore, (4.18) holds. Suppose that it is true for t with $d(t_0, t) = d$. Let t' be the vertex that is k-adjacent to t such that $d(t_0, t') = d + 1$. Then, we have

$$\begin{aligned} \tau_{t'}^{t_0}(x_{i;t'}) &= (\tau_t^{t_0} \circ \tau_{k;t})(x_{i;t'}) \\ &= \begin{cases} \tau_t^{t_0}\left(x_{k;t}^{-1} \prod_{j=1}^{n} x_{j;t}^{[-\varepsilon_{k;t} b_{jk;t}]_+} \right) = x^{-\mathbf{g}_{k;t} + \sum_{j=1}^{n} [-\varepsilon_{k;t} b_{jk;t}]_+ \mathbf{g}_{j;t}} & i = k, \\ \tau_t^{t_0}(x_{i;t}) = x^{\mathbf{g}_{i;t}} & i \neq k \end{cases} \\ &= x^{\mathbf{g}_{i;t'}}, \end{aligned} \tag{4.20}$$

where we used (2.13) in the last equality. Therefore, (4.18) holds for t'. Similarly, the equality (4.19) for y-variables follows from (2.12). □

Under these maps, $\hat{y}$-variables transform in the same way as y-variables as expected.

Proposition 4.4. *The following formulas hold:*

$$\tau_{k;t}(\hat{y}_{i;t'}) = \begin{cases} \hat{y}_{k;t}^{-1} & i = k, \\ \hat{y}_{i;t} \hat{y}_{k;t}^{[\varepsilon_{k;t} b_{ki;t}]_+} & i \neq k, \end{cases} \tag{4.21}$$

$$\rho_{k;t}(\hat{y}_{i;t}) = \hat{y}_{i;t}(1 + \hat{y}_{k;t}^{\varepsilon_{k;t}})^{-b_{ki;t}}, \tag{4.22}$$

$$\tau_t^{t_0}(\hat{y}_{i;t}) = \hat{y}^{\mathbf{c}_{i;t}} = x^{\hat{\mathbf{c}}_{i;t}}, \tag{4.23}$$

where $\hat{\mathbf{c}}_{i;t} = B_{t_0} \mathbf{c}_{i;t}$ *is a* $\hat{c}$*-vector in* (1.57).

Proof. The equality (4.21) follows from (4.6) and (1.12). The equality (4.22) follows from (4.9). The equality (4.23) follows from (4.18) and (1.56),

□

4.2 Nontropical parts and F-polynomials

Observing Proposition 4.3, it is clear that the nontropical parts $\rho_{k;t}$ are responsible to generate and mutate F-polynomials. Let us make this statement more manifest.

In addition to the inner product $(\ ,\ )_D$ in (2.45), we introduce a skew-symmetric form

$$\{\mathbf{u}, \mathbf{v}\}_{DB} := \mathbf{u}^T DB\mathbf{v} = (\mathbf{u}, B\mathbf{v})_D, \tag{4.24}$$

where $B = B_{t_0}$. Then, we introduce the following automorphisms $\mathfrak{q}_{k;t}$ for the initial x- and y-variables,

$$\mathfrak{q}_{k;t} : \mathbb{Q}(\mathbf{x}) \to \mathbb{Q}(\mathbf{x}),\ \mathbb{Q}(\mathbf{y}) \to \mathbb{Q}(\mathbf{y}), \tag{4.25}$$

$$\mathfrak{q}_{k;t}(x^{\mathbf{m}}) = x^{\mathbf{m}}(1 + \hat{y}^{\mathbf{c}^+_{k;t}})^{-(\delta_k \mathbf{c}_{k;t}, \mathbf{m})_D} \tag{4.26}$$

$$= x^{\mathbf{m}}(1 + x^{\hat{\mathbf{c}}^+_{k;t}})^{-(\delta_k \mathbf{c}_{k;t}, \mathbf{m})_D}, \tag{4.27}$$

$$\mathfrak{q}_{k;t}(y^{\mathbf{n}}) = y^{\mathbf{n}}(1 + y^{\mathbf{c}^+_{k;t}})^{-\{\delta_k \mathbf{c}_{k;t}, \mathbf{n}\}_{DB}}, \tag{4.28}$$

where $\mathbf{m}, \mathbf{n} \in \mathbb{Z}^n$, and $\mathbf{c}^+_{k;t} = \varepsilon_{k;t}\mathbf{c}_{k;t}$ and $\hat{\mathbf{c}}^+_{k;t} = \varepsilon_{k;t}\hat{\mathbf{c}}_{k;t}$ are a c^+-vector and a $\hat{c}^+$-vector, respectively, in Definition 2.30.

The seemingly asymmetric definitions in (4.26) and (4.28) are justified by the following properties.

Proposition 4.5. *The following facts hold:*

(a). We have the formulas

$$\mathfrak{q}_{k;t}(x^{\mathbf{g}_{i;t}}) = x^{\mathbf{g}_{i;t}}(1 + \hat{y}^{\mathbf{c}^+_{k;t}})^{-\delta_{ik}}, \tag{4.29}$$

$$\mathfrak{q}_{k;t}(y^{\mathbf{c}_{i;t}}) = y^{\mathbf{c}_{i;t}}(1 + y^{\mathbf{c}^+_{k;t}})^{-b_{ki;t}}. \tag{4.30}$$

(b). The following relation holds:

$$\tau_t^{t_0} \circ \rho_{k;t} = \mathfrak{q}_{k;t} \circ \tau_t^{t_0}. \tag{4.31}$$

(c). If t' and t are k-adjacent, we have

$$\mathfrak{q}_{k;t'} \circ \mathfrak{q}_{k;t} = \mathrm{id}. \tag{4.32}$$

Proof. (a). The claim follows from the following equalities:

$$(\delta_k \mathbf{c}_{k;t}, \mathbf{g}_{i;t})_D = \delta_{ki}, \tag{4.33}$$

$$\{\delta_k \mathbf{c}_{k;t}, \mathbf{c}_{i;t}\}_{DB} = b_{ki;t}. \tag{4.34}$$

The equality (4.33) is the duality in (2.47). The equality (4.34) follows from (2.14) as follows:

$$D^{-1}(C_t^T DB_{t_0} C_t) = D^{-1}(DB_t) = B_t. \tag{4.35}$$

(b). Thanks to (4.29) and Propositions 4.3 and 4.4, we have

$$x_{i;t} \overset{\rho_{k;t}}{\mapsto} x_{i;t}(1+\hat{y}_{k;t}^{\varepsilon_{k;t}})^{-\delta_{ik}} \overset{\tau_t^{t_0}}{\mapsto} x^{\mathbf{g}_{i;t}}(1+\hat{y}^{\mathbf{c}_{k;t}^+})^{-\delta_{ik}}, \tag{4.36}$$

$$x_{i;t} \overset{\tau_t^{t_0}}{\mapsto} x^{\mathbf{g}_{i;t}} \overset{\mathfrak{q}_{k;t}}{\mapsto} x^{\mathbf{g}_{i;t}}(1+\hat{y}^{\mathbf{c}_{k;t}^+})^{-\delta_{ik}}. \tag{4.37}$$

Therefore, the equality (4.31) holds for x-variables. Similarly, we have

$$y_{i;t} \overset{\rho_{k;t}}{\mapsto} y_{i;t}(1+y_{k;t}^{\varepsilon_{k;t}})^{-b_{ki;t}} \overset{\tau_t^{t_0}}{\mapsto} y^{\mathbf{c}_{i;t}}(1+y^{\mathbf{c}_{k;t}^+})^{-b_{ki;t}}, \tag{4.38}$$

$$y_{i;t} \overset{\tau_t^{t_0}}{\mapsto} y^{\mathbf{c}_{i;t}} \overset{\mathfrak{q}_{k;t}}{\mapsto} y^{\mathbf{c}_{i;t}}(1+y^{\mathbf{c}_{k;t}^+})^{-b_{ki;t}}. \tag{4.39}$$

Therefore, the equality (4.31) holds also for y-variables.

(c). This follows from $\mathbf{c}_{k;t'} = -\mathbf{c}_{k;t}$. □

For the same $t_0, t_1, \dots, t_{r+1} = t \in \mathbb{T}_n$ for $\mu_t^{t_0}$ and $\tau_t^{t_0}$ in (4.14) and (4.15), we define the automorphism

$$\mathfrak{q}_t^{t_0} := \mathfrak{q}_{k_0;t_0} \circ \mathfrak{q}_{k_1;t_1} \circ \cdots \circ \mathfrak{q}_{k_r;t_r} : \mathbb{Q}(\mathbf{x}) \to \mathbb{Q}(\mathbf{x}),\ \mathbb{Q}(\mathbf{y}) \to \mathbb{Q}(\mathbf{y}). \tag{4.40}$$

Again, by Proposition 4.5 (c), it depends only on t_0 and t.

One can separate the tropical and the nontropical parts of $\mu_t^{t_0}$ as follows:

Proposition 4.6. *The following decomposition holds:*

$$\mu_t^{t_0} = \mathfrak{q}_t^{t_0} \circ \tau_t^{t_0} : \mathbb{Q}(\mathbf{x}_t) \to \mathbb{Q}(\mathbf{x}),\ \mathbb{Q}(\mathbf{y}_t) \to \mathbb{Q}(\mathbf{y}). \tag{4.41}$$

Proof. First, we note that

$$\rho_{k_0;t_0}(x_i) = x_i(1+\hat{y}_{k_0})^{-\delta_{ik_0}} = \mathfrak{q}_{k_0;t_0}(x_i), \tag{4.42}$$

$$\rho_{k_0;t_0}(y_i) = y_i(1+y_{k_0})^{-b_{k_0i;t_0}} = \mathfrak{q}_{k_0;t_0}(y_i). \tag{4.43}$$

For x-variables, for example, by Proposition 4.5 (b), we have the following commutative diagram:

$$\begin{array}{ccccccccc}
\mathbb{Q}(\mathbf{x}_t) & \xrightarrow{\tau_{k_r;t_r}} & \mathbb{Q}(\mathbf{x}_{t_r}) & & & \xrightarrow{\qquad\tau_{t_r}^{t_0}\qquad} & & & \mathbb{Q}(\mathbf{x}_{t_0}) \\
 & \searrow^{\mu_{k_r;t_r}} & \downarrow{\scriptstyle\rho_{k_r;t_r}} & & & & & & \downarrow{\scriptstyle\mathfrak{q}_{k_r;t_r}} \\
 & & \mathbb{Q}(\mathbf{x}_{t_r}) & \xrightarrow{\tau_{k_{r-1};t_{r-1}}} & \mathbb{Q}(\mathbf{x}_{t_{r-1}}) & & \xrightarrow{\quad\tau_{t_{r-1}}^{t_0}\quad} & & \mathbb{Q}(\mathbf{x}_{t_0}) \\
 & & \ddots & & \vdots & & & & \vdots \\
 & & & & \mathbb{Q}(\mathbf{x}_{t_2}) & \xrightarrow{\tau_{k_1;t_1}} & \mathbb{Q}(\mathbf{x}_{t_1}) & \xrightarrow{\tau_{k_0;t_0}} & \mathbb{Q}(\mathbf{x}_{t_0}) \\
 & & & & & \searrow^{\mu_{k_1;t_1}} & \downarrow{\scriptstyle\rho_{k_1;t_1}} & & \downarrow{\scriptstyle\mathfrak{q}_{k_1;t_1}} \\
 & & & & & & \mathbb{Q}(\mathbf{x}_{t_1}) & \xrightarrow{\tau_{k_0;t_0}} & \mathbb{Q}(\mathbf{x}_{t_0}) \\
 & & & & & & & \searrow^{\mu_{k_0;t_0}} & \downarrow{\scriptstyle\rho_{k_0;t_0} = \mathfrak{q}_{k_0;t_0}} \\
 & & & & & & & & \mathbb{Q}(\mathbf{x}_{t_0})
\end{array} \tag{4.44}$$

Therefore, the claim holds. □

We conclude that the automorphisms $\mathfrak{q}_t^{t_0}$ generate the nontropical parts of x- and y-variables in the following manner.

Theorem 4.7. *The following formulas hold:*

$$\mathfrak{q}_t^{t_0}(x^{\mathbf{g}_{i;t}}) = x^{\mathbf{g}_{i;t}} F_{i;t}(\hat{\mathbf{y}}), \tag{4.45}$$

$$\mathfrak{q}_t^{t_0}(y^{\mathbf{c}_{i;t}}) = y^{\mathbf{c}_{i;t}} \prod_{j=1}^{n} F_{j;t}(\mathbf{y})^{b_{ji;t}}, \tag{4.46}$$

$$\mathfrak{q}_t^{t_0}(\hat{y}^{\mathbf{c}_{i;t}}) = \hat{y}^{\mathbf{c}_{i;t}} \prod_{j=1}^{n} F_{j;t}(\hat{\mathbf{y}})^{b_{ji;t}}. \tag{4.47}$$

Proof. The equalities (4.45) and (4.46) follow from Propositions 4.3 and 4.6. For example, for x-variables,

$$\mathfrak{q}_t^{t_0}(x^{\mathbf{g}_{i;t}}) = \mathfrak{q}_t^{t_0}(\tau_t^{t_0}(x_{i;t})) = \mu_t^{t_0}(x_{i;t}) = x^{\mathbf{g}_{i;t}} F_{i;t}(\hat{\mathbf{y}}). \tag{4.48}$$

Similarly, the equality (4.47) follows from Propositions 4.4 and 4.6. □

4.3 Detropicalization

Let $\mathbf{\Sigma}$ be any cluster pattern of rank n, and let t_0 be a given initial vertex. Let S_n be the symmetric group of degree n. We define the (left) action of a permutation $\nu \in S_n$ on x- and y-variables, etc., as follows.

$$\mathbf{x}' = \nu \mathbf{x}_t, \quad x'_i = x_{\nu^{-1}(i);t}, \tag{4.49}$$

$$\mathbf{y}' = \nu \mathbf{x}_t, \quad y'_i = y_{\nu^{-1}(i);t}, \tag{4.50}$$

$$B' = \nu B_t, \quad b'_{ij} = b_{\nu^{-1}(i)\nu^{-1}(j);t}, \tag{4.51}$$

$$C' = \nu C_t, \quad c'_{ij} = c_{i\nu^{-1}(j);t}, \tag{4.52}$$

$$G' = \nu G_t, \quad g'_{ij} = g_{i\nu^{-1}(j);t}. \tag{4.53}$$

Note that these actions are compatible with the separation formulas in Theorem 1.23.

For any $t, t' \in \mathbb{T}_n$ and any permutation $\nu \in S_n$, we define the following isomorphisms.

$$\nu_t^{t'} : \mathbb{Q}(\mathbf{x}_t) \to \mathbb{Q}(\mathbf{x}_{t'}),\ \mathbb{Q}(\mathbf{y}_t) \to \mathbb{Q}(\mathbf{y}_{t'}), \tag{4.54}$$

$$\nu_t^{t'}(x_{i;t}) = x_{\nu^{-1}(i);t'}, \tag{4.55}$$

$$\nu_t^{t'}(y_{i;t}) = y_{\nu^{-1}(i);t'}. \tag{4.56}$$

Then, by Proposition 4.3, one can rephrase the *periodicity condition* for G-matrices and x-variables as follows:

$$G_t^{t_0} = \nu G_{t'}^{t_0} \iff \tau_t^{t_0} = \tau_{t'}^{t_0} \circ \nu_t^{t'} \text{ (for } x\text{-variables)}, \tag{4.57}$$

$$\mathbf{x}_t = \nu \mathbf{x}_{t'} \quad \Longleftrightarrow \quad \mu_t^{t_0} = \mu_{t'}^{t_0} \circ \nu_t^{t'} \text{ (for } x\text{-variables).} \tag{4.58}$$

Also, the parallel statement holds for C-matrices and y-variables.

Proposition 4.8. *Let t_0, t, t' be any vertices in $\mathbb{T}_n$. Suppose that*

$$G_t^{t_0} = \nu G_{t'}^{t_0} \tag{4.59}$$

holds. Then, we have

$$\mathbf{x}_t = \nu \mathbf{x}_{t'} \quad \Longleftrightarrow \quad \mathfrak{q}_t^{t_0} = \mathfrak{q}_{t'}^{t_0} \text{ (for } x\text{-variables).} \tag{4.60}$$

Also, the parallel statement holds for C-matrices and y-variables.

Proof. The claim follows from (4.57), (4.58), and Proposition 4.6. □

Note that the permutation $\nu_t^{t'}$ drops off in the condition for $\mathfrak{q}_t^{t_0}$ in (4.60). The properties

$$G_t^{t_0} = \nu G_{t'}^{t_0} \quad \Longrightarrow \quad \mathbf{x}_t = \nu \mathbf{x}_{t'}, \tag{4.61}$$

$$C_t^{t_0} = \nu C_{t'}^{t_0} \quad \Longrightarrow \quad \mathbf{y}_t = \nu \mathbf{y}_{t'} \tag{4.62}$$

are called the *detropicalization* of x- and y-variables, respectively. They are the core of the *synchronicity phenomenon* occurring in cluster patterns systematically studied in [Nak21].

Remark 4.9. The implications (4.61) and (4.62) were proved for $\nu = \mathrm{id}$ by [CHL20, Lemma 2.4 & Theorem 2.5] and for general ν by [Nak21, Theorem 5.2]. However, the proofs therein depend on the Laurent positivity in Theorem 1.37, which we temporarily dismiss. We come back to the point later in Section 7.1.

5 Scattering diagrams

In this section we present a precise formulation of scattering diagrams following [GHK15, GHKK18]. To do this, we set the following guideline.

(a). Since in our approach we do not need scattering diagrams with *frozen variables* (corresponding to *coefficients of geometrical type* in [FZ07]), we skip the relevant formulation. This makes a considerable simplification of the formulation.

(b). We omit the description of the underlying Lie algebra $\mathfrak{g}$ and the group G in [GHKK18] for scattering diagrams. They are crucial for the construction of scattering diagrams, but we can skip them for our purpose.

(c). We clarify and establish the connection between cluster patterns and scattering diagrams in detail by step-by-step examples.

(d). We mostly follow the notations and the conventions in [GHKK18]. However, we take the *transpose* of the skew-symmetric forms therein to match the convention of the exchange matrices in [FZ07].

5.1 Fixed data and seed

Definition 5.1 (Fixed data). A *fixed data* Γ consists of the following:

- A lattice $N \simeq \mathbb{Z}^n$ of rank n with a skew-symmetric bilinear form
$$\{\cdot,\cdot\} : N \times N \to \mathbb{Q}. \tag{5.1}$$
- A sublattice $N^\circ \subset N$ of finite index (equivalently, of rank n) such that
$$\{N^\circ, N\} \subset \mathbb{Z}. \tag{5.2}$$
- Positive integers $\delta_1, \dots, \delta_n$ such that there is a basis $(e_1, \dots, e_n)$ of N, where $(\delta_1 e_1, \dots, \delta_n e_n)$ is a basis of N°.
- $M = \mathrm{Hom}(N, \mathbb{Z})$, $M^\circ = \mathrm{Hom}(N^\circ, \mathbb{Z})$, $M_\mathbb{R} = M \otimes_\mathbb{Z} \mathbb{R}$, and we regard
$$M \subset M^\circ \subset M_\mathbb{R}. \tag{5.3}$$

For a given $N^\circ \subset N$ of finite index, the above $\delta_1, \dots, \delta_n$ always exist; for example, take the elementary divisors of the embedding $N^\circ \hookrightarrow N$. However, other choices are equally good.

Let $\langle n, m\rangle : N^\circ \times M^\circ \to \mathbb{Z}$ be the canonical paring. We also write its linear extension $N \times M_\mathbb{R} \to \mathbb{R}$ by the same symbol $\langle n, m\rangle$. (Though it is a little confusing, the symbol n is used to denote both the rank and an element in N.)

Remark 5.2. (a). In (5.2) we already took the transpose of the form in [GHKK18] as mentioned in the above guideline (d).

(b). In [GHKK18] it is assumed that the numbers $\delta_1, \dots, \delta_n$ are coprime. Here we do not require it. See also Remark 5.29.

Definition 5.3 (Seed). A *seed* $\mathfrak{s} = (e_1, \dots, e_n)$ for a fixed data Γ is a basis of N such that $(\delta_1 e_1, \dots, \delta_n e_n)$ is a basis of N°.

Let $\mathfrak{s} = (e_1, \dots, e_n)$ be a seed for a fixed data Γ, and let $(e_1^*, \dots, e_n^*)$ be the dual basis of M. Let $f_i = \delta_i^{-1} e_i^*$ $(i = 1, \dots, n)$. Then, $(f_1, \dots, f_n)$ is a basis of M°.

Accordingly, we define isomorphisms, all of which are denoted by the same symbol for simplicity,

$$\begin{array}{rccc} \phi_{\mathfrak{s}}: & N & \rightarrow & \mathbb{Z}^n, \\ & e_i & \mapsto & \mathbf{e}_i, \end{array} \tag{5.4}$$

$$\begin{array}{rcccccc} \phi_{\mathfrak{s}}: & M^\circ & \rightarrow & \mathbb{Z}^n, & M_{\mathbb{R}} & \rightarrow & \mathbb{R}^n, \\ & f_i & \mapsto & \mathbf{e}_i. & & & \end{array} \tag{5.5}$$

The identifications under the above isomorphisms are written as, $N \simeq_{\mathfrak{s}} \mathbb{Z}^n$, $M^\circ \simeq_{\mathfrak{s}} \mathbb{Z}^n$, $M_{\mathbb{R}} \simeq_{\mathfrak{s}} \mathbb{R}^n$, respectively.

Let us identify $N \simeq_{\mathfrak{s}} \mathbb{Z}^n$ and $M^\circ \simeq_{\mathfrak{s}} \mathbb{Z}^n$. For $n \in N^\circ \subset N$ and $m \in M^\circ$, let $\mathbf{n} \in \mathbb{Z}^n$ and $\mathbf{m} \in \mathbb{Z}^n$ be the corresponding vectors. Then, the canonical paring $\langle n, m \rangle$ for $N^\circ \times M^\circ$ is given by

$$\langle n, m \rangle = (\mathbf{n}, \mathbf{m})_D := \mathbf{n}^T D \mathbf{m}, \quad D = \mathrm{diag}(\delta_1^{-1}, \dots, \delta_n^{-1}). \tag{5.6}$$

This symmetric bilinear form $(\cdot, \cdot)_D$ agrees with the one in (2.45). We also regard it as a pairing $N \times M_{\mathbb{R}} \to \mathbb{R}$, depending on the context.

Example 5.4. A seed $\mathfrak{s}$ for a fixed data Γ is identified with a seed $(\mathbf{x}, \mathbf{y}, B)$ (in the sense of Definition 1.3) in the following way: Let x^m $(m \in M^\circ)$ and y^n $(n \in N)$ be the monomial expressions with formal symbols x and y such that $x^{m+m'} = x^m x^{m'}$, $y^{n+n'} = y^n y^{n'}$. Let $\mathcal{F}_X$ and $\mathcal{F}_Y$ be the rational function fields generated by all formal exponentials x^m and y^n, respectively. Then, a seed $(\mathbf{x}_{\mathfrak{s}}, \mathbf{y}_{\mathfrak{s}}, B_{\mathfrak{s}}) = (\mathbf{x}, \mathbf{y}, B)$ in $(\mathcal{F}_X, \mathcal{F}_Y)$ is defined by

$$x_i = x^{f_i}, \quad y_i = y^{e_i}, \quad b_{ij} = \{\delta_i e_i, e_j\}. \tag{5.7}$$

In particular, the matrix D in (5.6) is a (left) skew-symmetrizer of $B = (b_{ij})$. Moreover, under the identification $N \simeq_{\mathfrak{s}} \mathbb{Z}^n$, the skew-symmetric bilinear form in (5.1) is given by

$$\{n, n'\} = \{\mathbf{n}, \mathbf{n}'\}_{DB} := \mathbf{n}^T DB \mathbf{n}'. \tag{5.8}$$

This skew-symmetric bilinear form agrees with the one in (4.24). This is the reason why we take D as (2.44) throughout. For a given seed $\mathfrak{s}$ (the *initial seed*) for a fixed data Γ and a given initial vertex $t_0 \in \mathbb{T}_n$, we associate a cluster pattern $\mathbf{\Sigma}_{\mathfrak{s}, t_0}$ whose initial seed Σ_{t_0} is given by the seed $(\mathbf{x}, \mathbf{y}, B)$ in (5.7). Below we consider the C-, G-, and F-patterns of $\mathbf{\Sigma}_{\mathfrak{s}, t_0}$ with the initial vertex t_0 unless otherwise mentioned.

Remark 5.5. In the scattering diagram formalism, it is natural to consider the diagonal matrix with positive integer diagonals $\Delta = \mathrm{diag}(\delta_1, \dots, \delta_n)$ and the skew-symmetric rational matrix $\Omega = (\omega_{ij})_{i,j=1}^n$, $\omega_{ij} = \{e_i, e_j\}$. Then, the matrices in (5.6)–(5.8) are given by

$$D = \Delta^{-1}, \quad B = \Delta\Omega, \quad DB = \Omega. \tag{5.9}$$

5.2 Walls

Let us introduce the homomorphism

$$\begin{array}{rccc} p^* : & N & \to & M^\circ \\ & n & \mapsto & \{\cdot, n\} \end{array}. \tag{5.10}$$

Example 5.6 (continued). Using the matrix B in (5.7), we have

$$p^*(e_j) = \sum_{i=1}^n b_{ij} f_i. \tag{5.11}$$

Therefore, under our identification $N \simeq_{\mathfrak{s}} \mathbb{Z}^n$ and $M^\circ \simeq_{\mathfrak{s}} \mathbb{Z}^n$, the matrix representation of p^* is just given by B. Also, again by (5.7) and (5.11),

$$x^{p^*(e_i)} = \prod_{j=1}^n x_j^{b_{ji}} = \hat{y}_i. \tag{5.12}$$

From now on we assume the following condition for a fixed data Γ:

Assumption 5.7 (Injectivity Assumption). The map p^* is injective.

Example 5.8 (continued.). Injectivity Assumption is equivalent to assume that the skew-symmetric form $\{\cdot,\cdot\}$ in (5.1) is nondegenerate. Thus, B in (5.7) is nonsingular, and $\hat{y}_1, \dots, \hat{y}_n$ in (5.12) are algebraically independent.

Define

$$N^+ = N_{\mathfrak{s}}^+ := \left\{ \sum_{i=1}^n a_i e_i \;\middle|\; a_i \in \mathbb{Z}_{\geq 0},\ \sum_{i=1}^n a_i \neq 0 \right\}, \tag{5.13}$$

$$P = P_{\mathfrak{s}} := \left\{ \sum_{i=1}^n a_i p^*(e_i) \;\middle|\; a_i \in \mathbb{Z}_{\geq 0} \right\} = p^*(N^+) \sqcup \{0\} \subset M^\circ. \tag{5.14}$$

We view P as a monoid.

Example 5.9 (continued.). Under the identification $M^\circ \simeq_{\mathfrak{s}} \mathbb{Z}^n$, we have

$$P = \left\{ \sum_{i=1}^n a_i \mathbf{b}_i \;\middle|\; a_i \in \mathbb{Z}_{\geq 0} \right\}, \tag{5.15}$$

where $\mathbf{b}_i$ is the ith column of B.

Let $\Bbbk$ be any field of characteristic zero, e.g., $\Bbbk = \mathbb{Q}$. Let $\Bbbk[P]$ be the monoid algebra of P over $\Bbbk$. Let J be the maximal ideal of $\Bbbk[P]$ generated by $P \setminus \{0\}$, and let

$$\Bbbk[[P]] = \varprojlim \Bbbk[P]/J^{\ell} \tag{5.16}$$

be the completion with respect to J. Using the formal exponentials x^m ($m \in P$), an element of $\Bbbk[P]$ (resp. $\Bbbk[[P]]$) is written as a polynomial (resp. a formal power series) in x as

$$f = \sum_{m \in P} c_m x^m, \quad (c_m \in \Bbbk). \tag{5.17}$$

Remark 5.10. In [GHKK18] the notation z^m is used. We use x^m because of the forthcoming identification with x-variables. This notation is also compatible with the one in Example 5.4.

Example 5.11 (continued.). Under the identification (5.12), $\Bbbk[P]$ is the polynomial algebra $\Bbbk[\hat{y}_1, \dots, \hat{y}_n]$. Also, $\Bbbk[[P]]$ is the algebra of formal power series $\Bbbk[[\hat{y}_1, \dots, \hat{y}_n]]$.

Definition 5.12 (Normalized automorphism). For $n \in N^+$ and $f = 1 + \sum_{k=1}^{\infty} c_k x^{kp^*(n)} \in \Bbbk[[P]]$, we define an automorphism $\mathfrak{p}_{n,f}$ of $\Bbbk[[P]]$ by

$$\mathfrak{p}_{n,f}(x^m) = x^m f^{\langle \delta(n)n, m \rangle}, \quad (m \in P), \tag{5.18}$$

where $\delta(n)$ is the smallest positive rational number such that $\delta(n)n \in N^{\circ}$. We call $\delta(n)$ the *normalization factor* of n and $\mathfrak{p}_{n,f}$ the *normalized automorphism* by n and f.

Example 5.13 (continued.). Let us identify $N \simeq_{\mathfrak{s}} \mathbb{Z}^n$ and $M^{\circ} \simeq_{\mathfrak{s}} \mathbb{Z}^n$.

(a). Let $n = \mathbf{e}_i$, where $p^*(n) = B\mathbf{e}_i = \mathbf{b}_i$. Then, we have $\delta(\mathbf{e}_i) = \delta_i$. Let $f = 1 + x^{\mathbf{b}_i} = 1 + \hat{y}_i$. Then, the automorphism $\mathfrak{p}_{n,f}$ is given by

$$\mathfrak{p}_{n,f}(x^{\mathbf{m}}) = x^{\mathbf{m}}(1 + \hat{y}_i)^{(\delta_i \mathbf{e}_i, \mathbf{m})_D}. \quad (\mathbf{m} \in P \subset M^{\circ}). \tag{5.19}$$

(b). Let $n = \mathbf{c}^+_{i;t}$, where $p^*(n) = B\mathbf{c}^+_{i;t} = \hat{\mathbf{c}}^+_{i;t}$. We first claim that

$$\delta(\mathbf{c}^+_{i;t}) = \delta_i. \tag{5.20}$$

We prove it by the induction on t along $\mathbb{T}_n$ starting from t_0. For $t = t_0$, this is the case (a) above. Assume that (5.20) holds for some t. Let t' be the vertex that is k-adjacent to t. By (2.6), we have

$$\begin{aligned} C_{t'} D^{-1} &= C_t (J_k + [\varepsilon_{k;t} B_t]_+^{k\bullet}) D^{-1} \\ &= C_t D^{-1} (J_k + [-\varepsilon_{k;t} (B_t)^T]_+^{k\bullet}), \end{aligned} \tag{5.21}$$

where (1.13) is used for the last equality. Since $|J_k + [-\varepsilon_{k;t}(B_t)^T]_+^{k\bullet}| = -1$ by Lemma 1.10, each column vector $\delta_i \mathbf{c}^+_{i;t'}$ belongs to N_0. Furthermore, δ_i is the smallest positive rational number satisfying this property because $\delta_1 \mathbf{c}^+_{1;t'}$, $\dots$, $\delta_n \mathbf{c}^+_{n;t'}$ are a $\mathbb{Z}$-basis of N°. This proves (5.20). Now, let $f = 1 + x^{\hat{\mathbf{c}}^+_{i;t}} = 1 + \hat{y}^{\mathbf{c}^+_{i;t}}$. Then, we have

$$\mathfrak{p}_{n,f}(x^{\mathbf{m}}) = x^{\mathbf{m}}(1 + \hat{y}^{\mathbf{c}^+_{i;t}})^{(\delta_i \mathbf{c}^+_{i;t}, \mathbf{m})_D}, \quad (\mathbf{m} \in P \subset M^\circ). \tag{5.22}$$

Observe that

$$\mathfrak{p}_{n,f}^{-\varepsilon_{i;t}} = \mathfrak{q}_{i;t}, \tag{5.23}$$

where $\mathfrak{p}_{n,f}^{-1}$ is the inverse of $\mathfrak{p}_{n,f}$, and $\mathfrak{q}_{i;t}$ is the one in (4.26) viewed as an automorphism of $\Bbbk[[P]]$. The factor $-\varepsilon_{i;t}$ in (5.23) is important for the integrity of the whole picture.

For any $n \in N^+$, we define a hyperplane in $M_\mathbb{R}$

$$n^\perp := \{m \in M_\mathbb{R} \mid \langle n, m \rangle = 0\} \subset M_\mathbb{R}. \tag{5.24}$$

We say that $n \in N^+$ is *primitive* if there is no pair $j \in \mathbb{Z}_{>1}$ and $n' \in N^+$ such that $n = jn'$. Let N^+_{pr} be the set of all primitive elements in N.

Now we introduce the fundamental ingredient of scattering diagrams.

Definition 5.14 (Wall). A *wall* for a seed $\mathfrak{s}$ is a triplet $\mathbf{w} = (\mathfrak{d}, f)_n$, where

- $n \in N^+_{\mathrm{pr}}$,
- $\mathfrak{d} \subset n^\perp$ is a (not necessarily strongly convex) cone in $M_\mathbb{R}$ of codimension one,
- f is an element of $\Bbbk[[x^{p^*(n)}]] \subset \Bbbk[[P]]$ with constant term 1,

$$f = 1 + \sum_{k=1}^{\infty} c_k x^{kp^*(n)}. \tag{5.25}$$

The cone $\mathfrak{d}$ is called the *support* of a wall. No names were explicitly given for f and n in [GHKK18]. Since this is inconvenient, here we call them the *wall function* and the *normal vector* of a wall. We say that a wall $\mathbf{w} = (\mathfrak{d}, f)_n$ is *incoming* if

$$p^*(n) \in \mathfrak{d}. \tag{5.26}$$

Otherwise, we say it is *outgoing*.

Remark 5.15. Since $\{n, n\} = 0$ by skew-symmetry, we have

$$p^*(n) \in n^\perp. \tag{5.27}$$

Example 5.16 (continued). Let us identify $N \simeq_{\mathfrak{s}} \mathbb{Z}^n$ and $M^\circ \simeq_{\mathfrak{s}} \mathbb{Z}^n$.

(a). Let $n = \mathbf{e}_i$, $\mathfrak{d} = \mathbf{e}_i^\perp$, and $f = 1 + \hat{y}_i$. Then, $\mathbf{b}_i = B\mathbf{e}_i \in \mathfrak{d}$, and $\mathbf{w}_{i;t_0} = (\mathbf{e}_i^\perp, 1 + \hat{y}_i)_{\mathbf{e}_i}$ is an *incoming* wall.

(b). Let $n = \mathbf{c}_{i;t}^+ = \varepsilon_{i;t}\mathbf{c}_{i;t}$, $\mathfrak{d} = \sigma_i(G_t)$, and $f = 1 + \hat{y}^{\mathbf{c}_{i;t}^+}$. Firstly, $\mathbf{c}_{i;t}^+$ is primitive due to the unimodularity (2.3) of C-matrices. Secondly, $\mathbf{c}_{i;t}^+$ is a normal vector of $\sigma_i(G_t)$ thanks to the duality in Proposition 2.16. Therefore, $\mathbf{w}_{i;t} = (\sigma_i(G_t), 1 + \hat{y}^{\mathbf{c}_{i;t}^+})_{\mathbf{c}_{i;t}^+}$ is a wall. Moreover, due to Proposition 2.32, if $\mathbf{c}_{i;t}^+ \neq \mathbf{e}_\ell$ for any ℓ, it is an *outgoing* wall. Note that, if $t, t' \in \mathbb{T}_n$ are k-adjacent, then two walls $\mathbf{w}_{k;t}$ and $\mathbf{w}_{k;t'}$ are identical.

5.3 Scattering diagrams

Let $\widehat{J}$ be the maximal ideal of $\Bbbk[[P]]$ generated by $P \setminus \{0\}$. We say that $f = 1 + \sum_{m \in P, m \neq 0} c_m x^m \in \Bbbk[[P]]$ is *trivial modulo* $\widehat{J}^\ell$ if its image in $\Bbbk[[P]]/\widehat{J}^\ell$ is 1.

Example 5.17 (continued). Under the identification $\Bbbk[[P]] = \Bbbk[[\hat{y}_1, \dots, \hat{y}_n]]$, f is trivial modulo $\widehat{J}^\ell$ if and only if $f - 1$ contains no monomial in $\hat{y}_1, \dots, \hat{y}_n$ whose total degree is less than ℓ.

Now we can give the definition of a scattering diagram.

Definition 5.18 (Scattering diagram). A *scattering diagram* $\mathfrak{D}$ for a seed $\mathfrak{s}$ is a collection of walls $\{\mathbf{w}_\lambda = (\mathfrak{d}_\lambda, f_\lambda)_{n_\lambda}\}_{\lambda \in \Lambda}$ with respect to a seed $\mathfrak{s}$, where Λ is a finite or countably infinite index set, satisfying the following *finiteness condition*:

- For each positive integer ℓ, there are only finitely many walls whose wall functions are not trivial modulo $\widehat{J}^\ell$.

Definition 5.19 (Support and singular locus). For a scattering diagram $\mathfrak{D} = \{(\mathfrak{d}_\lambda, f_\lambda)_{n_\lambda}\}_{\lambda \in \Lambda}$, the *support* and the *singular locus* of $\mathfrak{D}$ are defined by

$$\mathrm{Supp}(\mathfrak{D}) = \bigcup_{\lambda \in \Lambda} \mathfrak{d}_\lambda, \tag{5.28}$$

$$\mathrm{Sing}(\mathfrak{D}) = \bigcup_{\lambda \in \Lambda} \partial\mathfrak{d}_\lambda \cup \bigcup_{\substack{\lambda, \lambda' \in \Lambda \\ \dim \mathfrak{d}_\lambda \cap \mathfrak{d}_{\lambda'} = n-2}} \mathfrak{d}_\lambda \cap \mathfrak{d}_{\lambda'}. \tag{5.29}$$

Example 5.20 (continued). Let us identify $N \simeq_{\mathfrak{s}} \mathbb{Z}^n$ and $M^\circ \simeq_{\mathfrak{s}} \mathbb{Z}^n$. For the cluster pattern $\mathbf{\Sigma}_{\mathfrak{s},t_0}$, we define a set of walls

$$\mathfrak{D}(\mathbf{\Sigma}_{\mathfrak{s},t_0}) := \{\mathbf{w}_{i;t} = (\sigma_i(G_t), 1 + \hat{y}^{\mathbf{c}_{i;t}^+})_{\mathbf{c}_{i;t}^+} \mid i = 1, \dots, n;\ t \in \mathbb{T}_n\}, \tag{5.30}$$

where we discard all duplicate walls so that $\mathfrak{D}(\mathbf{\Sigma}_{\mathfrak{s},t_0})$ consists of mutually distinct walls. We may take a different label $(i;t)$ freely for identical walls.

The collection of walls $\mathfrak{D}(\mathbf{\Sigma}_{\mathfrak{s},t_0})$ is not necessarily a scattering diagram, because the finiteness condition is not guaranteed. In fact, there are examples where $\mathfrak{D}(\mathbf{\Sigma}_{\mathfrak{s},t_0})$ does not satisfy the finiteness condition. See Remark 6.15.

5.4 Wall-crossing automorphisms

Definition 5.21 (Admissible curve). A curve $\gamma : [0,1] \to M_{\mathbb{R}}$ is *admissible* for a scattering diagram $\mathfrak{D}$ if it satisfies the following properties:

- γ does not intersect $\mathrm{Sing}(\mathfrak{D})$.
- The end points of γ are in $M_{\mathbb{R}} \setminus \mathrm{Supp}(\mathfrak{D})$.
- It is a smooth curve, and it intersects $\mathrm{Supp}(\mathfrak{D})$ transversally.

Definition 5.22 (Wall-crossing automorphism). For any scattering diagram $\mathfrak{D}$ and any admissible curve γ, we define the associated *wall-crossing automorphism along* γ, $\mathfrak{p}_{\gamma,\mathfrak{D}} \in \mathrm{Aut}(\Bbbk[[P]])$ as follows: For each degree ℓ, take all walls whose wall functions are not trivial modulo $\widehat{J}^{\ell}$. There are only finitely many such walls. Let $(\mathfrak{d}_s, f_s)_{n_s}$ $(s = 1, \dots, r)$ be the walls among them which γ intersects. We may assume that the curve $\gamma(t)$ intersects $\mathfrak{d}_s$ at time $t = t_s$ such that

$$0 < t_1 \le t_2 \le \cdots \le t_r < 1. \tag{5.31}$$

For each s, we define the *intersection sign*

$$\epsilon_s = \begin{cases} 1 & \langle n_s, \gamma'(t_s)\rangle < 0, \\ -1 & \langle n_s, \gamma'(t_s)\rangle > 0, \end{cases} \tag{5.32}$$

where $\gamma'(t)$ is the velocity vector of $\gamma(t)$ at t. Then, we define

$$\mathfrak{p}^{\ell}_{\gamma,\mathfrak{D}} = \mathfrak{p}^{\epsilon_r}_{n_r,f_r} \circ \cdots \circ \mathfrak{p}^{\epsilon_2}_{n_2,f_2} \circ \mathfrak{p}^{\epsilon_1}_{n_1,f_1}, \tag{5.33}$$

$$\mathfrak{p}_{\gamma,\mathfrak{D}} = \lim_{\ell \to \infty} \mathfrak{p}^{\ell}_{\gamma,\mathfrak{D}}. \tag{5.34}$$

Note that $\mathfrak{p}_{\gamma,\mathfrak{D}}$ only depends on the homotopy class of γ in $M_{\mathbb{R}} \setminus \mathrm{Sing}(\mathfrak{D})$.

Remark 5.23. In [GHKK18], $\mathfrak{p}_{\gamma,\mathfrak{D}}$ is called a *path-ordered product* of automorphisms $\mathfrak{p}^{\epsilon_s}_{n_s,f_s}$.

Example 5.24 (continued). Let $\mathfrak{D}(\mathbf{\Sigma}_{\mathfrak{s},t_0})$ be the collection of walls in Example 5.20. Even though the finiteness condition does not hold in general, the wall-crossing automorphism is still well-defined for a curve crossing only finitely many walls. Let t_0, t_1, ..., $t_{r+1} = t \in \mathbb{T}_n$ be a sequence of vertices such that they are sequentially adjacent with edges labeled by k_0, k_1,..., k_r. Accordingly, we consider an admissible curve $\gamma^{t_0}_t$ further satisfying the following conditions:

- It is in the interior of the support $|\Delta(\mathbf{G}^{t_0})|$ of the G-fan $\Delta(\mathbf{G}^{t_0})$.

- It starts in $\sigma^\circ(G_{t_{r+1}})$.
- It sequentially intersects walls

$$(\sigma_{k_r}(G_{t_r}), 1 + \hat{y}^{\mathbf{c}^+_{k_r;t_r}})_{\mathbf{c}^+_{k_r;t_r}}, \dots, (\sigma_{k_0}(G_{t_0}), 1 + \hat{y}^{\mathbf{c}^+_{k_0;t_0}})_{\mathbf{c}^+_{k_0;t_0}}. \tag{5.35}$$

- It ends in $\sigma^\circ(G_{t_0})$.

By Proposition 2.16, for each $s = 0, \dots, r$, the c-vector $\mathbf{c}_{k_s;t_s}$ is inward for the cone $\sigma_{k_s}(G_{t_s})$. Thus, when $\gamma_t^{t_0}$ crosses $\sigma_{k_s}(G_{t_s})$ to *get into* $\sigma(G_{t_s})$, the velocity vector γ' and the c-vector $\mathbf{c}_{k_s;t_s}$ are always in the *same* direction with respect to $\sigma_{k_s}(G_{t_s})$. Since the normal vector of the wall is $n_s = \varepsilon_{k_s;t_s}\mathbf{c}_{k_s;t_s}$, the factor in (5.32) is given by $-\varepsilon_{k_s;t_s}$. Then, by (4.40) and (5.23), the wall-crossing automorphism along $\gamma_t^{t_0}$ is identified with $\mathfrak{q}_t^{t_0}$ in (4.40) as follows:

$$\begin{aligned}\tilde{\mathfrak{p}}_t^{t_0} := \mathfrak{p}_{\gamma_t^{t_0}, \mathfrak{D}(\mathbf{\Sigma}_{\mathfrak{s},t_0})} &= \mathfrak{p}^{-\varepsilon_{k_0,t_0}}_{\mathbf{c}^+_{k_0;t_0}, 1+\hat{y}^{\mathbf{c}^+_{k_0;t_0}}} \circ \cdots \circ \mathfrak{p}^{-\varepsilon_{k_r,t_r}}_{\mathbf{c}^+_{k_r;t_r}, 1+\hat{y}^{\mathbf{c}^+_{k_r;t_r}}} \\ &= \mathfrak{q}_{k_0;t_0} \circ \cdots \circ \mathfrak{q}_{k_r;t_r} \\ &= \mathfrak{q}_t^{t_0}.\end{aligned} \tag{5.36}$$

Note that the factor $-\varepsilon_{i;t}$ in (5.23) is absorbed in coordination with the factor in (5.32).

Definition 5.25 (Equivalence/Consistency).

- Two scattering diagrams $\mathfrak{D}$ and $\mathfrak{D}'$ with a common initial seed $\mathfrak{s}$ are *equivalent* if, for any curve γ that is admissible for both $\mathfrak{D}$ and $\mathfrak{D}'$, the equality $\mathfrak{p}_{\gamma,\mathfrak{D}} = \mathfrak{p}_{\gamma,\mathfrak{D}'}$ holds.
- A scattering diagram $\mathfrak{D}$ is *consistent* if for any admissible curve γ, the associated wall-crossing automorphism $\mathfrak{p}_{\gamma,\mathfrak{D}}$ depends only on the end points of γ.

Remark 5.26. For a given scattering diagram, one obtains equivalent scattering diagrams by repeating the following procedures:

- To join a pair of walls $(\mathfrak{d}_1, f)_n$ and $(\mathfrak{d}_2, f)_n$ intersecting in their common face $\mathfrak{d}_1 \cap \mathfrak{d}_2$ of codimension two into a wall $(\mathfrak{d}_1 \cup \mathfrak{d}_2, f)_n$, or to split it conversely.
- To join a pair of walls $(\mathfrak{d}, f_1)_n$ and $(\mathfrak{d}, f_2)_n$ into a wall $(\mathfrak{d}, f_1 f_2)_n$, or to split it conversely. One may join infinitely many walls with a common support as well by taking the infinite product of wall functions, or split it conversely.
- To add or to remove a trivial wall $(\mathfrak{d}, 1)$.

Applying these operations, and especially removing all trivial walls, one can obtain a scattering diagram whose support is minimal among other equivalent diagrams.

The following is the first fundamental theorem on scattering diagrams in view of the application to cluster algebra theory.

Theorem 5.27 ([GHKK18, Theorems 1.12 &1.28]). *Let $\mathfrak{s}$ be any initial seed for Γ. Then, there exists a scattering diagram $\mathfrak{D}_{\mathfrak{s}}$ satisfying the following properties:*

(a). $\mathfrak{D}_{\mathrm{in},\mathfrak{s}} := \{(e_i^{\perp}, 1+\hat{y}_i)_{e_i} \mid i=1,\dots,n\} \subset \mathfrak{D}_{\mathfrak{s}}$.

(b). $\mathfrak{D}_{\mathfrak{s}} \setminus \mathfrak{D}_{\mathrm{in},\mathfrak{s}}$ *consists only of outgoing walls, and their normal vectors are not equal to* e_i *for any* i.

(c). $\mathfrak{D}_{\mathfrak{s}}$ *is consistent.*

Moreover, such a scattering diagram is unique up to the equivalence.

We call $\mathfrak{D}_{\mathfrak{s}}$ a *cluster scattering diagram* (CSD for short) for $\mathfrak{s}$.

Remark 5.28. The second statement of (b) is taken from [GHKK18, Theorem 1.28, Remark 1.29], while all other statements are from [GHKK18, Theorem 1.12].

From now on we assume that $\mathfrak{D}_{\mathfrak{s}}$ has the minimal support among other equivalent diagrams. Under the identification $M_{\mathbb{R}} \simeq_{\mathfrak{s}} \mathbb{R}^n$, we define $\mathfrak{S}(B_{t_0}) := \mathrm{Supp}(\mathfrak{D}_{\mathfrak{s}}) \subset \mathbb{R}^n$. Now we see that the properties of the support diagram $\mathfrak{S}(B_{t_0})$ given in Proposition 3.1 follow from Definition 5.14 and Theorem 5.27.

Remark 5.29. There is some redundancy of fixed data, because we do not require the coprimeness of $\delta_1, \dots, \delta_n$ here. Suppose that the exchange matrix B in (5.7) is indecomposable. For a fixed data Γ, another fixed data Γ' is called a *rescaling* of Γ if there is some rational number $\lambda > 0$ such that the following relations hold:

$$N' = N, \tag{5.37}$$
$$\{n_1, n_2\}' = \lambda\{n_1, n_2\}, \tag{5.38}$$
$$(N^\circ)' = \lambda^{-1} N^\circ, \tag{5.39}$$
$$\delta_i' = \lambda^{-1}\delta_i, \tag{5.40}$$
$$(M^\circ)' = \lambda M^\circ. \tag{5.41}$$

For such a pair, a seed $\mathfrak{s}$ for Γ is also regarded as a seed for Γ'. Let $\phi_\lambda : \Bbbk[[P]] \to \Bbbk[[P']]$, $x^m \mapsto x^{\lambda m}$ be the induced algebra isomorphism. Then, we have, for $m \in P$,

$$\begin{aligned}(\mathfrak{p}'_{n,\phi_\lambda(f)} \circ \phi_\lambda)(x^m) &= \mathfrak{p}'_{n,\phi_\lambda(f)}(x^{\lambda m}) = x^{\lambda m}\phi_\lambda(f)^{\langle \delta_0'(n)n, \lambda m\rangle} \\ &= x^{\lambda m}\phi_\lambda(f)^{\langle \delta(n)n, m\rangle} = (\phi_\lambda \circ \mathfrak{p}_{n,f})(x^m).\end{aligned} \tag{5.42}$$

Thus, cluster scattering diagrams $\mathfrak{D}_{\mathfrak{s}}$ and $\mathfrak{D}'_{\mathfrak{s}}$ for Γ and Γ', respectively, are identified under the above correspondence. Moreover, for x_i, y_i, b_{ij} and x_i',

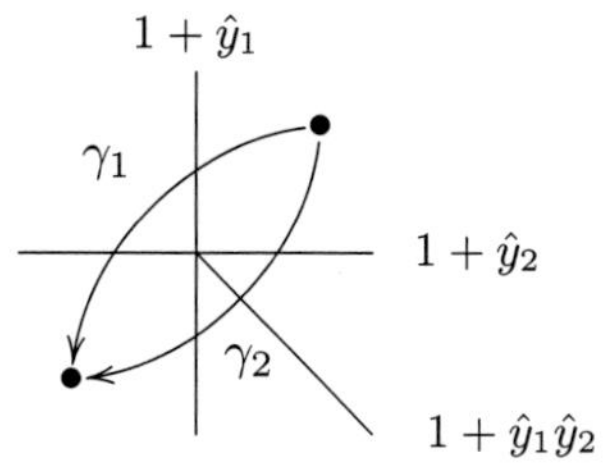

Figure 6: Consistent scattering diagram $\mathfrak{D}_\mathfrak{s}$ of Type A_2.

y'_i, b'_{ij} in (5.7) for Γ and Γ', respectively, the following relations hold:

$$x'_i = x_i^\lambda, \quad y'_i = y_i, \quad b'_{ij} = b_{ij}. \tag{5.43}$$

In particular, we see that a cluster scattering diagram $\mathfrak{D}_\mathfrak{s}$ depends only on the matrix $B = (b_{ij})$ as expected. In general, when B is not indecomposable, we may apply the above rescaling to each indecomposable component of B.

5.5 Rank 2 examples of consistent scattering diagrams $\mathfrak{D}_\mathfrak{s}$

Let us present the scattering diagrams $\mathfrak{D}_\mathfrak{s}$ in Theorem 5.27 in the rank 2 case under the identification $N \simeq_\mathfrak{s} \mathbb{Z}^n$ and $M_\mathbb{R} \simeq_\mathfrak{s} \mathbb{R}^n$, following [GHKK18]. Using the equivalence in Remark 5.29, we may assume that $\{e_2, e_1\} = 1$ without loosing generality. We use the same convention and notations in Section 2.5.

(I). Finite type. The scattering diagram $\mathfrak{D}_\mathfrak{s}$ is equivalent to $\mathfrak{D}(\mathbf{\Sigma}_{\mathfrak{s},t_0})$ constructed in Example 5.20. In other words, $\mathfrak{D}(\mathbf{\Sigma}_{\mathfrak{s},t_0})$ itself is a consistent scattering diagram.

(a). Type A_2. Let

$$\delta_1 = \delta_2 = 1; \ \hat{y}_1 = x_2, \ \hat{y}_2 = x_1^{-1}. \tag{5.44}$$

The scattering diagram $\mathfrak{D}(\mathbf{\Sigma}_{\mathfrak{s},t_0})$ consists of five distinct walls of the form in (5.30). They are rearranged up to the equivalence of scattering diagrams as two incoming walls and one outgoing walls as

$$(\mathbf{e}_1^\perp, 1+\hat{y}_1)_{\mathbf{e}_1}, \ (\mathbf{e}_2^\perp, 1+\hat{y}_2)_{\mathbf{e}_2}, \ (\sigma((1,-1)), 1+\hat{y}_1\hat{y}_2)_{(1,1)}. \tag{5.45}$$

This is the scattering diagram $\mathfrak{D}_\mathfrak{s}$, which is depicted in Figure 6. The only property to be checked is the following consistency condition

$$\mathfrak{p}_{\gamma_1,\mathfrak{D}_\mathfrak{s}} = \mathfrak{p}_{\gamma_2,\mathfrak{D}_\mathfrak{s}} \tag{5.46}$$

for the admissible curves γ_1 and γ_2 in Figure 6. One can check it easily by the direct calculation as follows:

$$\begin{aligned}\mathfrak{p}_{\gamma_1,\mathfrak{D}_s}(x_1) &= \mathfrak{p}_{\mathbf{e}_2,1+\hat{y}_2}(\mathfrak{p}_{\mathbf{e}_1,1+\hat{y}_1}(x_1))\\ &= \mathfrak{p}_{\mathbf{e}_2,1+\hat{y}_2}(x_1(1+\hat{y}_1))\\ &= x_1(1+\hat{y}_1(1+\hat{y}_2))\\ &= x_1(1+\hat{y}_1+\hat{y}_1\hat{y}_2).\end{aligned} \tag{5.47}$$

$$\begin{aligned}\mathfrak{p}_{\gamma_2,\mathfrak{D}_s}(x_1) &= \mathfrak{p}_{\mathbf{e}_1,1+\hat{y}_1}(\mathfrak{p}_{(1,1),1+\hat{y}_1\hat{y}_2}(\mathfrak{p}_{\mathbf{e}_2,1+\hat{y}_2}(x_1)))\\ &= \mathfrak{p}_{\mathbf{e}_1,1+\hat{y}_1}(\mathfrak{p}_{(1,1),1+\hat{y}_1\hat{y}_2}(x_1))\\ &= \mathfrak{p}_{\mathbf{e}_1,1+\hat{y}_1}(x_1(1+\hat{y}_1\hat{y}_2))\\ &= x_1(1+\hat{y}_1)(1+\hat{y}_1\hat{y}_2(1+\hat{y}_1)^{-1})\\ &= x_1(1+\hat{y}_1+\hat{y}_1\hat{y}_2).\end{aligned} \tag{5.48}$$

Similarly,

$$\begin{aligned}\mathfrak{p}_{\gamma_1,\mathfrak{D}_s}(x_2) &= \mathfrak{p}_{\mathbf{e}_2,1+\hat{y}_2}(\mathfrak{p}_{\mathbf{e}_1,1+\hat{y}_1}(x_2))\\ &= \mathfrak{p}_{\mathbf{e}_2,1+\hat{y}_2}(x_2)\\ &= x_2(1+\hat{y}_2).\end{aligned} \tag{5.49}$$

$$\begin{aligned}\mathfrak{p}_{\gamma_2,\mathfrak{D}_s}(x_2) &= \mathfrak{p}_{\mathbf{e}_1,1+\hat{y}_1}(\mathfrak{p}_{(1,1),1+\hat{y}_1\hat{y}_2}(\mathfrak{p}_{\mathbf{e}_2,1+\hat{y}_2}(x_2)))\\ &= \mathfrak{p}_{\mathbf{e}_1,1+\hat{y}_1}(\mathfrak{p}_{(1,1),1+\hat{y}_1\hat{y}_2}(x_2(1+\hat{y}_2)))\\ &= \mathfrak{p}_{\mathbf{e}_1,1+\hat{y}_1}(x_2(1+\hat{y}_1\hat{y}_2)(1+\hat{y}_2(1+\hat{y}_1\hat{y}_2)^{-1}))\\ &= \mathfrak{p}_{\mathbf{e}_1,1+\hat{y}_1}(x_2(1+\hat{y}_2+\hat{y}_1\hat{y}_2))\\ &= x_2(1+\hat{y}_2(1+\hat{y}_1)^{-1}+\hat{y}_1\hat{y}_2(1+\hat{y}_1)^{-1})\\ &= x_2(1+\hat{y}_2).\end{aligned} \tag{5.50}$$

It is also instructive to give an alternative derivation of (5.46) in view of Proposition 4.8. Recall the action of a permutation $\sigma \in S_n$ in (4.49)–(4.53). We know from (2.52) that

$$G_{t_2} = s_{12}G_{t_{-3}}, \tag{5.51}$$

where s_{12} is the transposition of 1 and 2. Also, the following property (the *pentagon periodicity*) is the first thing we learn in cluster algebra theory.

$$\mathbf{x}_{t_2} = s_{12}\mathbf{x}_{t_{-3}}. \tag{5.52}$$

Therefore, by Proposition 4.8, we have (for x-variables)

$$\mathfrak{q}^{t_0}_{t_2} = \mathfrak{q}^{t_0}_{t_{-3}}. \tag{5.53}$$

Thus, by (5.36),

$$\tilde{\mathfrak{p}}^{t_0}_{t_2} = \tilde{\mathfrak{p}}^{t_0}_{t_{-3}}. \tag{5.54}$$

Since $\mathfrak{p}_{\gamma_1,\mathfrak{D}_\mathfrak{s}} = (\tilde{\mathfrak{p}}^{t_0}_{t_2})^{-1}$ and $\mathfrak{p}_{\gamma_2,\mathfrak{D}_\mathfrak{s}} = (\tilde{\mathfrak{p}}^{t_0}_{t_{-3}})^{-1}$, we have (5.46). Moreover, it clarifies the meaning of the consistency condition (5.46); namely, *it guarantees the detropicalization of x-variables.*

(b). Type B_2. Let

$$\delta_1 = 1,\ \delta_2 = 2;\ \hat{y}_1 = x_2^2,\ \hat{y}_2 = x_1^{-1}. \tag{5.55}$$

The situation is the same as type A_2, and the outgoing walls are

$$(\sigma((1,-2)), 1 + \hat{y}_1\hat{y}_2)_{(1,1)},\ \left(\sigma((1,-1)), 1 + \hat{y}_1\hat{y}_2^2\right)_{(1,2)}. \tag{5.56}$$

(c). Type G_2. Let

$$\delta_1 = 1,\ \delta_2 = 3;\ \hat{y}_1 = x_2^3,\ \hat{y}_2 = x_1^{-1}. \tag{5.57}$$

The situation is the same as type A_2, and the outgoing walls are

$$\begin{aligned}&(\sigma((1,-3)), 1 + \hat{y}_1\hat{y}_2)_{(1,1)},\ \left(\sigma((1,-2)), 1 + \hat{y}_1^2\hat{y}_2^3\right)_{(2,3)},\\ &\left(\sigma((2,-3)), 1 + \hat{y}_1\hat{y}_2^2\right)_{(1,2)},\ \left(\sigma((1,-1)), 1 + \hat{y}_1\hat{y}_2^3\right)_{(1,3)}.\end{aligned} \tag{5.58}$$

(II). Infinite type. The consistent scattering diagram $\mathfrak{D}_\mathfrak{s}$ is an extension of $\mathfrak{D}(\mathbf{\Sigma}_{\mathfrak{s},t_0})$ such that the supports of additional walls are in the complement of $|\Delta(\mathbf{G}^{t_0})|$. In particular, $\mathfrak{D}(\mathbf{\Sigma}_{\mathfrak{s},t_0})$ is a scattering diagram but it is not consistent.

(d). Type $A_1^{(1)}$. Let

$$\delta_1 = \delta_2 = 2;\ \hat{y}_1 = x_2^2,\ \hat{y}_2 = x_1^{-2}. \tag{5.59}$$

There is one additional wall to $\mathfrak{D}(\mathbf{\Sigma}_{\mathfrak{s},t_0})$ [GHKK18, Example 1.15], which is

$$\left(\sigma((1,-1)), \left(\sum_{k=0}^{\infty} \hat{y}_1^k\hat{y}_2^k\right)^2\right)_{(1,1)}. \tag{5.60}$$

(e). Type $A_2^{(2)}$. Let

$$\delta_1 = 1,\ \delta_2 = 4;\ \hat{y}_1 = x_2^4,\ \hat{y}_2 = x_1^{-1}. \tag{5.61}$$

There is one additional wall to $\mathfrak{D}(\mathbf{\Sigma}_{\mathfrak{s},t_0})$ [Rea20a, Theorem 3.4.], which is

$$\left(\sigma((1,-2)), (1 + \hat{y}_1\hat{y}_2^2)\left(\sum_{k=0}^{\infty} \hat{y}_1^k\hat{y}_2^{2k}\right)^2\right)_{(1,2)}. \tag{5.62}$$

(f). Non-affine type. There are infinitely many additional walls inside the irrational cone spanned by $\mathbf{v}$ and $\mathbf{v}'$ in (2.68). The region is informally called the *Badlands.* We may also call it the *dark side* from the viewpoint of cluster patterns. The explicit description of walls therein is not yet known. However, for the skew-symmetric case $\delta_1 = \delta_2$, the existence of walls for every rational slope was proved in [DM21, Example 7.10].

6 Scattering diagrams and F-polynomials

In this section we show how F-polynomials and x-variables are related with the consistent scattering diagram $\mathfrak{D}_{\mathfrak{s}}$. The Laurent positivity follows from the positivity of theta functions in [GHKK18].

6.1 Mutations of scattering diagrams

We continue to assume that a fixed data Γ satisfies Injectivity Assumption throughout the section.

The following definition originates in [FG09a, §2.1].

Definition 6.1 (Seed mutation). Let $\mathfrak{s} = (e_1, \dots, e_k)$ be any seed for a fixed data Γ. For any $k = 1, \dots, n$, we define a new seed $\mu_k(\mathfrak{s}) = \mathfrak{s}' = (e'_1, \dots, e'_n)$ by

$$e'_i = \begin{cases} -e_k & i = k, \\ e_i + [b_{ki}]_+ e_k & i \neq k, \end{cases} \qquad b_{ij} = \{\delta_i e_i, e_j\}, \tag{6.1}$$

which is called the *mutation* of $\mathfrak{s}$ in direction k.

The transformation (6.1) is written by a matrix form P in Lemma 1.10. Thus, $e'_1, \dots, e'_n$ are certainly a basis of N. Also, under the mutation (6.1), The basis $\delta_1 e_1, \dots, \delta_n e_n$ of N° mutates as

$$\delta_i e'_i = \begin{cases} -\delta_k e_k & i = k, \\ \delta_i e_i + [-b_{ik}]_+ \delta_k e_k & i \neq k, \end{cases} \tag{6.2}$$

where we used the skew-symmetry $b_{ki}\delta_i = -b_{ik}\delta_k$. Thus, $\delta_1 e'_1, \dots, \delta_n e'_n$ are a basis of N° by the same reason as above. Therefore, $\mathfrak{s}'$ is certainly a seed in the sense of Definition 5.3.

On can easily confirm that the mutation (6.2) induces the following mutation for the basis $f_1, \dots, f_n$ of M°, which are dual to $\delta_1 e_1, \dots, \delta_n e_n$:

$$f'_i = \begin{cases} -f_k + \sum_{j=1}^{n} [-b_{jk}]_+ f_j & i = k, \\ f_i & i \neq k. \end{cases} \tag{6.3}$$

Also, we define the matrix $B' = (b'_{ij})$, $b'_{ij} = \{\delta_i e'_i, e'_j\}$ as in (5.7). Then, B' coincides with the matrix mutation of B in direction k by (1.12) with $\varepsilon = 1$.

Remark 6.2. The transformation (6.3) already appeared in (2.78) under the identification $M_{\mathbb{R}} \simeq_{\mathfrak{s}} \mathbb{R}^n$.

Remark 6.3. For $\mathfrak{s}' = \mu_k(\mathfrak{s})$, $\mu_k(\mathfrak{s}') = \mathfrak{s}$ does not hold, because $b'_{ki} = -b_{ki}$. Namely, μ_k is *not* an involution on the set of all seeds for a fixed data Γ. This causes a technical problem, for example, in the proof of the forthcoming

Theorem 6.13. To remedy the situation, we also define another mutation of a seed $\mathfrak{s}' = \mu_k^-(\mathfrak{s})$ by

$$e'_i = \begin{cases} -e_k & i = k, \\ e_i + [-b_{ki}]_+ e_k & i \neq k. \end{cases} \tag{6.4}$$

Again, we define the matrix $B' = (b'_{ij})$, $b'_{ij} = \{\delta_i e'_i, e'_j\}$. Then, B' coincides with the matrix mutation of B in direction k by (1.12) with $\varepsilon = -1$. Moreover, we have

$$\mu_k^-(\mu_k^+(\mathfrak{s})) = \mu_k^+(\mu_k^-(\mathfrak{s})) = \mathfrak{s}, \tag{6.5}$$

where $\mu_k^+ = \mu_k$. We call $\mu_k^\pm$ the *signed mutations* of $\mathfrak{s}$. They appeared, for example, in [IN14].

For any seed $\mathfrak{s} = (e_1, \dots, e_n)$ and $k = 1, \dots, n$, we define a linear map

$$\begin{array}{rccc} S_{k;\mathfrak{s}}: & M_\mathbb{R} & \to & M_\mathbb{R} \\ & m & \mapsto & m + \langle \delta_k e_k, m \rangle p^*(e_k). \end{array} \tag{6.6}$$

and its dual

$$\begin{array}{rccc} S^*_{k;\mathfrak{s}}: & N & \to & N \\ & n & \mapsto & n + \{\delta_k e_k, n\} e_k. \end{array} \tag{6.7}$$

The following equality can be easily verified:

$$\langle S^*_{k;\mathfrak{s}}(n), S_{k;\mathfrak{s}}(x) \rangle = \langle n, x \rangle. \tag{6.8}$$

Let

$$\mathcal{H}_{k,+} = \{m \in M_\mathbb{R} \mid \langle e_k, m \rangle \geq 0\}, \quad \mathcal{H}_{k,-} = \{m \in M_\mathbb{R} \mid \langle e_k, m \rangle \leq 0\}. \tag{6.9}$$

Then, following [GHKK18], we define a piecewise-linear map

$$\begin{array}{rccc} T_{k;\mathfrak{s}}: & M_\mathbb{R} & \to & M_\mathbb{R} \\ & m & \mapsto & \begin{cases} S_{k;\mathfrak{s}}(m) & m \in \mathcal{H}_{k,+}, \\ m & m \in \mathcal{H}_{k,-}. \end{cases} \end{array} \tag{6.10}$$

Remark 6.4. Under the identification $M_\mathbb{R} \simeq_\mathfrak{s} \mathbb{R}^n$, it coincides with the map $T_{k;t_0}$ in (2.81).

Definition 6.5 (Mutation of $\mathfrak{D}_\mathfrak{s}$)**.** For the consistent scattering diagram $\mathfrak{D}_\mathfrak{s}$ in Theorem 5.27 with minimal support, we define a collection of walls $T_k(\mathfrak{D}_\mathfrak{s})$, called the *mutation of* $\mathfrak{D}_\mathfrak{s}$ *in direction* k, by collecting all walls obtained below:

(a). For each wall $(\mathfrak{d}, f)_n$ of $\mathfrak{D}_\mathfrak{s}$ other than $(e_k^\perp, 1+\hat{y}_k)_{e_k}$, let $\mathfrak{d}_+ = \mathfrak{d} \cap \mathcal{H}_{k,+}$, and $\mathfrak{d}_- = \mathfrak{d} \cap \mathcal{H}_{k,-}$. We split the wall $(\mathfrak{d}, f)_n$ into two walls

$$(\mathfrak{d}_+, f)_n, \quad (\mathfrak{d}_-, f)_n, \tag{6.11}$$

where, if one of $\mathfrak{d}_+$ and $\mathfrak{d}_-$ is codimension more than 1, we throw that part away. Then, we replace $(\mathfrak{d}_+, f)_n$ with a new wall $(S_{k;\mathfrak{s}}(\mathfrak{d}_+), S_{k;\mathfrak{s}}(f))_{S^*_{k;\mathfrak{s}}(n)}$, where $S_{k;\mathfrak{s}}(f)$ is the formal power series obtained from f by replacing each term x^m $(m \in P)$ with $x^{S_{k;\mathfrak{s}}(m)}$, while we leave $(\mathfrak{d}_-, f)_n$ as it is.

(b). Replace $(e_k^\perp, 1+\hat{y}_k)_{e_k}$ with $(e_k'^\perp, 1+\hat{y}'_k)_{e'_k}$, where $e_k'^\perp = e_k{}^\perp$ and $\hat{y}'_k = \hat{y}_k^{-1}$ by (6.1).

To summarize, in both (a) and (b), each support of a wall (after splitting) is transformed by the piecewise-linear map $T_{k;\mathfrak{s}}$. On the other hand, each wall function in (a) obeys the same *linear map* applied to its support, while the case (b) is exceptional.

Remark 6.6. Here we encounter a temporal problem that the walls above are not walls for the initial seed $\mathfrak{s}$ by the following reasons:

- The normal vectors of walls $T_{k;\mathfrak{s}}(\mathfrak{d})$ may not belong to $N^+ = N^+_{\mathfrak{s}}$ in general, because the coefficients of e_k might be negative.
- Accordingly, the wall functions $T_{k;\mathfrak{s}}(f)$ may not belong to $\Bbbk[[P]]$ in general. In particular, the wall function $1+\hat{y}'_k = 1+\hat{y}_k^{-1}$ does not belong to $\Bbbk[[P]]$.

These problems will be harmonically solved in the next theorem.

Here is the second fundamental theorem on scattering diagrams in view of the application to cluster algebra theory.

Theorem 6.7 ([GHKK18, Theorem 1.24]). *The collection of walls $T_k(\mathfrak{D}_{\mathfrak{s}})$ is a scattering diagram for the seed $\mathfrak{s}' = \mu_k(\mathfrak{s})$. Moreover, $T_k(\mathfrak{D}_{\mathfrak{s}})$ is equivalent to the consistent scattering diagram $\mathfrak{D}_{\mathfrak{s}'}$.*

Let us explain how Proposition 3.2 follows from this theorem.

We do the identification $M_{\mathbb{R}} \simeq_{\mathfrak{s}} \mathbb{R}^n$ for $\mathfrak{D}_{\mathfrak{s}}$, and $M_{\mathbb{R}} \simeq_{\mathfrak{s}'} \mathbb{R}^n$ for $\mathfrak{D}_{\mathfrak{s}'}$, respectively. Under the change of a basis of $M_{\mathbb{R}}$ in (6.3), two identifications are related by the isomorphism $\eta^{t_1}_{t_0}$ in (2.80). All relevant maps are summarized in the following commutative diagram, where $\varphi^{t_1}_{t_0} = \eta^{t_1}_{t_0} \circ T_{k;t_0}$ as in Proposition 2.26:

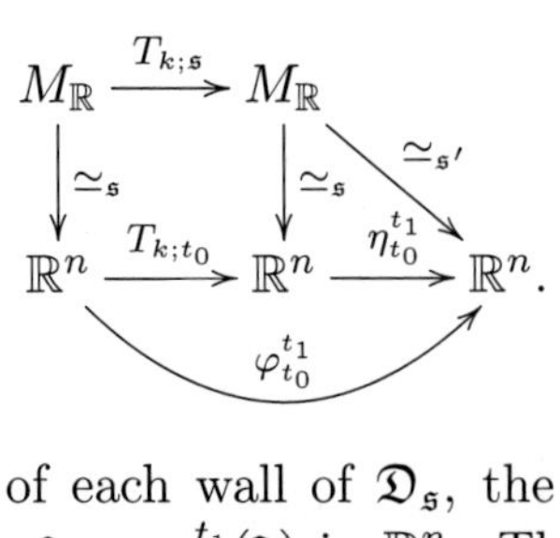

(6.12)

Then, for the support $\mathfrak{d}$ of each wall of $\mathfrak{D}_{\mathfrak{s}}$, the map $\mathfrak{d} \mapsto T_{k;\mathfrak{s}}(\mathfrak{d})$ in $M_{\mathbb{R}}$ is identified with the map $\mathfrak{d} \mapsto \varphi^{t_1}_{t_0}(\mathfrak{d})$ in $\mathbb{R}^n$. Thus, by forgetting the wall functions, we obtain Proposition 3.2.

6.2 Relation between $\mathfrak{D}(\mathbf{\Sigma}_{\mathfrak{s},t_0})$ and $\mathfrak{D}_\mathfrak{s}$

So far, what we know about the relation between $\mathfrak{D}(\mathbf{\Sigma}_{\mathfrak{s},t_0})$ and $\mathfrak{D}_\mathfrak{s}$ are the rank 2 examples and the relations of the supports in Theorem 3.3. Below we show that the walls of $\mathfrak{D}(\mathbf{\Sigma}_{\mathfrak{s},t_0})$ are integrated into a part of the consistent scattering diagram $\mathfrak{D}_\mathfrak{s}$ in an essential way. To do that, we apply a parallel construction of the mutation of $\mathfrak{D}_\mathfrak{s}$ to $\mathfrak{D}(\mathbf{\Sigma}_{\mathfrak{s},t_0})$.

Throughout this subsection, we omit the normal vector $n \in N^+$ of a wall $(\mathfrak{d}, f)_n$ for simplicity, because it is recovered from $\mathfrak{d}$ or f without ambiguity.

For $t_0 \in \mathbb{T}_n$ and $k = 1, \dots, n$, we define a linear map

$$\begin{array}{rccc} S_{k;t_0}: & \mathbb{R}^n & \to & \mathbb{R}^n \\ & \mathbf{v} & \mapsto & (I + (B_{t_0})^{\bullet k})\mathbf{v}, \end{array} \tag{6.13}$$

which correspond to the linear map $S_{k;\mathfrak{s}}$ in (6.6) under the identification $M_\mathbb{R} \simeq_\mathfrak{s} \mathbb{R}^n$. Let $\mathbb{R}^n_{k,+}$ and $\mathbb{R}^n_{k,-}$ be the ones in (2.70). Then, the piecewise-linear map (2.81), which corresponds to (6.10), is written as

$$\begin{array}{rccc} T_{k;t_0}: & \mathbb{R}^n & \to & \mathbb{R}^n \\ & \mathbf{v} & \mapsto & \begin{cases} S_{k;t_0}(\mathbf{v}) & \mathbf{v} \in \mathbb{R}^n_{k,+}, \\ \mathbf{v} & \mathbf{v} \in \mathbb{R}^n_{k,-}. \end{cases} \end{array} \tag{6.14}$$

Let us explicitly denote the dependence of $\hat{y}$-variables to the initial vertex t_0 as

$$\hat{y}^{\mathbf{v}}_{t_0} := x^{B_{t_0}\mathbf{v}}. \tag{6.15}$$

Definition 6.8 (Mutation of $\mathfrak{D}(\mathbf{\Sigma}_{\mathfrak{s},t_0})$). For the collection of walls $\mathfrak{D}(\mathbf{\Sigma}_{\mathfrak{s},t_0})$, let $T_k(\mathfrak{D}(\mathbf{\Sigma}_{\mathfrak{s},t_0}))$ be the collection of walls $T_k(\mathfrak{d}, f)$ obtained from each wall $(\mathfrak{d}, f)$ of $\mathfrak{D}(\mathbf{\Sigma}_{\mathfrak{s},t_0})$ as below.

(a). For each wall $(\mathfrak{d}, f) = (\sigma_i(G^{t_0}_t), 1 + \hat{y}^{\mathbf{c}^{t_0+}_{i;t}}_{t_0})$ such that $\mathbf{c}^{t_0+}_{i;t} \neq \mathbf{e}_k$, we define

$$T_k(\mathfrak{d}, f) = \begin{cases} (S_{k;t_0}(\mathfrak{d}), S_{k;t_0}(f)) & \mathfrak{d} \subset \mathbb{R}^n_{k,+}, \\ (\mathfrak{d}, f) & \mathfrak{d} \subset \mathbb{R}^n_{k,-}, \end{cases} \tag{6.16}$$

where

$$S_{k;t_0}(1 + \hat{y}^{\mathbf{c}^{t_0+}_{i;t}}_{t_0}) = S_{k;t_0}(1 + x^{\hat{\mathbf{c}}^{t_0+}_{i;t}}) := 1 + x^{S_{k;t_0}(\hat{\mathbf{c}}^{t_0+}_{i;t})}. \tag{6.17}$$

Note that by Proposition 2.20 either $\mathfrak{d} \subset \mathbb{R}^n_{k,+}$ or $\mathfrak{d} \subset \mathbb{R}^n_{k,-}$ occurs. Thus, we do not have to divide a wall.

(b). For each wall $(\mathfrak{d}, f) = (\sigma_i(G^{t_0}_t), 1 + \hat{y}^{\mathbf{e}_k}_{t_0})$ we define

$$T_k(\mathfrak{d}, f) = (\sigma_i(G^{t_0}_t), 1 + \hat{y}^{-\mathbf{e}_k}_{t_0}). \tag{6.18}$$

To summarize, in both (a) and (b) each support $\mathfrak{d}$ is transformed by the piecewise-linear map $T_{k;t_0}$. On the other hand, each wall function f in (a) obeys the same *linear map* applied to its support, while the case (b) is exceptional.

Remark 6.9. The same remark as Remark 6.6 is applied to walls in the above.

Let $t_1 \in \mathbb{T}_n$ be the vertex that is k-adjacent to the initial vertex t_0. Let $\eta_{t_0}^{t_1}$ be the linear map in (2.80). For any wall $(\mathfrak{d}, f) \in T_k(\mathfrak{D}(\boldsymbol{\Sigma}_{\mathfrak{s},t_0}))$, we define a new wall by

$$\eta_{t_0}^{t_1}(\mathfrak{d}, f = 1 + x^{\mathbf{v}}) := (\eta_{t_0}^{t_1}(\mathfrak{d}), 1 + x^{\eta_{t_0}^{t_1}(\mathbf{v})}). \tag{6.19}$$

We have a parallel result to Theorem 6.7. Also, this is an upgrade of Proposition 2.22 including wall functions.

Proposition 6.10. *The walls $\eta_{t_0}^{t_1}(\mathfrak{d}, f)$ in (6.19) are walls of $\mathfrak{D}(\boldsymbol{\Sigma}_{\mathfrak{s}',t_1})$ with $\mathfrak{s}' = \mu_k(\mathfrak{s})$. Moreover, it induces the following bijections of walls:*

$$\begin{array}{ccccc} \mathfrak{D}(\boldsymbol{\Sigma}_{\mathfrak{s},t_0}) & \overset{T_k}{\to} & T_k(\mathfrak{D}(\boldsymbol{\Sigma}_{\mathfrak{s},t_0})) & \overset{\eta_{t_0}^{t_1}}{\to} & \mathfrak{D}(\boldsymbol{\Sigma}_{\mathfrak{s}',t_1}) \\ (\sigma_i(G_t^{t_0}), 1 + \hat{y}_{t_0}^{\mathbf{c}_{i;t}^{t_0+}}) & \mapsto & T_k(\sigma_i(G_t^{t_0}), 1 + \hat{y}_{t_0}^{\mathbf{c}_{i;t}^{t_0+}}) & \mapsto & (\sigma_i(G_t^{t_1}), 1 + \hat{y}_{t_1}^{\mathbf{c}_{i;t}^{t_1+}}), \end{array} \tag{6.20}$$

where

$$\hat{y}_{t_0}^{\mathbf{c}_{i;t}^{t_0+}} = x^{\hat{\mathbf{c}}_{i;t}^{t_0+}}, \quad \hat{y}_{t_1}^{\mathbf{c}_{i;t}^{t_1+}} = x^{\hat{\mathbf{c}}_{i;t}^{t_1+}}. \tag{6.21}$$

Proof. We only need to show the correspondence by $\eta_{t_0}^{t_1}$ in (6.20). First we prove the claim for the supports. Under the map $\eta_{t_0}^{t_1} \circ T_k$, the support $\sigma_i(G_t^{t_0})$ maps to $\eta_{t_0}^{t_1}(T_{k;t_0}(\sigma_i(G_t^{t_0})))$. By Propositions 2.22 and 2.26, it equals to $\sigma_i(G_t^{t_1})$.

Next we prove the claim for the wall functions. Let us consider the case (a) in Definition 6.8, where $\mathbf{c}_{i;t}^{t_0} \neq \pm\mathbf{e}_k$ is assumed. Under the map $\eta_{t_0}^{t_1} \circ T_k$, the $\hat{c}^+$-vector $\hat{\mathbf{c}}_{i;t}^{t_0+} = \varepsilon_{i;t}^{t_0}\hat{\mathbf{c}}_{i;t}^{t_0}$ obeys the same linear transformations as its support $\sigma_i(G_t^{t_0})$. Therefore, by comparing (2.20) and (2.37), its image equals to $\varepsilon_{i;t}^{t_0}\hat{\mathbf{c}}_{i;t}^{t_1}$. Now we need to compare $\varepsilon_{i;t}^{t_0}$ and $\varepsilon_{i;t}^{t_1}$. Here we repeat a similar argument after (2.27) in the proof of Proposition 2.7. By (2.19), $C_t^{t_0}$ and $C_t^{t_1}$ differ only in their kth rows. Since we assume $\mathbf{c}_{i;t}^{t_0} \neq \pm\mathbf{e}_k$, there is at least one nonzero element in the ith column of $C_t^{t_0}$ other than at the k-th row. Thus, we have $\varepsilon_{i;t}^{t_0} = \varepsilon_{i;t}^{t_1}$, and we conclude that $\hat{\mathbf{c}}_{i;t}^{t_0+}$ maps to $\hat{\mathbf{c}}_{i;t}^{t_1+}$ as desired. Now we consider the case (b) in Definition 6.8, where $\mathbf{c}_{i;t}^{t_0} = \pm\mathbf{e}_k$ is assumed. It is enough to prove the equality

$$\eta_{t_0}^{t_1}(-\hat{\mathbf{c}}_{i;t}^{t_0+}) = \hat{\mathbf{c}}_{i;t}^{t_1+}. \tag{6.22}$$

Again by (2.19), we have $\mathbf{c}_{i;t}^{t_1} = \mp \mathbf{e}_k$. Therefore, $\mathbf{c}_{i;t}^{t_0+} = \mathbf{c}_{i;t}^{t_1+} = \mathbf{e}_k$. Thus, the equality (6.22) is equivalent to

$$\eta_{t_0}^{t_1}(-\mathbf{b}_{k;t_0}) = \mathbf{b}_{k;t_1}, \tag{6.23}$$

which can be proved as

$$\eta_{t_0}^{t_1}(-\mathbf{b}_{k;t_0}) = -(J_k + [-B_{t_0}]_+^{\bullet k})\mathbf{b}_{k;t_0} = -\mathbf{b}_{k;t_0} = \mathbf{b}_{k;t_1}. \tag{6.24}$$

(This proof also explains the necessity of the exception in the case (b) in view of the exceptional change of tropical signs $\varepsilon_{i;t}^{t_0} = -\varepsilon_{i;t}^{t_1}$ therein.) □

Example 6.11 (cf. Example 2.24)**.** Let us clarify Definition 6.8 and Proposition 6.10 explicitly for type A_2 based on the convention in Section 2.5. The relevant data are as follows:

$$B_{t_0} = \begin{pmatrix} 0 & -1 \\ 1 & 0 \end{pmatrix}, \quad B_{t_1} = B_{t_{-1}} = \begin{pmatrix} 0 & 1 \\ -1 & 0 \end{pmatrix}, \tag{6.25}$$

$$\hat{y}_{1;t_0} = x_2,\ \hat{y}_{2;t_0} = x_1^{-1};\ \hat{y}_{1;t_1} = \hat{y}_{1;t_{-1}} = x_2^{-1},\ \hat{y}_{2;t_1} = \hat{y}_{2;t_{-1}} = x_1, \tag{6.26}$$

$$S_{1;t_0} = \begin{pmatrix} 1 & 0 \\ 1 & 1 \end{pmatrix}, \quad S_{2;t_0} = \begin{pmatrix} 1 & -1 \\ 0 & 1 \end{pmatrix}, \tag{6.27}$$

$$\eta_{t_0}^{t_1} = \begin{pmatrix} -1 & 0 \\ 0 & 1 \end{pmatrix}, \quad \eta_{t_0}^{t_{-1}} = \begin{pmatrix} 1 & 1 \\ 0 & -1 \end{pmatrix}. \tag{6.28}$$

The results are depicted in Figure 7, where $\dot{G}_{t_0}^{t} := T_{k;t_0}(G_{t_0}^{t})$ therein. See also Figure 5.

To describe the relation between $\mathfrak{D}(\Sigma_{\mathfrak{s},t_0})$ and $\mathfrak{D}_{\mathfrak{s}}$, we introduce the following notion.

Definition 6.12. For any wall $(\mathfrak{d}, f)$ and a scattering diagram $\mathfrak{D}$ for a common seed $\mathfrak{s}$, we say that $(\mathfrak{d}, f)$ is *essential to* $\mathfrak{D}$ if there is a scattering diagram $\mathfrak{D}'$ that is equivalent to $\mathfrak{D}$ satisfying the following condition:

- The wall $(\mathfrak{d}, f)$ is a wall of $\mathfrak{D}'$; moreover, for any wall $(\mathfrak{d}', f')$ of $\mathfrak{D}'$ other than $(\mathfrak{d}, f)$, it holds that $\dim \mathfrak{d} \cap \mathfrak{d}' < n - 1$.

Combining Theorem 6.7 and Proposition 6.10, we obtain the following relation between $\mathfrak{D}(\Sigma_{\mathfrak{s},t_0})$ and $\mathfrak{D}_{\mathfrak{s}}$. (The claim (b) was already given in Theorem 3.3 (a), but it is included here again for the summary.)

Theorem 6.13. *Under the identification $M_{\mathbb{R}} \simeq_{\mathfrak{s}} \mathbb{R}^n$, the following facts hold:*

(a). Every wall $(\sigma_i(G_t^{t_0}), 1 + \hat{y}_{t_0}^{\mathbf{c}_{i;t}^{t_0+}})$ of $\mathfrak{D}(\Sigma_{\mathfrak{s},t_0})$ is essential to $\mathfrak{D}_{\mathfrak{s}}$.

(b). For any $t \in \mathbb{T}_n$, the set $\sigma^{\circ}(G_t^{t_0})$, which is a chamber of $\mathrm{Supp}(\mathfrak{D}(\Sigma_{\mathfrak{s},t_0}))$, *is also a chamber of* $\mathrm{Supp}(\mathfrak{D}_{\mathfrak{s}})$.

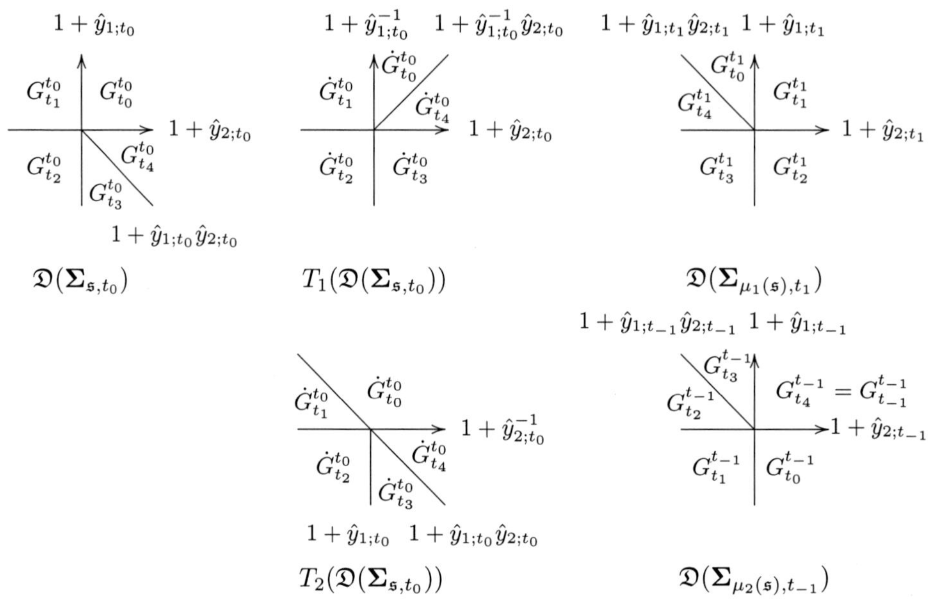

Figure 7: Mutations of $\mathfrak{D}(\mathbf{\Sigma}_{\mathfrak{s},t_0})$ for type A_2.

Proof. We only need to prove (a). Basically, we prove the claim by the induction on t_0. However, as explained in Remark 6.3, the mutation $\mu_k(\mathfrak{s})$ in (6.1) is not involutive. So, we pay special attention to it. Let $t_0, t \in \mathbb{T}_n$ be any pair of vertices. Let $d = d(t_0, t)$ be their distance in $\mathbb{T}_n$. Then, there is a unique sequence of vertices $t_0, t_1, \dots, t_d = t$ that are sequentially adjacent with edges labeled by $k_1, \dots, k_d$ such that $d(t_0, t_r) = r$. First, we assign a seed $\mathfrak{s} = \mathfrak{s}_0$ at the vertex t_0, which is the target of the proof. Then, we assign a seed $\mathfrak{s}_r$ at the vertex t_r recursively as $\mathfrak{s}_{r+1} = \mu_{k_{r+1}}^{-}(\mathfrak{s}_r)$ by the signed mutation in (6.4). Then, by (6.5), we have $\mathfrak{s}_r = \mu_{k_{r+1}}(\mathfrak{s}_{r+1})$. Now, the claim (a) follows from the following claim.

Claim 6.14. Let t be as above. Then, for any $r = 0, \dots, d$, each wall $(\sigma_i(G_t^{t_r}), 1 + \hat{y}_{t_r}^{\mathbf{c}_{i;t}^{t_r+}})$ of $\mathfrak{D}(\mathbf{\Sigma}_{\mathfrak{s}_r,t_r})$ is essential to $\mathfrak{D}_{\mathfrak{s}_r}$.

For each $i = 1, \dots, n$, we prove the claim by the finite induction on r from d to 0.

First, consider the case $r = d$, namely, $t_d = t$. We split the support of the wall $(e_i^{\perp}, 1 + \hat{y}_{t_d}^{\mathbf{e}_i})$ of $\mathfrak{D}_{\mathfrak{s}_d}$ in Theorem 5.27 into the orthants

$$e_i^{\perp} = \bigcup_{\kappa_1,\dots,\kappa_{i-1},\kappa_{i+1},\dots,\kappa_n=\pm 1} \sigma(\kappa_1 e_1, \dots, \kappa_{i-1} e_{i-1}, \kappa_{i+1} e_{i+1}, \dots, \kappa_n e_n), \quad (6.29)$$

so that the one with $\kappa_1 = \cdots = \kappa_n = 1$ is $\sigma_i(G_t^{t_d}) = \sigma_i(I)$. By Theorem 5.27, there is no other wall of $\mathfrak{D}_{\mathfrak{s}_d}$ whose normal vector is $\mathbf{e}_i$. Thus, $(\sigma_i(G_t^{t_d}), 1+\hat{y}_{t_d}^{\mathbf{e}_i})$ is essential to $\mathfrak{D}_{\mathfrak{s}_d}$.

Next, suppose that the claim is true for some r. Namely, for the wall $(\mathfrak{d}, f) := (\sigma_i(G_t^{t_r}), 1 + \hat{y}_{t_r}^{\mathbf{c}_{i;t}^{t_r+}})$ of $\mathfrak{D}(\mathbf{\Sigma}_{\mathfrak{s}_r,t_r})$, there is a scattering diagram $\mathfrak{D}'$ that is equivalent to $\mathfrak{D}_{\mathfrak{s}_r}$ such that the following holds:

- Under the identification $M_{\mathbb{R}} \simeq_{\mathfrak{s}_r} \mathbb{R}^n$, the wall $(\mathfrak{d}, f)$ is a wall of $\mathfrak{D}'$; moreover, for any wall $(\mathfrak{d}', f')$ of $\mathfrak{D}'$ other than $(\mathfrak{d}, f)$, it holds that $\dim \mathfrak{d} \cap \mathfrak{d}' < n - 1$.

We apply the same construction of $T_{k_r}(\mathfrak{D}_{\mathfrak{s}_r})$ to $\mathfrak{D}'$ and obtain $T_{k_r}(\mathfrak{D}')$. Since the construction of $T_{k_r}(\mathfrak{D}')$ is compatible with that of $T_{k_r}(\mathfrak{D}(\mathbf{\Sigma}_{\mathfrak{s}_r,t_r}))$, we have the property:

- Under the identification $M_{\mathbb{R}} \simeq_{\mathfrak{s}_r} \mathbb{R}^n$, the wall $T_{k_r}(\mathfrak{d}, f)$ is a wall of $T_{k_r}(\mathfrak{D}')$; moreover, for any wall $(\mathfrak{d}', f')$ of $T_{k_r}(\mathfrak{D}')$ other than $T_{k_r}(\mathfrak{d}, f)$, it holds that $\dim T_{k_r;t_r}(\mathfrak{d}) \cap \mathfrak{d}' < n - 1$.

By construction, $T_{k_r}(\mathfrak{D}')$ is equivalent to $T_{k_r}(\mathfrak{D}_{\mathfrak{s}_r})$. Also, by Theorem 6.7, $T_{k_r}(\mathfrak{D}_{\mathfrak{s}_r})$ is equivalent to $\mathfrak{D}_{\mathfrak{s}_{r-1}}$, where we recall $\mu_{k_r}(\mathfrak{s}_r) = \mathfrak{s}_{r-1}$. Thus, $T_{k_r}(\mathfrak{D}')$ is also equivalent to $\mathfrak{D}_{\mathfrak{s}_{r-1}}$. Therefore, the wall $T_{k_r}(\mathfrak{d}, f)$ is essential to $\mathfrak{D}_{\mathfrak{s}_{r-1}}$. Now, we change the identification from $M_{\mathbb{R}} \simeq_{\mathfrak{s}_r} \mathbb{R}^n$ to $M_{\mathbb{R}} \simeq_{\mathfrak{s}_{r-1}} \mathbb{R}^n$ by the linear isomorphism $\eta_{t_r}^{t_{r-1}}$. Then, by Proposition 6.10, the wall $T_{k_r}(\mathfrak{d}, f)$ is identified with $(\sigma_i(G_t^{t_{r-1}}), 1+\hat{y}_{t_{r-1}}^{\mathbf{c}_{i;t}^{t_{r-1}+}})$. Therefore, we conclude that the wall $(\sigma_i(G_t^{t_{r-1}}), 1 + \hat{y}_{t_{r-1}}^{\mathbf{c}_{i;t}^{t_{r-1}+}})$ is essential to $\mathfrak{D}_{\mathfrak{s}_{r-1}}$. This completes the proof of the claim. □

Remark 6.15. The following remark is due to Nathan Reading. Theorem 6.13 does not mean that $\mathfrak{D}(\mathbf{\Sigma}_{\mathfrak{s},t_0})$ satisfies the finiteness condition. For example, consider the initial exchange matrix

$$B_{t_0} = \begin{pmatrix} 0 & -1 & -1 \\ 1 & 0 & -1 \\ 1 & 1 & 0 \end{pmatrix}. \tag{6.30}$$

(This matrix is singular, but it is irrelevant to the point here.) Then, there are infinitely many distinct faces $\sigma_i(G_t)$ sharing the common normal vector $(0, 1, 0)$, thus, also with the common wall function. Therefore, in this case $\mathfrak{D}(\mathbf{\Sigma}_{\mathfrak{s},t_0})$ is not a scattering diagram in the sense of Definition 5.18. However, Theorem 6.13 guarantees that one can combine them into finitely many walls to satisfy the finiteness condition. After this, $\mathfrak{D}(\mathbf{\Sigma}_{\mathfrak{s},t_0})$ becomes a scattering diagram.

6.3 F-polynomials and wall-crossing automorphisms

Based on Theorem 6.13, now we can identify F-polynomials with wall-crossing automorphisms of $\mathfrak{D}_{\mathfrak{s}}$.

So far, a wall-crossing automorphism $\mathfrak{p}_{\gamma,\mathfrak{D}}$ is defined to act on $\Bbbk[[P]]$ by (5.18) as a ring automorphism. Now we consider a $\Bbbk[[P]]$-module $x^m\Bbbk[[P]]$ for any $m \in M^\circ$ such that the action $\mathfrak{p}_{\gamma,\mathfrak{D}}(x^m)$ is defined by the same formula (5.18).

Let us recall the situation in Example 5.24. Consider a sequence of vertices $t_0, t_1, \dots, t_{r+1} = t \in \mathbb{T}_n$ such that they are sequentially adjacent with edges labeled by $k_0, \dots, k_r$. Let $\gamma_t^{t_0}$ be an admissible curve therein. Namely, $\gamma_t^{t_0}$ starts in $\sigma^\circ(G_t)$, passing through G-cones in $\mathbb{R}^n$, and it ends in $\sigma^\circ(G_{t_0})$. Under the identification $M_{\mathbb{R}} \simeq_{\mathfrak{s}} \mathbb{R}^n$, we define

$$\mathfrak{p}_t^{t_0} := \mathfrak{p}_{\gamma_t^{t_0},\mathfrak{D}_{\mathfrak{s}}}. \tag{6.31}$$

The following theorem clarifies the relation between F-polynomials and wall-crossing automorphisms in the scattering diagram $\mathfrak{D}_{\mathfrak{s}}$.

Theorem 6.16 ([Rea20b, Theorem 5.6], [Rea20a, Theorem 2.9]). *Under the identification $M_{\mathbb{R}} \simeq_{\mathfrak{s}} \mathbb{R}^n$, the following facts hold.*

(a). The wall-crossing automorphism $\mathfrak{p}_t^{t_0}$ in (6.31) for $\mathfrak{D}_{\mathfrak{s}}$ coincides with the wall-crossing automorphism $\tilde{\mathfrak{p}}_t^{t_0}$ in (5.36) for $\mathfrak{D}(\mathbf{\Sigma}_{\mathfrak{s},t_0})$.

(b). The following formula holds in $x^{\mathbf{g}_{i;t}}\Bbbk[[P]]$:

$$\mathfrak{p}_t^{t_0}(x^{\mathbf{g}_{i;t}}) = x^{\mathbf{g}_{i;t}} F_{i;t}(\hat{\mathbf{y}}). \tag{6.32}$$

Proof. (a). By Theorems 6.13, one may assume that, up to the equivalence, the walls of $\mathfrak{D}_{\mathfrak{s}}$ passed by the above curve $\gamma_t^{t_0}$ all belong to $\mathfrak{D}(\mathbf{\Sigma}_{\mathfrak{s},t_0})$. Thus, we have

$$\tilde{\mathfrak{p}}_t^{t_0} = \mathfrak{p}_{\gamma_t^{t_0},\mathfrak{D}(\mathbf{\Sigma}_{\mathfrak{s},t_0})} = \mathfrak{p}_{\gamma_t^{t_0},\mathfrak{D}_{\mathfrak{s}}} = \mathfrak{p}_t^{t_0}. \tag{6.33}$$

(b). Recall that $\tilde{\mathfrak{p}}_t^{t_0} = \mathfrak{q}_t^{t_0}$ by (5.36). Therefore, by (4.45), we have the equality. □

Remark 6.17. By Injectivity Assumption, the initial exchange matrix B is nonsingular. Thus, $\hat{y}_1, \dots, \hat{y}_n$ are algebraically independent. Therefore, (6.32) uniquely determines F-polynomials $F_{i;t}(\mathbf{y})$ themselves.

6.4 Theta functions and Laurent positivity

Let us present the outline of the proof of the Laurent positivity by [GHKK18].

The following object was introduced in [GHKK18] to provide a combinatorial description of wall-crossing automorphisms.

Definition 6.18 (Broken line). Let $\mathfrak{D}$ be any scattering diagram with respect to a given initial seed $\mathfrak{s}$. Let $m_0 \in M^\circ \setminus \{0\}$, and let $Q \in M_{\mathbb{R}} \setminus \mathrm{Supp}(\mathfrak{D})$. A *broken line for m_0 with endpoint Q* is a piecewise-linear curve $\gamma : (-\infty, 0] \to M_{\mathbb{R}} \setminus \mathrm{Sing}(\mathfrak{D})$ satisfying the following properties:

(1) The endpoint $\gamma(0)$ is Q.

(2) There are numbers $-\infty = t_0 < t_1 < t_2 < \cdots < t_{r+1} = 0$ $(r \geq 0)$ such that γ is linear in each interval $I_j = (t_j, t_{j+1})$, while it breaks at each t_j for $j = 1, \dots, r$. Moreover, γ passes some walls of $\mathfrak{D}$ at each break point t_j. (It is possible that γ passes some walls in an interval I_j without breaking.)

(3) To each I_j, a monomial $c_j x^{m_j} \in \Bbbk[M^\circ]$ is attached. In particular, to I_0, a monomial x^{m_0} is attached, where m_0 is the given data.

(4) The velocity γ' is $-m_j$ in the interval I_j.

(5) Let γ_j be a segment of γ for the interval $(t_j - \delta, t_j + \delta)$ with sufficiently small $\delta > 0$ such that γ_j passes walls only at t_j. Then, $c_j x^{m_j}$ is a monomial in $\mathfrak{p}_{\gamma_j, \mathfrak{D}}(c_{j-1} x^{m_{j-1}})$.

Finally, we define

$$\mathrm{Mono}(\gamma) := c_r x^{m_r}. \tag{6.34}$$

Below we implicitly assume that Q is in a general position so that it has the maximal set of broken lines in the neighborhood of Q. Namely, no broken line for Q' near Q drops out accidentally by intersecting $\mathrm{Sing}(\mathfrak{D})$ when Q' approaches to Q.

Remark 6.19. At t_j, the velocity changes from $-m_j$ to $-m_{j+1} = -m_j - k_j p^*(n_j)$, where n_j is the common normal vector of the walls at t_j, and k_j is some positive integer. Since $p^*(n_j) \in n_j^{\perp}$, the broken line always *crosses* walls, *not reflects nor stops* at them as stated in Condition (2).

Remark 6.20 ([GHKK18, Remark 3.2])**.** In Condition (5) of Definition 6.18, suppose that γ_j crosses possibly multiple walls $(\mathfrak{d}_\lambda, f_\lambda)_{n_j}$ $(\lambda \in \Lambda_j)$ at t_j. Since γ does not intersect $\mathrm{Sing}(\mathfrak{D})$, these walls share the common normal vector $n_j \in N_{\mathrm{pr}}^+$. Note that $\gamma' = -m_{j-1}$ just before γ crosses these walls. Then, the intersection signs in (5.32) are given by

$$\epsilon_j = \begin{cases} 1 & \langle n_j, m_{j-1} \rangle > 0, \\ -1 & \langle n_j, m_{j-1} \rangle < 0. \end{cases} \tag{6.35}$$

Thus, $\epsilon_j \langle n_j, m_{j-1} \rangle = |\langle n_j, m_{j-1} \rangle|$. Therefore, by (5.18), (5.33), and (5.34), we have

$$\mathfrak{p}_{\gamma_j, \mathfrak{D}}(c_{j-1} x^{m_{j-1}}) = c_{j-1} x^{m_{j-1}} \prod_{\lambda \in \Lambda_j} f_\lambda^{|\langle \delta(n_j) n_j, m_{j-1} \rangle|}. \tag{6.36}$$

It is crucial that there is *no division* in this expression for the forthcoming positivity of theta functions.

Definition 6.21 (Theta function)**.** Under the same assumption and notations in Definition 6.18, the *theta function* ϑ_{Q,m_0} for m_0 with endpoint Q is defined by

$$\vartheta_{Q,m_0} = \sum_{\gamma} \mathrm{Mono}(\gamma), \tag{6.37}$$

where the sum is over all broken lines for m_0 with endpoint Q. We also define

$$\vartheta_{Q,0} = 1. \tag{6.38}$$

Let us quote the basic properties of theta functions from [GHKK18].

Proposition 6.22 ([GHKK18, Theorem 3.4])**.** *For any scattering diagram $\mathfrak{D}$, we have*

$$\vartheta_{Q,m_0} \in x^{m_0}\Bbbk[[P]]. \tag{6.39}$$

Proposition 6.23 ([CPS, Lemma 4.8], [GHKK18, Theorem 3.5])**.** *Let $\mathfrak{D}$ be a consistent scattering diagram. Let $m_0 \in M^\circ$ and $Q, Q' \in M_{\mathbb{R}} \setminus \mathrm{Supp}(\mathfrak{D})$. Then, for any admissible curve γ from Q to Q', we have*

$$\vartheta_{Q',m_0} = \mathfrak{p}_{\gamma,\mathfrak{D}}(\vartheta_{Q,m_0}). \tag{6.40}$$

Now let us specialize to the scattering diagram $\mathfrak{D}_{\mathfrak{s}}$ in Theorem 5.27.

Proposition 6.24 ([GHKK18, Corollary 3.9])**.** *Let $\mathfrak{D}_{\mathfrak{s}}$ be the scattering diagram in Theorem 5.27. Let us identify $M_{\mathbb{R}} \simeq_{\mathfrak{s}} \mathbb{R}^n$. For any G-cone $\sigma(G_t)$, let $\mathbf{m}_0 \in \sigma(G_t) \cap \mathbb{Z}^n$ and $Q \in \sigma^\circ(G_t)$. Then, we have*

$$\vartheta_{Q,\mathbf{m}_0} = x^{\mathbf{m}_0}. \tag{6.41}$$

From Propositions 6.23, 6.24, we have the third fundamental theorem on scattering diagrams in view of the application to cluster algebra theory.

Theorem 6.25. *Let $\mathfrak{D}_{\mathfrak{s}}$ be the scattering diagram in Theorem 5.27. Let us identify $M_{\mathbb{R}} \simeq_{\mathfrak{s}} \mathbb{R}^n$. For any G-cone $\sigma(G_t)$, let $\mathbf{m}_0 \in \sigma(G_t) \cap \mathbb{Z}^n$ and $Q \in \sigma^\circ(G_{t_0})$. Then, for any admissible curve γ from any point in $\sigma^\circ(G_t)$ to Q, we have*

$$\vartheta_{Q,\mathbf{m}_0} = \mathfrak{p}_{\gamma,\mathfrak{D}}(x^{\mathbf{m}_0}). \tag{6.42}$$

From Theorems 6.16 and 6.25, we have the following identification of x-variables in $\mathbf{\Sigma}_{\mathfrak{s},t_0}$ with theta functions.

Theorem 6.26 ([GHKK18, Theorem 4.9])**.** *Let $\mathfrak{D}_{\mathfrak{s}}$ be the scattering diagram in Theorem 5.27. Let us identify $M_{\mathbb{R}} \simeq_{\mathfrak{s}} \mathbb{R}^n$. Then, for any $t \in \mathbb{T}_n$ and $Q \in \sigma^\circ(G_{t_0})$, we have*

$$\vartheta_{Q,\mathbf{g}_{i;t}} = x^{\mathbf{g}_{i;t}} F_{i;t}(\hat{\mathbf{y}}) = x_{i;t}. \tag{6.43}$$

Proof. We set $\mathbf{m}_0 = \mathbf{g}_{i;t}$ in (6.42). Then, comparing it with (6.32), we obtain the equality. □

The following positivity result holds.

Proposition 6.27 ([GHKK18, Theorem 1.13]). *The scattering diagram $\mathfrak{D}_\mathfrak{s}$ in Theorem 5.27 is equivalent to a scattering diagram such that every wall function has a form*

$$f = (1 + x^{p^*(n)})^c \tag{6.44}$$

for some (not necessarily primitive) $n \in N^+$ and a positive integer c.

Remark 6.28. By Theorem 6.13, the complexity of wall functions with non-primitive n or $c \neq 1$ occurs only outside $|\Delta(\mathbf{G})|$, i.e., in the dark side.

Example 6.29. The wall function in (5.60) can be split into the form (6.44) by

$$\left(\sum_{k=0}^{\infty} \hat{y}_1^k \hat{y}_2^k \right)^2 = \prod_{j=0}^{\infty} \left(1 + \hat{y}_1^{2^j} \hat{y}_2^{2^j} \right)^2. \tag{6.45}$$

By the definition of broken lines, Remark 6.20, and Proposition 6.27, we have the fourth fundamental theorem on scattering diagrams in view of the application to cluster algebra theory.

Theorem 6.30 ([GHKK18, Theorem 1.13 & Remark 3.2]). *For the scattering diagram $\mathfrak{D}_\mathfrak{s}$ in Theorem 5.27, every theta function $\vartheta_{Q,m_0} \in x^{m_0}\Bbbk[[P]]$ has only nonnegative integer coefficients.*

Finally, we obtain the Laurent positivity in Theorem 1.37.

Theorem 6.31 (Laurent positivity, [GHKK18, Theorem 4.10]). *For any cluster patterns $\mathbf{\Sigma}$ and a given initial vertex t_0, every F-polynomial $F_{i;t}(\mathbf{y}) \in \mathbb{Z}[\mathbf{y}]$ has no negative coefficients.*

Proof. Recall that we have assumed Injectivity Assumption for Γ. For a nonsingular B-pattern, it is a corollary of Theorems 6.30, 6.26, and Remark 6.17. The singular case reduces to the nonsingular case by Proposition 2.11. □

6.5 Linear independence of cluster monomials

Let us present another important consequence of Theorem 6.26.

Following [FZ03b], we introduce the notion of cluster monomials.

Definition 6.32 (Cluster monomials). For a given cluster pattern $\mathbf{\Sigma}$, a *cluster monomial* at $t \in \mathbb{T}_n$ is a monomial (including 1) in the cluster variables $x_{1;t}, \dots, x_{n;t}$ at t.

Two cluster variables taken from different t and t' in $\mathbb{T}_n$ may coincide. For example, if t and t' are k-adjacent, a cluster monomial at t that does not contain $x_{k;t}$ as a factor is also a cluster monomial at t'. We regard them as the same cluster monomial. We are interested in the collection of all *distinct* cluster monomials taken from all $t \in \mathbb{T}_n$.

The following property was conjectured by [FZ03b, Conj. 4.16], then partially proved by [CK08, DWZ10, GLS12, Pla11, CILF11, CIKLFP13] for various classes, and proved by [GHKK18] in full generality. See also Corollary 6.37.

Theorem 6.33 ([GHKK18, Theorem 7.20]). *For a given cluster pattern $\mathbf{\Sigma}$, all distinct cluster monomials are linearly independent over $\mathbb{Z}$.*

Below we present a proof of Theorem 6.33 following [GHKK18].

First, we express cluster monomials with theta functions in the same way as cluster variables given in Theorem 6.26. A cluster monomial at t has the following form

$$x_t^{\mathbf{a}} = \prod_{i=1}^{n} x_{i;t}^{a_i} \quad (\mathbf{a} = (a_i) \in \mathbb{Z}_{\geq 0}^n). \tag{6.46}$$

We define the g-vector $\mathbf{g}_t^{\mathbf{a}}$ of $x_t^{\mathbf{a}}$ by

$$\mathbf{g}_t^{\mathbf{a}} = \sum_{i=1}^{n} a_i \mathbf{g}_{i;t}. \tag{6.47}$$

By definition, this is in the G-cone $\sigma(G_t)$. Under the situation of Theorem 6.26, we have

$$\vartheta_{Q,\mathbf{g}_t^{\mathbf{a}}} = \mathfrak{p}_{\gamma,\mathfrak{D}}(x^{\mathbf{g}_t^{\mathbf{a}}}) = \mathfrak{p}_{\gamma,\mathfrak{D}}\left(\prod_{i=1}^{n} (x^{\mathbf{g}_{i;t}})^{a_i}\right) = \prod_{i=1}^{n} \mathfrak{p}_{\gamma,\mathfrak{D}}(x^{\mathbf{g}_{i;t}})^{a_i} = x_t^{\mathbf{a}}, \tag{6.48}$$

where we used (6.43) in the last equality.

By the Laurent phenomenon, we know that $\vartheta_{Q,\mathbf{g}_t^{\mathbf{a}}} = x_t^{\mathbf{a}}$ has a Laurent polynomial expression in $\mathbf{x}$. We say that a Laurent monomial in $\mathbf{x}$ is *proper* if it has a negative exponent for some x_i.

The following lemma is a key to prove Theorem 6.33.

Lemma 6.34 (In the proof of [GHKK18, Theorem 7.20]). *In the Laurent polynomial expression of the theta function $\vartheta_{Q,\mathbf{g}_t^{\mathbf{a}}}$ in (6.48), every Laurent monomial in $\mathbf{x}$ is proper if $\mathbf{g}_t^{\mathbf{a}} \notin \sigma(G_{t_0}) \cap \mathbb{Z}^n = \mathbb{Z}_{\geq 0}^n$.*

Proof. We may assume that $\mathbf{g}_t^{\mathbf{a}} \neq \mathbf{0}$. Suppose that $\vartheta_{Q,\mathbf{g}_t^{\mathbf{a}}}$ contains a non-proper Laurent monomial M in $\mathbf{x}$. Then, there is a broken line γ for $\mathbf{g}_t^{\mathbf{a}}$ with endpoint Q such that the final velocity $\gamma'(0)$ does not have any positive component. Since $\gamma'(0)$ is not zero, it is a negative vector. Such a broken line has no break point; therefore, it is a ray ending at Q with constant velocity $-\mathbf{g}_t^{\mathbf{a}}$. Thus, $\mathbf{g}_t^{\mathbf{a}}$ is a positive vector. □

Let us rephrase the above result in terms of cluster monomials.

Lemma 6.35. *Let $\mathbf{\Sigma}$ be a given cluster pattern. In the Laurent polynomial expression of a cluster monomial $x_t^{\mathbf{a}}$ for $\mathbf{\Sigma}$ at t, every Laurent monomial in $\mathbf{x}$ is proper if $x_t^{\mathbf{a}}$ does not coincide with any cluster monomial at t_0.*

Proof. First, assume that the underlying B-pattern $\mathbf{B}$ is nonsingular. Suppose that $x_t^{\mathbf{a}}$ contains a non-proper Laurent monomial M in $\mathbf{x}$. Then, by (6.48) and Lemma 6.34, $x_t^{\mathbf{a}} = \vartheta_{Q,\mathbf{g}_t^{\mathbf{a}}}$ and $\mathbf{g}_t^{\mathbf{a}}$ belongs to $\sigma(G_{t_0})$. Thus, we can write $\mathbf{g}_t^{\mathbf{a}} = \sum_{i=1}^n b_i \mathbf{e}_i$ ($b_i \in \mathbb{Z}_{\geq 0}$). Then, we obtain

$$x_t^{\mathbf{a}} = \vartheta_{Q,\mathbf{g}_t^{\mathbf{a}}} = \mathfrak{p}_{\gamma,\mathfrak{D}}(x^{\mathbf{g}_t^{\mathbf{a}}}) = \mathfrak{p}_{\gamma,\mathfrak{D}}\left(\prod_{i=1}^n x_i^{b_i}\right) = \prod_{i=1}^n (\mathfrak{p}_{\gamma,\mathfrak{D}}(x_i))^{b_i} = x^{\mathbf{g}_t^{\mathbf{a}}}. \tag{6.49}$$

Thus, $x_t^{\mathbf{a}}$ coincides with a cluster monomial at t_0.

Next, assume that $\mathbf{B}$ is singular. Consider the principal extension $\overline{\mathbf{B}}$. By Proposition 2.11 and the separation formula (1.46), $x_t^{\mathbf{a}}$ for $\overline{\mathbf{B}}$ is given by

$$x_t^{\mathbf{a}} = \prod_{i=1}^n \left(\prod_{j=1}^n x_j^{g_{ji;t}}\right)^{a_i} F_{i;t}(\hat{\mathbf{y}})^{a_i} \quad (i = 1, \dots, n), \tag{6.50}$$

where G_t and $F_{i;t}$ are a G-matrix and an F-polynomial for $\mathbf{B}$. Thus, this expression formally coincides with the one for $\mathbf{B}$. However, in the formula, for $\overline{\mathbf{B}}$, we have

$$\hat{y}_i = \sum_{j=1}^{2n} x_j^{b_{ji}} = x_{n+i} \sum_{j=1}^{n} x_j^{b_{ji}} \quad (i = 1, \dots, n), \tag{6.51}$$

and, for $\mathbf{B}$, we have $\hat{y}_i = \sum_{j=1}^n x_j^{b_{ji}}$. To extract the result for $\mathbf{B}$, we note the following facts:

(a) Any cluster monomial $x_t^{\mathbf{a}}$ for $\mathbf{B}$ is obtained from the corresponding $x_t^{\mathbf{a}}$ for $\overline{\mathbf{B}}$ by the specialization $x_{n+1} = \cdots = x_{2n} = 1$.
(b) Since F-polynomials are polynomials in $\hat{\mathbf{y}}$, the exponents of x_{n+i} ($i = 1, \dots, n$) are nonnegative for every Laurent monomial in the RHS of (6.50) for $\overline{\mathbf{B}}$. Therefore, all the above proper Laurent monomials are still proper under the specialization $x_{n+1} = \cdots = x_{2n} = 1$.

Now, suppose that $x_t^{\mathbf{a}}$ for B contains a non-proper Laurent monomial in $\mathbf{x} = (x_1, \dots, x_n)$. Then, by (b), the corresponding $x_t^{\mathbf{a}}$ for $\overline{B}$ also contains a non-proper Laurent monomial in $\tilde{\mathbf{x}} = (x_1, \dots, x_{2n})$. Thus, by the first part of the proof, $x_t^{\mathbf{a}}$ for $\overline{B}$ coincides with a cluster monomial for $\overline{B}$ at t_0. Therefore, by (a), $x_t^{\mathbf{a}}$ for B coincides with a cluster monomial for B at t_0. □

The property presented in Lemma 6.35 is called the *proper Laurent monomial property* in [CILF11]. It was shown in [CILF11, Theorem 6.4] that the linear independence in Theorem 6.33 immediately follows form this property. For the reader's convenience, we reproduce the proof here.

Proof of Theorem 6.33. Consider a $\mathbb{Z}$-linear relation among distinct cluster monomials. One can organize it in the following form:

$$\sum_i c_i x_{t_i}^{\mathbf{a}_i} = \sum_i c'_i x^{\mathbf{a}'_i}, \tag{6.52}$$

where each cluster monomial $x_{t_i}^{\mathbf{a}_i}$ in the LHS does not coincide with any cluster monomial at t_0. Also, $t_i = t_j$ may happen for $i \neq j$. Then, by Lemma 6.35, the LHS is a sum of proper Laurent monomials. Thus, to have the equality, the both sides of (6.52) vanish. Then, it follows that $c'_i = 0$ for any i by the linear independence of the cluster monomials at t_0, which is obvious. One can repeat this argument by changing the initial vertex t_0 to other vertices in $\mathbb{T}_n$ and show that $c_i = 0$ for any i. □

Remark 6.36. Using the same idea, the linear independence of a wider class of theta functions including cluster monomials (called the *theta basis*) was proved in [GHKK18, Theorem 7.20].

One can immediately extend Theorem 6.33 to a cluster pattern with coefficients in the sense of [FZ07]. This was the original conjecture given by [FZ07].

Corollary 6.37. *For any cluster pattern $\mathbf{\Sigma}$ with coefficients in a given semifield $\mathbb{P}$, all distinct cluster monomials are linearly independent over $\mathbb{ZP}$.*

Proof. Suppose that there is a linear relation among cluster monomials over $\mathbb{ZP}$ for some cluster pattern with coefficients. Then, by trivializing the coefficients, we obtain a linear relation over $\mathbb{Z}$ for the cluster pattern without coefficients. This is a contradiction. □

6.6 Remarks on singular case

One can actually remove Injectivity Assumption of Γ, or equivalently, the assumption of the nonsingularity of the corresponding B-pattern, and still obtain an analogous result to Theorem 6.26 directly. Since this is beyond the scope of Part II, we only give a sketch of the outline here.

On the F-polynomial side, we need to modify the result (4.45). We consider a cluster pattern with *principal coefficients* at t_0 in [FZ07]. Let

$\mathbb{P} = \mathrm{Trop}(\mathbf{y})$ be the tropical semifield of $\mathbf{y} = \mathbf{y}_{t_0}$. Let $\mathbb{ZP}$ be the group ring of $\mathbb{P}$, and let $\mathbb{QP}$ be the fraction field of $\mathbb{ZP}$. The $\hat{y}$-variables are defined by

$$\hat{y}_{i;t} = y_{i;t} \prod_{j=1}^{n} x_{j;t}^{b_{ji;t}}. \tag{6.53}$$

In particular, let $\hat{\mathbf{y}} = (\hat{y}_i)$ be the initial $\hat{y}$-variables,

$$\hat{y}_i = \hat{y}_{i;t_0} = y_i \prod_{j=1}^{n} x_j^{b_{ji}}, \tag{6.54}$$

which are algebraically independent in the ambient field $\mathcal{F}_X = (\mathbb{QP})(\mathbf{x})$ whether B is nonsingular or singular. Let $t, t' \in \mathbb{T}_n$ be vertices that are k-adjacent. The y-variables transform in $\mathbb{P}$ by

$$y_{i;t'} = \begin{cases} y_{k;t}^{-1} & i = k, \\ y_{i;t} y_{k;t}^{[\varepsilon b_{ki;t}]_+} (1 \oplus y_{k;t}^{\varepsilon})^{-b_{ki;t}} & i \neq k. \end{cases} \tag{6.55}$$

Meanwhile, one can regard mutations of x-variables as isomorphisms of fields as follows, where $\mathbf{x}_t$, $\mathbf{x}_{t'}$ are n-tuple of formal variables:

$$\mu_{k;t} : (\mathbb{QP})(\mathbf{x}_{t'}) \to (\mathbb{QP})(\mathbf{x}_t), \tag{6.56}$$

$$\mu_{k;t}(x_{i;t'}) = \begin{cases} x_{k;t}^{-1} \left(\prod_{j=1}^{n} x_{j;t}^{[-\varepsilon b_{jk;t}]_+} \right) \dfrac{1 + \hat{y}_{k;t}^{\varepsilon}}{1 \oplus \hat{y}_{k;t}^{\varepsilon}} & i = k, \\ x_{i;t} & i \neq k. \end{cases} \tag{6.57}$$

Compare it with (4.2). Note that, for the tropical sign $\varepsilon_{k;t}$, we have

$$1 \oplus y_{k;t}^{\varepsilon_{k;t}} = 1, \tag{6.58}$$

thus, the factor $1 \oplus \hat{y}_{k;t}^{\varepsilon}$ in (6.55) and (6.57) disappears by setting $\varepsilon = \varepsilon_{k;t}$. Then, we just repeat the same decomposition (4.4) and so on, and we obtain the same formula as (4.45),

$$\mathfrak{q}_t^{t_0}(x^{\mathbf{g}_{i;t}}) = x^{\mathbf{g}_{i;t}} F_{i;t}(\hat{\mathbf{y}}), \tag{6.59}$$

but now $\hat{y}_1, \dots, \hat{y}_n$ are algebraically independent as mentioned.

Meanwhile, on the scattering diagram side, following [GHKK18, Appendix B], we consider the extension of the lattice M° of a fixed data Γ,

$$\tilde{M}^\circ = M^\circ \oplus N. \tag{6.60}$$

Then, we introduce a parallel map to p^* in (5.10),

$$\tilde{p}^* : N \to \tilde{M}^\circ, \tag{6.61}$$

$$(\tilde{p}^*(n))(n'+m') = \{n',n\} + \langle n,m\rangle, \quad (n' \in N^\circ, m' \in M). \tag{6.62}$$

We have

$$\tilde{p}^*(e_j) = \sum_{i=1}^{n} b_{ij} f_i + e_j, \tag{6.63}$$

so that the *map $\tilde{p}^*$ is injective whether B is singular or not.* Also, under the identification (5.7), we have

$$x^{\tilde{p}^*(e_i)} = y_i \prod_{j=1}^{n} x_j^{b_{ji}} =: \hat{y}_i, \tag{6.64}$$

which coincides with (6.53). We also need to extend the monoid P in (5.14) with $\tilde{P} \subset \tilde{M}^\circ$ as follows:

(i). $\tilde{P} = \sigma \cap \tilde{M}^\circ$, where σ is a $2r$-dimensional strongly convex cone in $\tilde{M}^\circ_{\mathbb{R}}$.
(ii). $\tilde{p}^*(e_1), \cdots, \tilde{p}^*(e_r) \in \tilde{P}$.

Such $\tilde{P}$ is not unique at all, and we choose one arbitrarily. The result does not depend on the choice of $\tilde{P}$. Then, for a wall $(\mathfrak{d}, f)_n$, we replace the wall function f in (5.25) with

$$f = 1 + \sum_{k=1}^{\infty} c_k x^{k\tilde{p}^*(n)} \in \Bbbk[[\tilde{P}]], \tag{6.65}$$

while n and $\mathfrak{d}$ remain the same. Then, under this modification, we have a parallel result for Theorems 6.25 and 6.26 without assuming Injectivity Assumption. More details will be found in Part III.

7 Some applications

Let us give some applications of the results and the techniques presented so far.

7.1 Detropicalization revisited

We return to the situation in Section 4.3.

As mentioned in Remark 4.9, the following result was given for $\sigma = \mathrm{id}$ by [CHL20], and for general σ by [Nak21], where both rely on the Laurent positivity. Here, we give an alternative proof based on Theorem 6.16 and the consistency of the scattering diagram $\mathfrak{D}_{\mathfrak{s}}$, without relying on the Laurent positivity.

Theorem 7.1 (Detropicalization [CHL20, Lemma 2.4 & Theorem 2.5], [Nak21, Theorem 5.2]). *Let $\mathbf{\Sigma}$ be any cluster pattern of rank n, and let t_0 be a given initial vertex. Then, for any $t, t' \in \mathbb{T}_n$ and any permutation $\nu \in S_n$, the following facts hold:*

(a). $G_t = \nu G_{t'} \implies \mathbf{x}_t = \nu \mathbf{x}_{t'}$.

(b). $C_t = \nu C_{t'} \implies \mathbf{y}_t = \nu \mathbf{y}_{t'}$.

Proof. First, assume that the underlying B-pattern $\mathbf{B}$ is nonsingular.

(a). Assume that $G_t = \nu G_{t'}$. Then, $\sigma(G_t) = \sigma(G_{t'})$. Therefore, for $\mathfrak{p}_t^{t_0}$ in (6.31), we have

$$\mathfrak{p}_t^{t_0} = \mathfrak{p}_{t'}^{t_0} \tag{7.1}$$

by the consistency of the scattering diagram $\mathfrak{D}_{\mathfrak{s}}$. Then, we have $\mathfrak{q}_t^{t_0} = \mathfrak{q}_{t'}^{t_0}$ (for x-variables) by Theorem 6.16 (a) and (5.36). Therefore, $\mathbf{x}_t = \nu \mathbf{x}_{t'}$ by Proposition 4.8. Alternatively, by Theorem 6.16 (b) and (7.1), we have

$$\mathfrak{p}_t^{t_0}(x^{\mathbf{g}_{i;t}}) = x^{\mathbf{g}_{i;t}} F_{i;t}(\hat{\mathbf{y}}) = x_{i;t}, \tag{7.2}$$

$$\mathfrak{p}_t^{t_0}(x^{\mathbf{g}_{i;t}}) = \mathfrak{p}_{t'}^{t_0}(x^{\mathbf{g}_{\nu^{-1}(i);t'}}) = x^{\mathbf{g}_{\nu^{-1}(i);t'}} F_{\nu^{-1}(i);t'}(\hat{\mathbf{y}}) = x_{\nu^{-1}(i);t'}. \tag{7.3}$$

Thus, we have $\mathbf{x}_t = \nu \mathbf{x}_{t'}$.

(b). Assume that $C_t = \nu C_{t'}$. Then, by the duality (2.4) and (2.14), we obtain

$$\delta_{\nu(i)} = \delta_i, \ G_t = \nu G_{t'}, \ B_t = \nu B_{t'}. \tag{7.4}$$

See [Nak21, Prop. 4.4 & Cor. 4.5] for details. Then, from (7.2) and (7.3), we have

$$F_{i;t}(\hat{\mathbf{y}}) = F_{\nu^{-1}(i);t'}(\hat{\mathbf{y}}). \tag{7.5}$$

By Remark 6.17, we have

$$F_{i;t}(\mathbf{y}) = F_{\nu^{-1}(i);t'}(\mathbf{y}). \tag{7.6}$$

Thus, by the separation formula (1.47), we obtain $\mathbf{y}_t = \nu\mathbf{y}_{t'}$.

Next, we consider the case when $\mathbf{B}$ is singular. Thanks to Proposition 2.11, it reduces to the nonsingular case as follows:

$$G_t = \nu G_{t'} \implies \overline{G}_t = \nu\overline{G}_{t'} \implies \overline{\mathbf{x}}_t = \nu\overline{\mathbf{x}}_{t'} \implies \mathbf{x}_t = \nu\mathbf{x}_{t'}, \tag{7.7}$$

$$C_t = \nu C_{t'} \implies \overline{C}_t = \nu\overline{C}_{t'} \implies \overline{\mathbf{y}}_t = \nu\overline{\mathbf{y}}_{t'} \implies \mathbf{y}_t = \nu\mathbf{y}_{t'}, \tag{7.8}$$

where $\overline{\mathbf{\Sigma}} = \{\overline{\Sigma}_t = (\overline{\mathbf{x}}_t, \overline{\mathbf{y}}_t, \overline{B}_t)\}_{t\in\mathbb{T}_n}$ is a cluster pattern in Proposition 2.11. □

7.2 Bijection between g-vectors and x-variables

Here we prove a statement, which sharpens Theorem 7.1 (a). This was proved by [CIKLFP13] in the skew-symmetric case with the representation/categorical method. Also, the implication $\Longrightarrow$ was proved by [CL20, Theorem 3.2] in the skew-symmetrizable case with principal coefficients. In contrast to the proof of Theorem 7.1, here we rely on the Laurent positivity.

Theorem 7.2. *Let $\mathbf{\Sigma}$ be any cluster pattern, and let t_0 be a given initial vertex. Then, we have*

$$\mathbf{g}_{i;t}^{t_0} = \mathbf{g}_{i';t'}^{t_0} \iff x_{i;t} = x_{i';t'}. \tag{7.9}$$

Proof. ($\Longleftarrow$) First we apply the separation formula (1.46) to $x_{i';t'}$ with respect to the initial vertex t. Then, by assumption, we have

$$x_{i';t'} = \left(\prod_{j=1} x_{j;t}^{g_{ji';t'}^t}\right) F_{i';t'}^t(\hat{\mathbf{y}}_t) = x_{i;t}. \tag{7.10}$$

By the equivalent statement of the sign-coherence in Conjecture 1.32, which is now proved, the constant term of $F_{i';t'}^t(\mathbf{y}_t)$ is 1. Then, by the Laurent positivity, we have $F_{i;t'}^t(\mathbf{y}_t) = 1$ and also $\mathbf{g}_{i';t'}^t = \mathbf{e}_i = \mathbf{g}_{i;t}^t$. Then, applying the bijection in Proposition 2.25 repeatedly, we obtain $\mathbf{g}_{i';t'}^{t_0} = \mathbf{g}_{i;t}^{t_0}$.

($\Longrightarrow$) From the assumption, we obtain $\mathbf{g}_{i';t'}^t = \mathbf{e}_i$ by the opposite procedure to the above. Thus, we have

$$x_{i';t'} = x_{i;t}F_{i';t'}^t(\hat{\mathbf{y}}_t). \tag{7.11}$$

Changing the role of t and t', we also have

$$x_{i;t} = x_{i';t'}F_{i;t}^{t'}(\hat{\mathbf{y}}_{t'}). \tag{7.12}$$

From two equalities, we obtain

$$F^{t}_{i';t'}(\hat{\mathbf{y}}_t)F^{t'}_{i;t}(\hat{\mathbf{y}}'_t) = 1. \tag{7.13}$$

Again, by the constant term 1 of F-polynomials and the Laurent positivity, we have $F^{t}_{i';t'}(\mathbf{y}_t) = F^{t}_{i;t}(\mathbf{y}_t) = 1$. Therefore, $x_{i;t} = x_{i';t'}$.

($\Longrightarrow$) Let us give an alternative proof using the consistency of the scattering diagram $\mathfrak{D}_{\mathfrak{s}}$ without using the Laurent positivity. Assume that $\mathbf{B}$ is nonsingular. Let us identify $M_{\mathbb{R}} \simeq_{\mathfrak{s}} \mathbb{R}^n$. Then, we have

$$x_{i;t} = \mathfrak{p}^{t_0}_{t}(x^{\mathbf{g}_{i;t}}), \quad x_{i';t'} = \mathfrak{p}^{t_0}_{t'}(x^{\mathbf{g}_{i';t'}}) = \mathfrak{p}^{t_0}_{t'}(x^{\mathbf{g}_{i;t}}). \tag{7.14}$$

By assumption, two cones $\sigma(G_t)$, $\sigma(G_{t'})$ have the ray $\sigma(\mathbf{g}_{i;t}) = \sigma(\mathbf{g}_{i';t'})$ as a common face. Let $\gamma^{t'}_{t}$ be an admissible curve from $\sigma^{\circ}(G_t)$ to $\sigma^{\circ}(G_{t'})$ for $\mathfrak{D}_{\mathfrak{s}}$. (It may not be completely contained in $|\Delta(\mathbf{G}^{t_0})|$.) For each degree ℓ, by ignoring all walls of $\mathfrak{D}_{\mathfrak{s}}$ whose wall functions are trivial modulo $\widehat{J}^{\ell}$, one can take $\gamma^{t'}_{t}$ close enough to the ray $\sigma(\mathbf{g}_{i;t})$ so that $\gamma^{t'}_{t}$ passes through only walls of $\mathfrak{D}_{\mathfrak{s}}$ that contain $\sigma(\mathbf{g}_{i;t})$. Since the normal vectors of these walls are orthogonal to $\mathbf{g}_{i;t}$, the contribution to the wall-crossing automorphism $\mathfrak{p}_{\gamma^{t'}_{t},\mathfrak{D}_{\mathfrak{s}}}$ from each wall is trivial. Therefore, we have

$$\mathfrak{p}_{\gamma^{t'}_{t},\mathfrak{D}_{\mathfrak{s}}}(x^{\mathbf{g}_{i;t}}) \equiv x^{\mathbf{g}_{i;t}} \mod \widehat{J}^{\ell}. \tag{7.15}$$

Since ℓ is arbitrary, this implies that

$$\mathfrak{p}_{\gamma^{t'}_{t},\mathfrak{D}_{\mathfrak{s}}}(x^{\mathbf{g}_{i;t}}) = x^{\mathbf{g}_{i;t}}. \tag{7.16}$$

By the consistency of $\mathfrak{D}_{\mathfrak{s}}$, we have

$$\mathfrak{p}^{t_0}_{t'} \circ \mathfrak{p}_{\gamma^{t'}_{t}} = \mathfrak{p}^{t_0}_{t}. \tag{7.17}$$

By combining (7.14)–(7.17), we obtain $x_{i;t} = x_{i';t'}$. Next, we consider the case when $\mathbf{B}$ is singular. Thanks to Proposition 2.11, it reduces to the nonsingular case as follows:

$$\mathbf{g}^{t_0}_{i;t} = \mathbf{g}^{t_0}_{i';t'} \Longrightarrow \overline{\mathbf{g}}^{t_0}_{i;t} = \overline{\mathbf{g}}^{t_0}_{i';t'} \Longrightarrow \overline{x}_{i;t} = \overline{x}_{i';t'} \Longrightarrow x_{i;t} = x_{i';t'}. \tag{7.18}$$

□

Remark 7.3. The first proof of the implication $\Longrightarrow$ is a little simplified version of the proof of [CL20, Theorem 3.2].

Remark 7.4. The above proof (as for the implication $\Longrightarrow$, the first one) is applicable to any cluster pattern *with arbitrary coefficients* in the sense of [FZ07], with a little care of coefficients. Originally in [FZ07], a g-vector is uniquely associated with each x-variable only for cluster patterns with geometric type satisfying a certain condition in [FZ07, Eq. (7.10)]. Now we can safely associate a g-vector with each x-variable for any cluster pattern *with arbitrary coefficients.*

Remark 7.5. As for y-variables and c-vectors, the implication

$$\mathbf{c}_{i;t}^{t_0} = \mathbf{c}_{i';t'}^{t_0} \quad \Longleftarrow \quad y_{i;t} = y_{i';t'} \tag{7.19}$$

holds by Proposition 1.30. However, the opposite implication does not hold. For example, for type A_2 case in Section 2.5, we have

$$y_{2;t_0} = y_2, \quad y_{2;t_1} = y_2(1 + y_1), \tag{7.20}$$

which is already a counterexample.

One can rephrase Theorem 7.2 using the notion of the cluster complex in [FZ03a].

Definition 7.6 (Cluster complex). The *cluster complex* $\Delta(\mathbf{\Sigma})$ of a cluster pattern $\mathbf{\Sigma}$ is the simplicial complex whose vertices are x-variables (cluster variables) and whose simplices are nonempty subsets of clusters.

We define a parallel notion for a G-pattern $\mathbf{G}^{t_0}$ of $\mathbf{\Sigma}$, where we encounter the conflict of notation with a G-fan $\Delta(\mathbf{G}^{t_0})$. However, two notions are essentially equivalent, so that we abuse the notation.

Definition 7.7 (G-complex). The *G-complex* $\Delta(\mathbf{G}^{t_0})$ of a G-pattern $\mathbf{G}^{t_0}$ is the simplicial complex whose vertices are g-vectors in $\mathbf{G}^{t_0}$ and whose simplices are nonempty sets of g-vectors belonging to a common G-matrix.

One can geometrically identify a simplex in the G-complex $\Delta(\mathbf{G}^{t_0})$ with the intersection $\sigma \cap S^{n-1}$ of a cone σ in the G-fan $\Delta(\mathbf{G}^{t_0})$ and the unit sphere S^{n-1} in $\mathbb{R}^n$.

We have a corollary of Theorem 7.2.

Corollary 7.8. *For any cluster pattern $\mathbf{\Sigma}$ and any vertex t_0, we have an isomorphism of simplicial complexes*

$$\begin{array}{ccc} \Delta(\mathbf{\Sigma}) & \simeq & \Delta(\mathbf{G}^{t_0}) \\ x_{i;t} & \longleftrightarrow & \mathbf{g}_{i;t}. \end{array} \tag{7.21}$$

Due to Remark 7.4, we can extend the result to any cluster pattern with arbitrary coefficients in the sense of [FZ07]. This extends the result of [CIKLFP13, Cor. 5.6] in the skew-symmetric case.

Corollary 7.9. *For any cluster pattern $\mathbf{\Sigma}$ with arbitrary coefficients in the sense of [FZ07] and any vertex t_0, we have the same isomorphism as* (7.21).

We have a further corollary of Corollary 7.9 and Theorem 2.17. The equivalence $(a) \Longleftrightarrow (d)$ below is a part of a conjecture by [FZ03b, Conj. 4.14(b)], and it was proved by [GSV08] with algebraic arguments. Here we replace the proof with geometric arguments.

Corollary 7.10 (cf. [GSV08, Theorem 5]). . *For any cluster pattern* $\mathbf{\Sigma}$ *with arbitrary coefficients in the sense of [FZ07] and any vertex* t_0*, the following conditions for* $t, t' \in \mathbb{T}_n$ *are equivalent:*

(a). $\mathbf{x}_t$ *and* $\mathbf{x}_{t'}$ *contain exactly* $n-1$ *common elements (as a set), where* n *is the rank of* $\mathbf{\Sigma}$.

(b). $\sigma(G_t)$ *and* $\sigma(G_{t'})$ *intersects in their common face of codimension one.*

(c). There are some $t'' \in \mathbb{T}_n$ *that is adjacent to* t *and some permutation* $\nu \in S_n$ *such that* $G_{t'} = \nu G_{t''}$.

(d). There are some $t'' \in \mathbb{T}_n$ *that is adjacent to* t *and some permutation* $\nu \in S_n$ *such that* $\mathbf{x}_{t'} = \nu \mathbf{x}_{t''}$.

Proof. The equivalences $(a) \Longleftrightarrow (b)$ and $(c) \Longleftrightarrow (d)$ are due to Corollary 7.9, while $(b) \Longleftrightarrow (c)$ is due to Theorem 2.17. $\square$

Part III
Cluster Scattering Diagrams

0 Introduction to Part III

This is a self-contained exposition of several fundamental properties of *cluster scattering diagrams* (CSDs for short) introduced and studied by Gross, Hacking, Keel, and Kontsevich [GHKK18].

Scattering diagrams originated in the study of homological mirror symmetry, and they are developed in the works of Kontsevich-Soibelman [KS06, KS08, KS14], Gross-Siebert [GS11], Gross-Pandharipande-Siebert [GPS10], Gross [Gro11], Carl-Pumperla-Siebert [CPS], etc. The connection to cluster algebras was established by Gross-Hacking-Keel-Kontsevich [GHKK18], where CSDs were introduced. Then, with a scattering diagram technique, several important conjectures on cluster algebras were proved.

Here is a brief description of the contents.

- In the first half (Sections 1–3) we present a construction of CSDs without assuming any knowledge on scattering diagrams. The main result here is the existence and the uniqueness of a CSD for a given seed (Theorem 3.12).
- In the second half (Sections 4–7) we present the fundamental properties of CSDs, which are necessary to prove several conjectures in cluster algebras, including
  - the mutation invariance (Theorem 6.9),
  - the positivity of theta functions (Theorems 5.2 and 7.9).
- We present detailed proofs for all propositions in a self-contained way.

The connection between CSDs and cluster patterns are given in Part II based on the above results. Thus, this part is regarded as a supplement to Part II. Also, it may be read independently as an introductory text to CSDs.

While the most results are taken from [GHKK18, KS14], there are several added or modified proofs and new results due to us. Most notably, we employ the following approach, which adds novelty to our presentation:

- We clearly separate the structure group of walls and its faithful representation (the *principal x-representation*) employed for the construction of CSDs in [GHKK18]. In the first half we only use the structure group itself. This enables us to lift the assumption of the nondegeneracy for the fixed data in [GHKK18], thereby makes the construction of CSDs more transparent. In the second half we apply the principal x-representation to the wall elements when needed.
- Throughout we highlight the fundamental role of the *dilogarithm elements* and the *pentagon relation* in CSDs, which is implicit in [GHKK18].

More recently, the *quantum CSDs* were studied by Davison-Mandel [DM21], where the counterparts of many results presented here were given.

1 Underlying algebraic structure

In this section we introduce some underlying algebraic structure behind scattering diagrams we are going to study.

1.1 Fixed data and seed

Definition 1.1 (Fixed data). A *fixed data* Γ consists of the following:

- A lattice $N \simeq \mathbb{Z}^r$ of rank r with a skew-symmetric bilinear form
$$\{\cdot,\cdot\} : N \times N \to \mathbb{Q}. \tag{1.1}$$
- A sublattice $N^\circ \subset N$ of finite index (equivalently, of rank r) such that
$$\{N^\circ, N\} \subset \mathbb{Z}. \tag{1.2}$$
- Positive integers $\delta_1, \dots, \delta_r$ such that there is a basis $(e_1, \dots, e_r)$ of N, where $(\delta_1 e_1, \dots, \delta_r e_r)$ is a basis of N°.
- $M = \mathrm{Hom}(N, \mathbb{Z})$, $M^\circ = \mathrm{Hom}(N^\circ, \mathbb{Z})$.

We say that Γ is *nondegenerate* (resp. *degenerate*) if $\{\cdot,\cdot\}$ is nondegenerate (resp. degenerate).

For a given $N^\circ \subset N$ of finite index, the above $\delta_1, \dots, \delta_r$ always exist; for example, take the elementary divisors of the embedding $N^\circ \hookrightarrow N$. However, other choices are equally good.

Let $M_\mathbb{R} = M \otimes_\mathbb{Z} \mathbb{R} \simeq \mathbb{R}^r$, and we regard
$$M \subset M^\circ \subset M_\mathbb{R}. \tag{1.3}$$
Let $\langle \cdot,\cdot \rangle : N^\circ \times M^\circ \to \mathbb{Z}$ be the canonical paring. We also write its linear extension $N \times M_\mathbb{R} \to \mathbb{R}$ by the same symbol.

Definition 1.2 (Seed). A *seed* $\mathfrak{s} = (e_1, \dots, e_r)$ for a fixed data Γ is a basis of N such that $(\delta_1 e_1, \dots, \delta_r e_r)$ is a basis of N°.

For a seed $\mathfrak{s}$ for Γ, we have a basis $(e_1^*, \dots, e_r^*)$ of M that is dual to $(e_1, \dots, e_r)$. Let $f_i = e_i^*/\delta_i$. Then, we also have a basis $(f_1, \dots, f_r)$ of M° that is dual to the basis $(\delta_1 e_1, \dots, \delta_r e_r)$ of N°.

Let $\Omega = \Omega_{\Gamma,\mathfrak{s}} = (\omega_{ij})$ be the $r \times r$ skew-symmetric rational matrix defined by $\omega_{ij} = \{e_i, e_j\}$, and let $\Delta = \Delta_{\Gamma,\mathfrak{s}} = \mathrm{diag}(\delta_1, \dots, \delta_r)$ be the diagonal matrix with diagonal entries $\delta_1, \dots, \delta_r$. Let $B = B_{\Gamma,\mathfrak{s}} = (b_{ij})$ be the $r \times r$ integer matrix defined by
$$B = \Delta\Omega, \quad b_{ij} = \{\delta_i e_i, e_j\}. \tag{1.4}$$
Since $\Omega = \Delta^{-1}B$ is skew-symmetric, B is *skew-symmetrizable* with a (left) skew-symmetrizer $D = \Delta^{-1}$ in the sense of [FZ02a].

Conversely, for any $r \times r$ skew-symmetrizable integer matrix $B = (b_{ij})$, there is a (not unique) decomposition

$$B = \Delta\Omega, \tag{1.5}$$

where $\Delta = \mathrm{diag}(\delta_1, \dots, \delta_r)$ with positive integers $\delta_1, \dots, \delta_r$ and $\Omega = (\omega_{ij})$ is an $r \times r$ skew-symmetric rational matrix. Then, we set

$$N = \bigoplus_{i=1}^{r} \mathbb{Z}e_i, \quad \{e_i, e_j\} = \omega_{ij}, \quad N^\circ = \bigoplus_{i=1}^{r} \mathbb{Z}\delta_i e_i. \tag{1.6}$$

This yields a (not unique) pair of a fixed data Γ and a seed $\mathfrak{s}$ such that B is given by (1.4).

The seed $\mathfrak{s}$ here is identified with the initial seed $(\mathbf{x}, \mathbf{y}, B)$ for a cluster pattern or a cluster algebra in the sense of [FZ07] with

$$x_i = x^{f_i}, \quad y_i = y^{e_i}, \quad b_{ij} = \{\delta_i e_i, e_j\}, \tag{1.7}$$

where $\mathbf{x} = (x_i)_{i=1}^r$, $\mathbf{y} = (y_i)_{i=1}^r$, and $B = (b_{ij})_{i,j=1}^r$ are *cluster variables (x-variables)*, *coefficients (y-variables)*, and an *exchange matrix* in [FZ07], and x and y are symbols for the formal exponential. See Part II for more details.

Remark 1.3. In (1.2) we took the transpose of the bilinear form $\{\cdot, \cdot\}$ in [GHKK18] to match the convention of exchange matrices in [FZ07] in the above correspondence. As a consequence, every expression concerning the bilinear form is the transpose of the one in [GHKK18] throughout Part III.

Let us introduce a group homomorphism

$$\begin{array}{rccc} p^*: & N & \to & M^\circ \\ & n & \mapsto & \{\cdot, n\}. \end{array} \tag{1.8}$$

By definition, we have

$$\langle n', p^*(n) \rangle = \{n', n\} \in \mathbb{Z} \quad (n' \in N^\circ, n \in N). \tag{1.9}$$

Lemma 1.4. *Let e_i, f_i, $B = (b_{ij})$ be as above.*

(a). We have

$$p^*(e_j) = \sum_{i=1}^{r} b_{ij} f_i. \tag{1.10}$$

Thus, the matrix representation of p^* with respect to the above bases is given by B.

(b). Γ is nondegenerate if and only if the matrix B is nonsingular.

(c). The map p^* is injective if and only if Γ is nondegenerate.

Proof. (a). One can verify it as

$$\langle \delta_i e_i, p^*(e_j) \rangle = \{\delta_i e_i, e_j\} = b_{ij} = \left\langle \delta_i e_i, \sum_{k=1}^{r} b_{kj} f_k \right\rangle. \tag{1.11}$$

The properties (b) and (c) follow from (a). □

Throughout the section, we fix a seed $\mathfrak{s} = (e_1, \dots, e_r)$ of a given fixed data Γ unless otherwise mentioned.

Let N^+ be the set of *positive elements* in N with respect to $\mathfrak{s}$, namely,

$$N^+ = N^+_{\mathfrak{s}} := \left\{ \sum_{i=1}^{r} a_i e_i \;\middle|\; a_i \in \mathbb{Z}_{\geq 0}, \sum_{i=1}^{r} a_i > 0 \right\}. \tag{1.12}$$

We say that $n \in N$ is *primitive* if there is no pair $j \in \mathbb{Z}_{>1}$ and $n' \in N$ such that $n = jn'$. Let N^+_{pr} be the set of all primitive elements in N^+.

We have a natural *degree function* $\deg : N^+ \to \mathbb{Z}_{>0}$ given by

$$\deg : \sum_{i=1}^{r} a_i e_i \mapsto \sum_{i=1}^{r} a_i. \tag{1.13}$$

For each integer $\ell > 0$, we set

$$(N^+)^{\leq \ell} = \{ n \in N^+ \mid \deg(n) \leq \ell \}, \quad (N^+_{\mathrm{pr}})^{\leq \ell} = (N^+)^{\leq \ell} \cap N^+_{\mathrm{pr}}, \tag{1.14}$$

$$(N^+)^{> \ell} = \{ n \in N^+ \mid \deg(n) > \ell \}, \quad (N^+_{\mathrm{pr}})^{> \ell} = (N^+)^{> \ell} \cap N^+_{\mathrm{pr}}, \tag{1.15}$$

$$(N^+)^{\ell} = \{ n \in N^+ \mid \deg(n) = \ell \}, \quad (N^+_{\mathrm{pr}})^{\ell} = (N^+)^{\ell} \cap N^+_{\mathrm{pr}}. \tag{1.16}$$

In particular, $(N^+)^{\leq \ell}$ is a finite set.

1.2 Structure group G

Let us fix a field $\Bbbk$ of characteristic zero arbitrarily, e.g., $\Bbbk = \mathbb{Q}$.

For a given fixed data Γ and a seed $\mathfrak{s}$, we introduce an N^+-graded Lie algebra $\mathfrak{g}$ over $\Bbbk$ with generators X_n $(n \in N^+)$ as follows:

$$\mathfrak{g} = \mathfrak{g}_{\mathfrak{s}} := \bigoplus_{n \in N^+} \mathfrak{g}_n, \quad \mathfrak{g}_n = \Bbbk X_n, \tag{1.17}$$

$$[X_n, X_{n'}] = \{n, n'\} X_{n+n'}. \tag{1.18}$$

The Jacobi identity is easily seen by the following cyclic expression:

$$[X_{n_1}, [X_{n_2}, X_{n_3}]] = (\{n_1, n_2\}\{n_2, n_3\} - \{n_2, n_3\}\{n_3, n_1\}) X_{n_1+n_2+n_3}. \tag{1.19}$$

For each integer $\ell > 0$, we define a Lie algebra ideal and a quotient of $\mathfrak{g}$,

$$\mathfrak{g}^{>\ell} := \bigoplus_{n\in(N^+)^{>\ell}} \mathfrak{g}_n, \quad \mathfrak{g}^{\leq\ell} := \mathfrak{g}/\mathfrak{g}^{>\ell} \simeq \bigoplus_{n\in(N^+)^{\leq\ell}} \mathfrak{g}_n \text{ (as a vector space).} \tag{1.20}$$

It is important that $\mathfrak{g}^{\leq\ell}$ is a *nilpotent* Lie algebra. We have the canonical projections

$$\pi_{\ell',\ell} : \mathfrak{g}^{\leq\ell'} \to \mathfrak{g}^{\leq\ell} \quad (\ell' > \ell). \tag{1.21}$$

Thus, we have the completion of $\mathfrak{g}$,

$$\widehat{\mathfrak{g}} = \widehat{\mathfrak{g}}_{\mathfrak{s}} := \lim_{\longleftarrow} \mathfrak{g}^{\leq\ell} \simeq \prod_{n\in N^+} \mathfrak{g}_n \quad \text{(as a vector space)} \tag{1.22}$$

with the canonical projections $\pi_\ell : \widehat{\mathfrak{g}} \to \mathfrak{g}^{\leq\ell}$ $(\ell > 0)$.

For any Lie subalgebra $\mathfrak{h}$ of $\mathfrak{g}$, its completion $\widehat{\mathfrak{h}}$ is defined by

$$\mathfrak{h}^{>\ell} = \mathfrak{h} \cap \mathfrak{g}^{>\ell}, \quad \mathfrak{h}^{\leq\ell} = \mathfrak{h}/\mathfrak{h}^{>\ell}, \quad \widehat{\mathfrak{h}} = \lim_{\longleftarrow} \mathfrak{h}^{\leq\ell}. \tag{1.23}$$

Next, we define the exponential group G for the Lie algebra $\widehat{\mathfrak{g}}$ in (1.22). In fact, there are two equivalent ways to define it.

(a). For each nilpotent Lie algebra $\mathfrak{g}^{\leq\ell}$ in (1.20), we define the associated exponential group

$$G^{\leq\ell} = \exp(\mathfrak{g}^{\leq\ell}) \tag{1.24}$$

as follows: As a set, $G^{\leq\ell}$ is identified with $\mathfrak{g}^{\leq\ell}$ with $\exp(X) \leftrightarrow X$, and its product is defined by the *Baker-Campbell-Hausdorff (BCH) formula* (e.g., [Jac79, §V.5]),

$$\begin{aligned}&\exp(X)\exp(Y)\\ &= \exp\Bigl(X + Y + \frac{1}{2}[X,Y] + \frac{1}{12}[X,[X,Y]] - \frac{1}{12}[Y,[X,Y]] + \cdots\Bigr),\end{aligned} \tag{1.25}$$

where we do not need the explicit form of the higher order commutators here. Originally, the BCH formula is a product formula for the exponential element $e^X = \sum_{k=0}^{\infty} X^k/k!$ in the situation where e^X makes sense. Here, the product is well-defined due to the nilpotency of $\mathfrak{g}^{\leq\ell}$. Moreover, it is compatible with the canonical projections

$$\pi_{\ell',\ell} : G^{\leq\ell'} \to G^{\leq\ell} \quad (\ell' > \ell). \tag{1.26}$$

Thus, by the inverse limit we obtain a group

$$G = G_{\mathfrak{s}} = G_{\Gamma,\mathfrak{s}} := \lim_{\longleftarrow} G^{\leq\ell} \tag{1.27}$$

with the canonical projections $\pi_\ell : G \to G^{\leq \ell}$ ($\ell > 0$). Let $G^{>\ell} := \operatorname{Ker} \pi_\ell$. When we discuss $G^{\leq \ell}$, we often work with elements in G modulo $G^{>\ell}$.

(b). Alternatively, we may directly define

$$G := \exp(\widehat{\mathfrak{g}}), \tag{1.28}$$

where the infinite sum in the BCH formula (1.25) makes sense for $\widehat{\mathfrak{g}}$ again due to the nilpotency of $\mathfrak{g}^{\leq \ell}$. An element $\exp(X)$ ($X \in \widehat{\mathfrak{g}}$) is identified with an element $(\exp(\pi_\ell(X)))_{\ell=1}^{\infty}$ of G in (1.27).

Here we call G the *structure group* for the forthcoming scattering diagrams. We remark that the group G depends only on N, $\{\cdot,\cdot\}$, and $e_1, \dots, e_r$ among the data from Γ and $\mathfrak{s}$. In view of the decomposition (1.5), they corresponds to the matrix Ω. Thus, one can take a common G for those $(\Gamma, \mathfrak{s})$ where N, $\{\cdot,\cdot\}$, and $e_1, \dots, e_r$ (and Ω) are fixed but N° and $\delta_1, \dots, \delta_r$ (and Δ) are taken differently.

For any $g = \exp(X) \in G$ and any $c \in \Bbbk$, a power g^c of g is defined by

$$g^c := \exp(cX). \tag{1.29}$$

In particular, for any $c \in \mathbb{Q}$, the fractional power g^c of g is unambiguously defined.

We may consider an *infinite* product in G as follows. Suppose that there is a countable and linearly ordered sequence $(g_i)_{i \in I}$ ($g_i \in G$). Suppose that for any $\ell \in \mathbb{Z}_{>0}$, there are only finitely many $i \in I$ such that $\pi_\ell(g_i) \neq \mathrm{id}$. Let p_ℓ be the product of those g_i such that $\pi_\ell(g_i) \neq \mathrm{id}$ along its linear order, say, from left to right. Then, $\lim_{\ell \to \infty} p_\ell$ yields an element in G, which is the infinite product $\prod_{i \in I} g_i$.

Let $\mathfrak{h}$ be any Lie subalgebra of $\mathfrak{g}$, and let $\mathfrak{h}^{\leq \ell}$ and $\widehat{\mathfrak{h}}$ be the ones in (1.23). Then, the corresponding subgroup H of G is defined by

$$H^{\leq \ell} := \exp(\mathfrak{h}^{\leq \ell}), \quad H := \varprojlim H^{\leq \ell} = \exp(\widehat{\mathfrak{h}}). \tag{1.30}$$

The following subgroups of G are especially important.

Definition 1.5 (Parallel subgroup). For each $n \in N_{\mathrm{pr}}^{+}$, we define a Lie subalgebra of $\mathfrak{g}$,

$$\mathfrak{g}_n^{\parallel} := \bigoplus_{j>0} \mathfrak{g}_{jn}. \tag{1.31}$$

This is abelian by (1.18) and the skew-symmetry of (1.1). Then, we have an abelian subgroup $G_n^{\parallel}$ of G corresponding to $\widehat{\mathfrak{g}}_n^{\parallel}$ as defined in (1.30). We call $G_n^{\parallel}$ the *parallel subgroup* of n.

From the BCH formula (1.25), we obtain the *Zassenhaus formula* [Mag54], which is regarded as the dual of the BCH formula:

$$\begin{aligned}&\exp(X+Y)\\&=\exp(X)\exp(Y)\exp\Big(-\frac{1}{2}[X,Y]\Big)\exp\Big(\frac{1}{3}[Y,[X,Y]]+\frac{1}{6}[X,[X,Y]]\Big)\cdots,\end{aligned} \tag{1.32}$$

where again the explicit form of higher commutators are not necessary, and the infinite product makes sense in G. Together with the BCH formula, it also implies the useful formulas

$$\exp(X)\exp(Y)=\exp(Y)\exp(X)\exp([X,Y])C \tag{1.33}$$

$$=\exp(Y)\exp([X,Y])C'\exp(X). \tag{1.34}$$

where C and C' are (possibly infinite) products of the exponentials of higher commutators of X and Y.

1.3 Representations of G

Let us introduce two kinds of representations of G, which build a bridge between scattering diagrams and cluster algebras.

Let us recall some basic facts on derivations [Jac79, §I.2]. Let A be a (not necessarily associative) algebra over $\Bbbk$. A linear map $D: A \to A$ is a *derivation* of A if $D(ab)=D(a)b+aD(b)$ holds for any $a,b\in A$. For any derivations D_1 and D_2 of A, $[D_1,D_2]:=D_1D_2-D_2D_1$ is also a derivation. Thus, we have the Lie algebra $\mathrm{Der}(A)$ consisting of all derivations of A called the *derivation (Lie) algebra*. For a derivation D, suppose that the infinite sum $\exp(D)=\sum_{k=0}^{\infty}D^k/k!$ is well-defined as a linear map on A. Then, $\exp(D)$ is an algebra automorphism of A.

(a). *y-representation.* We consider a monoid

$$Q=Q_{\mathfrak{s}}:=\Big\{\sum_{i=1}^{r}a_ie_i\ \Big|\ a_i\in\mathbb{Z}_{\geq 0}\Big\}=N^+\sqcup\{0\}\subset N. \tag{1.35}$$

Let $\Bbbk[Q]$ be the monoid algebra of Q over $\Bbbk$. Let J be the maximal ideal of $\Bbbk[Q]$ generated by $N^+=Q\setminus\{0\}$, and let

$$\Bbbk[[Q]]=\varprojlim \Bbbk[Q]/J^\ell \tag{1.36}$$

be the completion with respect to J. We express any element $f\in\Bbbk[[Q]]$ as a formal power series in a symbol y as

$$f=\sum_{n\in Q}c_ny^n. \tag{1.37}$$

Let $\mathrm{Der}(\Bbbk[[Q]])$ be the derivation algebra of $\Bbbk[[Q]]$.

For each $n \in N^+$, we define a derivation $\tilde{X}_n$ on $\Bbbk[[Q]]$ by

$$\tilde{X}_n(y^{n'}) := \{n, n'\} y^{n'+n} \quad (n' \in Q). \tag{1.38}$$

Proposition 1.6 (cf. [GHKK18, §1.1]). *The map*

$$\begin{array}{rccc} \rho_y : & \widehat{\mathfrak{g}} & \to & \mathrm{Der}(\Bbbk[[Q]]) \\ & X = \displaystyle\sum_{n \in N^+} c_n X_n & \mapsto & \tilde{X} = \displaystyle\sum_{n \in N^+} c_n \tilde{X}_n, \end{array} \tag{1.39}$$

yields a Lie algebra homomorphism, where the sum could be infinite. Moreover, if Γ is nondegenerate, it is injective.

Proof. First, we regard ρ_y in (1.39) with finite sum as a map $\mathfrak{g} \to \mathrm{Der}(\Bbbk[Q])$. Then, we have

$$\begin{aligned} \tilde{X}_{n_1} \tilde{X}_{n_2}(y^n) &= \tilde{X}_{n_1}(\{n_2, n\} y^{n+n_2}) \\ &= (\{n_1, n\} + \{n_1, n_2\})\{n_2, n\} y^{n+n_2+n_1}. \end{aligned} \tag{1.40}$$

Thus, we obtain

$$\begin{aligned} [\tilde{X}_{n_1}, \tilde{X}_{n_2}](y^n) &= \{n_1, n_2\}\{n_1 + n_2, n\} y^{n+n_1+n_2} \\ &= \{n_1, n_2\} \tilde{X}_{n_1+n_2}(y^n). \end{aligned} \tag{1.41}$$

Therefore, $\rho_y : \mathfrak{g} \to \mathrm{Der}(\Bbbk[Q])$ is a Lie algebra homomorphism. We show that it is extended to the one in (1.39). First, we note that $\mathfrak{g}(J^{\ell+1}) \subset J^{\ell+2}$. Thus, $\mathfrak{g}$ acts on $\Bbbk[Q]^{\leq \ell} := \Bbbk[Q]/J^{\ell+1}$. Next, we have $\mathfrak{g}^{>\ell}(\Bbbk[Q]) \subset J^{\ell+1}$. Thus, it is extended to the action of $\widehat{\mathfrak{g}}$. Moreover, this action is compatible with projections $\pi_{\ell',\ell} : \Bbbk[Q]^{\leq \ell'} \to \Bbbk[Q]^{\leq \ell}$ for $\ell' > \ell$. Therefore, $\widehat{\mathfrak{g}}$ acts on $\Bbbk[[Q]]$. If $\{\cdot, \cdot\}$ is non-degenerate, $\tilde{X}_n$ $(n \in N^+)$ in (1.38) are linearly independent. Thus, ρ_y is injective. □

Remark 1.7. The above representation ρ_y is equivalent to the direct sum of the adjoint and the trivial representations of $\widehat{\mathfrak{g}}$.

For any $X \in \widehat{\mathfrak{g}}$, let $\tilde{X} := \rho_y(X)$. We define

$$\mathrm{Exp}(\tilde{X}) := \sum_{k=0}^{\infty} \frac{1}{k!} \tilde{X}^k \in \mathrm{GL}(\Bbbk[[Q]]), \tag{1.42}$$

which is well-defined by the same reason in the proof of Proposition 1.6. Since $\tilde{X}$ is a derivation, $\mathrm{Exp}(\tilde{X})$ is an algebra automorphism of $\Bbbk[[Q]]$ as we mentioned. Then, we have the following group homomorphism, where we abuse the symbol ρ_y, for simplicity.

Proposition 1.8. *We have a group homomorphism*

$$\begin{array}{rccc} \rho_y : & G & \to & \mathrm{Aut}(\Bbbk[[Q]]) \\ & \exp(X) & \mapsto & \mathrm{Exp}(\tilde{X}). \end{array} \tag{1.43}$$

Moreover, if Γ *is nondegenerate, it is injective.*

Proof. The map ρ_y is a group homomorphism, because the exponential (1.42) also satisfies the same BCH formula as mentioned. Suppose that Γ is nondegenerate. Let us show that ρ_y is injective. Assume that $\mathrm{Exp}(\tilde{X}) = \mathrm{id}$. Then, by the action (1.38), we have $\tilde{X} = 0$. Then, $X = 0$ in $\widehat{\mathfrak{g}}$ by the injectivity of ρ_y in Proposition 1.6. Thus, $\exp(X) = \mathrm{id}$. □

The representation ρ_y is closely related with the mutations of y-variables in (1.7) as we will see soon. Thus, we call it the *y-representation* of G.

(b). *x-representation.* Here we assume that Γ is nondegenerate from the beginning. Then, thanks to Lemma 1.4, the map p^* in (1.8) is injective. In particular, $p^*(e_1)$, …, $p^*(e_r)$ are $\mathbb{Z}$-linearly independent. We define a monoid

$$P = P_{\mathfrak{s}} := \left\{ \sum_{i=1}^{r} a_i p^*(e_i) \;\middle|\; a_i \in \mathbb{Z}_{\geq 0} \right\} = p^*(Q) \subset M^{\circ}. \tag{1.44}$$

Let $\Bbbk[P]$ the monoid algebra of P over $\Bbbk$, and $\Bbbk[[P]]$ be the completion by the maximal ideal of $\Bbbk[P]$ generated by $P \setminus \{0\}$. (Note that, if Γ is degenerate, we cannot define $\Bbbk[[P]]$.) We express any element $f \in \Bbbk[[P]]$ as a formal power series in a symbol x as

$$f = \sum_{m \in P} c_m x^m. \tag{1.45}$$

Then, in parallel to (1.38), for each $n \in N^+$, we define a derivation $\tilde{X}_n$ on $\Bbbk[[P]]$ by

$$\tilde{X}_n(x^m) := \langle n, m \rangle x^{m + p^*(n)} \quad (m \in P). \tag{1.46}$$

Then, repeating the same argument for ρ_y, we obtain the following group homomorphism.

Proposition 1.9. *Suppose that* Γ *is nondegenerate. Then, we have an injective group homomorphism*

$$\begin{array}{rccc} \rho_x : & G & \to & \mathrm{Aut}(\Bbbk[[P]]) \\ & \exp(X) & \mapsto & \mathrm{Exp}(\tilde{X}). \end{array} \tag{1.47}$$

The representation ρ_x is closely related with the mutations of x-variables in (1.7) as we will see soon. Thus, we call it the *x-representation* of G.

Even though a faithful representation of G is desirable in some situation, it is too restrictive to assume that Γ is nondegenerate in view of cluster algebra theory. We will treat this problem later in Section 4.

1.4 Dilogarithm elements and pentagon relation

As an application of the actions of G in the previous subsection, we show a remarkable relation among certain distinguished elements in G.

Let $\mathrm{Li}_2(x)$ be the *Euler dilogarithm* [Lew81] defined by

$$\mathrm{Li}_2(x) := \sum_{j=1}^{\infty} \frac{x^j}{j^2} = -\int_0^x \frac{\log(1-y)}{y}\,dy. \tag{1.48}$$

It has the following key property:

$$x\frac{d}{dx}(-\mathrm{Li}_2(-x)) = \sum_{j=1}^{\infty} \frac{(-1)^{j+1}}{j} x^j = \log(1+x). \tag{1.49}$$

Having this in mind, we introduce the following elements of G.

Definition 1.10 (Dilogarithm element)**.** For any $n \in N^+$, let

$$\Psi[n] := \exp\left(\sum_{j=1}^{\infty} \frac{(-1)^{j+1}}{j^2} X_{jn} \right) \in G^{\parallel}_{n_0}, \tag{1.50}$$

where $n_0 \in N^+_{\mathrm{pr}}$ is the unique one such that $n = tn_0$ for some integer $t > 0$. We call $\Psi[n]$ the *dilogarithm element* for n.

Remark 1.11. It is "only formally" expressed as

$$\Psi[n] = \exp(-\mathrm{Li}_2(-X_n)), \tag{1.51}$$

if we interpret X_n^j as X_{jn}. (Caution: X_n^j and X_{jn} act differently on $\Bbbk[[Q]]$ and $\Bbbk[[P]]$.)

The elements $\Psi[n]$ are introduced because of the following property.

Proposition 1.12 (Cf. [GHKK18, Lemma 1.3])**.** *(a). Under the y-representation ρ_y, $\Psi[n]$ acts on $\Bbbk[[Q]]$ as*

$$\Psi[n](y^{n'}) = y^{n'}(1+y^n)^{\{n,n'\}} \quad (n' \in Q). \tag{1.52}$$

(b). Suppose that Γ is nondegenerate. Then, under the x-representation ρ_x, $\Psi[n]$ acts on $\Bbbk[[P]]$ as

$$\Psi[n](x^m) = x^m(1+x^{p^*(n)})^{\langle n,m\rangle} \quad (m \in P). \tag{1.53}$$

Proof. Since the calculations are parallel, we concentrate on the case (b).

$$\begin{aligned}\Psi[n](x^m) &= x^m \exp\left(\sum_{j>0} \frac{(-1)^{j+1}}{j^2} x^{jp^*(n)} \langle jn, m\rangle\right)\\ &= x^m \exp\left(\sum_{j>0} \frac{(-1)^{j+1}}{j} x^{jp^*(n)}\right)^{\langle n,m\rangle}\\ &= x^m(1 + x^{p^*(n)})^{\langle n,m\rangle}.\end{aligned} \tag{1.54}$$

□

Observe that the calculation in (1.54) is parallel to the derivation of the formula (1.49) by the power series calculation. The above automorphisms (1.52) and (1.53) are identified with the automorphism part of the Fock-Goncharov decomposition of mutations of *y-variables (coefficients)* and x-*variables (cluster variables)* in cluster algebras, respectively, in [FG09a, §2.1].

The dilogarithm elements are generators of G in the following sense.

Proposition 1.13. *The group G is generated by $\Psi[tn]^c$ ($n \in N^+_{\mathrm{pr}}, t \in \mathbb{Z}_{>0}, c \in \Bbbk$) admitting the infinite product.*

Proof. By inverting the expression (1.50), one can express $\exp(X_n)$ ($n \in N^+$) as an infinite product of $\Psi[tn]^c$ ($t \in \mathbb{Z}_{>0}$, $c \in \mathbb{Q}$). □

The dilogarithm elements satisfy the following remarkable relations.

Proposition 1.14. *Let $n, n' \in N^+$. The following relations hold in G.*

(a). If $\{n', n\} = 0$, for any $c, c' \in \Bbbk$,

$$\Psi[n']^{c'}\Psi[n]^c = \Psi[n]^c\Psi[n']^{c'}. \tag{1.55}$$

(b). (Pentagon relation [GHKK18, Example 1.14] for $c = 1$.) If $\{n', n\} = c$ ($c \in \mathbb{Q}, c \neq 0$),

$$\Psi[n']^{1/c}\Psi[n]^{1/c} = \Psi[n]^{1/c}\Psi[n + n']^{1/c}\Psi[n']^{1/c}. \tag{1.56}$$

Proof. (a). This is clear from (1.18). (b). Let us consider the rank 2 sublattice N' of N generated by n and n'. By the assumption $\{n', n\} \neq 0$, the form $\{\cdot, \cdot\}$ restricted on N' is nondegenerate. Let G' be the subgroup of G corresponding to N'. Similarly, we consider a monoid

$$P' = \{ap^*(n) + a'p^*(n') \mid a, a' \in \mathbb{Z}_{\geq 0}\} \subset M^\circ. \tag{1.57}$$

Since the relation (1.56) involves only elements in G', one can prove it by the x-representation of G' on $\Bbbk[[P']]$, which is faithful. The left hand side is given by

$$\begin{aligned}
&\Psi[n']^{1/c}\Psi[n]^{1/c}(x^m) \\
&= \Psi[n']^{1/c}(x^m(1+x^{p^*(n)})^{\langle n,m\rangle/c}) \\
&= x^m(1+x^{p^*(n')})^{\langle n',m\rangle/c}(1+x^{p^*(n)}(1+x^{p^*(n')}))^{\langle n,m\rangle/c}.
\end{aligned} \tag{1.58}$$

The right hand side is given by

$$\begin{aligned}
&\Psi[n]^{1/c}\Psi[n+n']^{1/c}\Psi[n']^{1/c}(x^m) \\
&= \Psi[n]^{1/c}\Psi[n+n']^{1/c}(x^m(1+x^{p^*(n')})^{\langle n',m\rangle/c}) \\
&= \Psi[n]^{1/c}(x^m(1+x^{p^*(n+n')})^{\langle n+n',m\rangle/c} \\
&\qquad \times (1+x^{p^*(n')}(1+x^{p^*(n+n')})^{-1})^{\langle n',m\rangle/c}) \\
&= \Psi[n]^{1/c}(x^m(1+x^{p^*(n+n')})^{\langle n,m\rangle/c}(1+x^{p^*(n+n')}+x^{p^*(n')})^{\langle n',m\rangle/c}) \\
&= x^m(1+x^{p^*(n)})^{\langle n,m\rangle/c}(1+x^{p^*(n+n')}(1+x^{p^*(n)})^{-1})^{\langle n,m\rangle/c} \\
&\qquad \times (1+x^{p^*(n+n')}(1+x^{p^*(n)})^{-1}+x^{p^*(n')}(1+x^{p^*(n)})^{-1})^{\langle n',m\rangle/c} \\
&= x^m(1+x^{p^*(n)}+x^{p^*(n+n')})^{\langle n,m\rangle/c}(1+x^{p^*(n')})^{\langle n',m\rangle/c}.
\end{aligned} \tag{1.59}$$

Thus, two expressions agree. □

Remark 1.15. (a). The above calculation for $c = 1$ is essentially the same for the *pentagon periodicity* of x-variables for a cluster algebra of type A_2 in [FZ02a]. Alternatively, one can also work with the y-representation. Then, the calculation for $c = 1$ is essentially the same for the pentagon periodicity of y-variables for a cluster algebra of type A_2 in [FZ02a].

(b). The case $c \neq 1$ follows from the case $c = 1$ by Proposition 1.17.

1.5 Equivalence of fixed data and normalization factor

There is some redundancy for choosing a fixed data. Let us start with a simple situation.

Definition 1.16 (Rescaling of fixed data)**.** Let Γ and Γ' be a pair of fixed data. We say that Γ' is a *rescaling of* Γ if there is some rational number $\lambda > 0$ such that the following relations hold:

$$N' = N, \tag{1.60}$$

$$\{n_1, n_2\}' = \lambda\{n_1, n_2\}, \tag{1.61}$$

$$(N^\circ)' = \lambda^{-1}N^\circ, \tag{1.62}$$

$$\delta'_i = \lambda^{-1}\delta_i, \tag{1.63}$$

$$(M^\circ)' = \lambda M^\circ. \tag{1.64}$$

The exchange matrix B associated with $(\Gamma, \mathfrak{s})$ in (1.4) is invariant under the above rescaling, namely,

$$b'_{ij} = \{\delta'_i e_i, e_j\}' = \lambda\{\lambda^{-1}\delta_i e_i, e_j\} = b_{ij}. \tag{1.65}$$

Let $\mathfrak{s} = (e_1, \dots, e_r)$ be a seed for Γ. Then, $\mathfrak{s}$ is also regarded as a seed for Γ'. Let $\widehat{\mathfrak{g}}'$ and G' be the corresponding Lie algebra and group for Γ' and $\mathfrak{s}$.

Proposition 1.17. *We have a Lie algebra isomorphism*

$$\begin{array}{cccc} \phi_\lambda : & \widehat{\mathfrak{g}}' & \to & \widehat{\mathfrak{g}} \\ & X'_n & \mapsto & \lambda X_n \end{array} \tag{1.66}$$

and a group isomorphism

$$\begin{array}{cccc} \phi_\lambda : & G' & \to & G \\ & \exp(X'_n) & \mapsto & \exp(\lambda X_n). \end{array} \tag{1.67}$$

Proof. This is due to the relation (1.61) and the commutation relation (1.18). □

Definition 1.18 (Normalization factor)**.** For any $n \in N^+$, let $\delta(n)$ be the smallest positive rational number such that $\delta(n)n \in N^\circ$. We call $\delta(n)$ the *normalization factor* of n with respect to N°.

Example 1.19. (a). For the seed $\mathfrak{s} = (e_1, \dots, e_r)$ under consideration, we have $\delta(e_i) = \delta_i$.

(b). For any $n \in N^+_{\mathrm{pr}}$ and any integer $t > 0$, we have

$$\delta(tn) = \delta(n)/t, \tag{1.68}$$

because $\delta(tn)tn = \delta(n)n$. Also, $\delta(n)$ is an integer. Otherwise, let $\delta(n) = r/s$ be the irreducible rational expression with $s \neq 1$. Then, we have $n, (r/s)n \in N$. Thus, $(1/s)n \in N$. This contradicts that n is primitive.

Proposition 1.20. *Let G and G' be as above. For any $n \in N^+_{\mathrm{pr}}$, let $g \in G_n^{\parallel}$ and $g' \in (G')_n^{\parallel}$ with*

$$g = \exp\left(\delta(n)\sum_{j>0} c_j X_{jn}\right), \quad g' = \exp\left(\delta'(n)\sum_{j>0} c'_j X'_{jn}\right), \tag{1.69}$$

where $\delta(n)$ and $\delta'(n)$ are the normalization factors with respect to N° and $(N^\circ)'$. Then,

$$\phi_\lambda(g') = g. \tag{1.70}$$

if and only $c_j = c'_j$ for any $j > 0$.

Proof. This follows from (1.66) and the transformation of the normalization factors $\delta'(n) = \lambda^{-1}\delta(n)$ by (1.62). □

In view of Proposition 1.20, it is natural to express an element $g \in G_n^{\parallel}$ as

$$g = g_0^{\delta(n)}, \tag{1.71}$$

where $g_0 = g^{1/\delta(n)}$. We call (1.71) the *normalized form* of g. Then, the coefficients c_j of the expansion $g_0 = \exp(\sum_{j>0} c_j X_{jn})$ are *invariant* under the rescaling of Γ.

Example 1.21. For $\Psi[n]$ in (1.50) with $n = tn_0$ therein, we have

$$\phi_\lambda(\Psi'[n]^{\delta'(n_0)}) = \Psi[n]^{\delta(n_0)}. \tag{1.72}$$

Thus, we also have

$$\phi_\lambda(\Psi'[n]^{\delta'(n)}) = \Psi[n]^{\delta(n)}. \tag{1.73}$$

These elements play the main role in the forthcoming CSDs.

Now we consider a more general situation.

Definition 1.22. Let $(\Gamma, \mathfrak{s})$ be a pair of a fixed data and a seed, and let $B_{\Gamma,\mathfrak{s}}$ be the associated matrix in (1.4). We say that $(\Gamma, \mathfrak{s})$ and $(\Gamma', \mathfrak{s}')$ are *B-equivalent* if $B_{\Gamma,\mathfrak{s}} = B_{\Gamma',\mathfrak{s}'}$.

In Section 1.1 we have seen that, for a given skew-symmetrizable matrix B, there is a (not unique) pair $(\Gamma, \mathfrak{s})$ such that $B = B_{\Gamma,\mathfrak{s}}$. Let us show that the associated group $G_{\Gamma,\mathfrak{s}}$ is uniquely determined from B up to isomorphism. This fact is natural in view of the correspondence to cluster algebras.

Proposition 1.23. *If $(\Gamma, \mathfrak{s})$ and $(\Gamma', \mathfrak{s}')$ are B-equivalent, then the associated groups $G_{\Gamma,\mathfrak{s}}$ and $G_{\Gamma',\mathfrak{s}'}$ are isomorphic.*

Proof. Let $B = B_{\Gamma,\mathfrak{s}} = B_{\Gamma',\mathfrak{s}'}$. We say that an $n \times n$ matrix A is *indecomposable* if there is no pair $I, J \subset \{1, \dots, r\}$, $I \sqcup J = \{1, \dots, r\}$, $I, J \neq \emptyset$ such that $a_{ij} = 0$ for any $i \in I$ and $j \in J$. Suppose that B is decomposed into $B_1 \oplus \cdots \oplus B_p$ up to simultaneous permutation of rows and columns, where B_i are indecomposable submatrices of B. Accordingly, the groups $G = G_{\Gamma,\mathfrak{s}}$ and $G' = G_{\Gamma',\mathfrak{s}'}$ are factorized into the corresponding subgroups $G = \prod_{i=1}^p G_i$ and $G' = \prod_{i=1}^p G'_i$. Thus, we only need to prove the claim when B is indecomposable. Then, it is well known and easy to show that a (left) skew-symmetrizer of B is unique up to a positive rational number. This implies that δ_i and δ'_i are related by $\delta'_i = \lambda^{-1}\delta_i$ with a common multiple λ^{-1}. Let us identify e_i with e'_i $(i = 1, \dots, r)$ and N with N'. Now we are in the situation of Definition 1.16. Therefore, G and G' are isomorphic by Proposition 1.17. □

The above proof also tells that the parallel results to Proposition 1.20 and Example 1.21 hold in the situation of Proposition 1.23.

1.6 Formulas for dilogarithm elements

Let us present some useful formulas for dilogarithm elements.

Proposition 1.24. *Let $n, n' \in N^+$.*

(a). ([GHKK18, Lemma C.8].) If $\{\delta(n')n', n\} = \pm\alpha$ ($\alpha \in \mathbb{Z}_{>0}$), then the following equality holds for the x-representation on $\Bbbk[[P']]$ for P' in (1.57):

$$\Psi[n']^{\pm\delta(n')} X_n \Psi[n']^{\mp\delta(n')} = \sum_{j=0}^{\alpha} \binom{\alpha}{j} X_{n+jn'}. \tag{1.74}$$

(b). If $\{\delta(n')n', n\} = \alpha$ and $\{\delta(n)n, n'\} = -\beta$ ($\alpha, \beta \in \mathbb{Z}_{>0}$), then

$$\Psi[n']^{\delta(n')}\Psi[n]^{\delta(n)} = \Psi[n]^{\delta(n)}\Psi[\beta n + n']^{\delta(n')} C \Psi[n + \alpha n']^{\delta(n)}\Psi[n']^{\delta(n')}, \tag{1.75}$$

where C is a (possibly infinite) product of $\Psi[n'']^{c''}$ ($c'' \in \mathbb{Q}$) such that $n'' = j_1 n + j_2 n'$ ($j_1, j_2 \in \mathbb{Z}_{>0}$) with $\alpha^{-1} < j_1/j_2 < \beta$.

Proof. (a). We have

$$\begin{aligned} &X_n \Psi[n']^{\mp\delta(n')}(x^m) \\ &= X_n(x^m(1 + x^{p^*(n')})^{\mp\langle\delta(n')n', m\rangle}) \\ &= \langle n, m\rangle x^{m+p^*(n)}(1 + x^{p^*(n')})^{\mp\langle\delta(n')n', m\rangle} \\ &\quad + \alpha\langle n', m\rangle x^{m+p^*(n'+n)}(1 + x^{p^*(n')})^{\mp\langle\delta(n')n', m\rangle - 1}, \end{aligned} \tag{1.76}$$

$$\begin{aligned} &\Psi[n']^{\mp\delta(n')} \sum_{j=0}^{\alpha} \binom{\alpha}{j} X_{n+jn'}(x^m) \\ &= \Psi[n']^{\mp\delta(n')} \sum_{j=0}^{\alpha} \binom{\alpha}{j} \langle n + jn', m\rangle x^{m+p^*(n+jn')} \\ &= \sum_{j=0}^{\alpha} \binom{\alpha}{j} \langle n + jn', m\rangle x^{m+p^*(n+jn')}(1 + x^{p^*(n')})^{\mp\langle\delta(n')n', m\rangle - \alpha} \\ &= \sum_{j=0}^{\alpha} \binom{\alpha}{j} \langle n, m\rangle x^{m+p^*(n+jn')}(1 + x^{p^*(n')})^{\mp\langle\delta(n')n', m\rangle - \alpha} \\ &\quad + \sum_{j=1}^{\alpha} \alpha \binom{\alpha - 1}{j - 1} \langle n', m\rangle x^{m+p^*(n+jn')}(1 + x^{p^*(n')})^{\mp\langle\delta(n')n', m\rangle - \alpha} \\ &= \langle n, m\rangle x^{m+p^*(n)}(1 + x^{p^*(n')})^{\mp\langle\delta(n')n', m\rangle} \\ &\quad + \alpha\langle n', m\rangle x^{m+p^*(n'+n)}(1 + x^{p^*(n')})^{\mp\langle\delta(n')n', m\rangle - 1}. \end{aligned} \tag{1.77}$$

Thus, they coincide.

(b). As in the proof of Proposition 1.14, it is enough to prove it for the x-representation on $\Bbbk[[P']]$ for P' in (1.57). For any $t \in \mathbb{Z}_{>0}$, we obtain from (1.74) that

$$\Psi[n']^{\delta(n')} X_{tn} \Psi[n']^{-\delta(n')} = X_{tn} + \sum_{j=1}^{t\alpha-1} \binom{t\alpha}{j} X_{tn+jn'} + X_{t(n+\alpha n')}. \quad (1.78)$$

Taking the sum and exponential in accordance with (1.50), then using the Zassenhaus formula (1.32) and Proposition 1.13, we obtain

$$\Psi[n']^{\delta(n')} \Psi[n]^{\delta(n)} \Psi[n']^{-\delta(n')} = \Psi[n]^{\delta(n)} C_1 \Psi[n+\alpha n']^{\delta(n)}, \quad (1.79)$$

where C_1 is a (possibly infinite) product of $\Psi[n'']^{c''}$ ($c'' \in \mathbb{Q}$) such that $n'' = j_1 n + j_2 n'$ ($j_1, j_2 \in \mathbb{Z}_{>0}$) with $0 < j_2/j_1 < \alpha$. Similarly, by exchange the role of n and n', we obtain

$$\Psi[n]^{-\delta(n)} \Psi[n']^{\delta(n')} \Psi[n]^{\delta(n)} = \Psi[\beta n + n']^{\delta(n')} C_2 \Psi[n']^{\delta(n')}, \quad (1.80)$$

where C_2 is a (possibly infinite) product of $\Psi[n'']^{c''}$ ($c'' \in \mathbb{Q}$) such that $n'' = j_1 n + j_2 n'$ ($j_1, j_2 \in \mathbb{Z}_{>0}$) with $0 < j_1/j_2 < \beta$. Combining (1.79) and (1.80), we obtain (1.75). □

Notes

The contents are mostly taken from [GHKK18, §1.1] and partly from [KS14, §2.1]. The y-representation of G in Section 1.4 appeared in [GHK15, §2]. The pentagon relation (1.56) generalizes the one in [GHKK18, Example 1.14] for $c = 1$. The equivalence of fixed data in Section 1.5 is distilled from the *change of lattice trick* in [GHKK18, Appendix C.3]. Proposition 1.24 (b) is due to us.

2 Consistent scattering diagrams

In this section we introduce and study consistent scattering diagrams. In particular, we show that there is a natural one-to-one correspondence between consistent scattering diagrams and elements in the structure group G.

2.1 Scattering diagrams

We continue to fix a seed $\mathfrak{s}$ for a given (possibly degenerate) fixed data Γ.

For any $n \in N$, $n \neq 0$, the rational hyperplane $n^\perp$ in $M_\mathbb{R}$ that is orthogonal to n is defined by

$$n^\perp := \{z \in M_\mathbb{R} \mid \langle n, z \rangle = 0\}. \tag{2.1}$$

Definition 2.1 (Cone). For $m_1, \dots, m_s \in M^\circ$, let

$$\sigma = \sigma(m_1, \dots, m_s) := \mathbb{R}_{\geq 0} m_1 + \cdots + \mathbb{R}_{\geq 0} m_s \subset M_\mathbb{R}. \tag{2.2}$$

We also set $\sigma(\emptyset) = \{0\}$. We call σ a *convex rational polyhedral cone*, or simply, a *cone* in $M_\mathbb{R}$. We say that a cone σ is *strongly convex* if $\sigma \cap (-\sigma) = \{0\}$.

Definition 2.2 (Wall). A *wall* for a seed $\mathfrak{s}$ is a triplet $\mathbf{w} = (\mathfrak{d}, g)_n$, where

- $n \in N^+_{\mathrm{pr}}$,
- $\mathfrak{d} \subset n^\perp$ is a (not necessarily strongly convex) cone of codimension 1 in $M_\mathbb{R}$,
- $g \in G_n^{\parallel}$.

We call n, $\mathfrak{d}$, g the *normal vector*, the *support*, the *wall element* of a wall $\mathbf{w}$, respectively.

Informally, the support $\mathfrak{d}$ is also referred to as a *wall*; for example, we say "A curve in $M_\mathbb{R}$ crosses a wall." The normal vector n is uniquely specified from the data $\mathfrak{d}$. So, we may omit it as $\mathbf{w} = (\mathfrak{d}, g)$ when it is not necessary to indicate it explicitly.

Definition 2.3 (Scattering diagram). A *scattering diagram* $\mathfrak{D}$ for a seed $\mathfrak{s}$ is a collection of walls $\{\mathbf{w}_\lambda = (\mathfrak{d}_\lambda, g_\lambda)_{n_\lambda}\}_{\lambda \in \Lambda}$ for $\mathfrak{s}$, where Λ is a finite or countably infinite index set, satisfying the following *finiteness condition*:

- For each integer $\ell > 0$, there are only finitely many walls such that $\pi_\ell(g_\lambda) \neq \mathrm{id}$, where $\pi_\ell : G \to G^{\leq \ell}$ is the canonical projection.

For any scattering diagram $\mathfrak{D}$ and any integer $\ell > 0$, let $\mathfrak{D}_\ell$ be the collection of walls obtained from $\mathfrak{D}$ by removing all walls such that $\pi_\ell(g_\lambda) = \mathrm{id}$. We call $\mathfrak{D}_\ell$ the *reduction* of $\mathfrak{D}$ at degree ℓ. By the finiteness condition, $\mathfrak{D}_\ell$ has only finitely many walls.

We give some related notions.

Definition 2.4 (Support/Singular locus). For a scattering diagram $\mathfrak{D} = \{\mathbf{w}_\lambda = (\mathfrak{d}_\lambda, g_\lambda)_{n_\lambda}\}_{\lambda \in \Lambda}$, the *support* and the *singular locus* of $\mathfrak{D}$ are defined by

$$\mathrm{Supp}(\mathfrak{D}) = \bigcup_{\lambda \in \Lambda} \mathfrak{d}_\lambda, \tag{2.3}$$

$$\mathrm{Sing}(\mathfrak{D}) = \bigcup_{\lambda \in \Lambda} \partial \mathfrak{d}_\lambda \cup \bigcup_{\substack{\lambda, \lambda' \in \Lambda \\ \dim \mathfrak{d}_\lambda \cap \mathfrak{d}_{\lambda'} = r-2}} \mathfrak{d}_\lambda \cap \mathfrak{d}_{\lambda'}. \tag{2.4}$$

We note that, if $\dim \mathfrak{d}_\lambda \cap \mathfrak{d}_{\lambda'} < r - 2$, then it is contained in $\partial \mathfrak{d}_\lambda \cup \partial \mathfrak{d}_{\lambda'}$.

Definition 2.5 (Admissible curve). A curve $\gamma : [0, 1] \to M_{\mathbb{R}}$ is *admissible* for a scattering diagram $\mathfrak{D}$ if it satisfies the following conditions:

- The end points of γ are in $M_{\mathbb{R}} \setminus \mathrm{Supp}(\mathfrak{D})$.
- It is a smooth curve, and it intersects $\mathrm{Supp}(\mathfrak{D})$ transversally.
- γ does not intersect $\mathrm{Sing}(\mathfrak{D})$.

Definition 2.6 (Path-ordered product). Let $\mathfrak{D}$ be any scattering diagram. For any admissible curve γ, we define an element $\mathfrak{p}_{\gamma,\mathfrak{D}} \in G$ as follows: For each integer $\ell > 0$, the reduction $\mathfrak{D}_\ell$ at ℓ has only finitely many walls. Suppose that $\gamma = \gamma(t)$ crosses walls $\mathbf{w}_i = (\mathfrak{d}_i, g_i)_{n_i}$ $(i = 1, \dots, s)$ of $\mathfrak{D}_\ell$ in this order at $t = t_i$ with

$$0 < t_1 \le t_2 \le \cdots \le t_s < 1. \tag{2.5}$$

Since $\gamma(t_i) \notin \mathrm{Sing}(\mathfrak{D})$, when γ crosses multiple walls at a time, these walls have a common normal vector. We define the *intersection sign* ϵ_i $(i = 1, \dots, s)$ by

$$\epsilon_i = \begin{cases} 1 & \langle n_i, \gamma'(t_i) \rangle < 0, \\ -1 & \langle n_i, \gamma'(t_i) \rangle > 0, \end{cases} \tag{2.6}$$

where $\gamma'(t_i)$ is the velocity vector of $\gamma(t)$ at t_i. Now, we define

$$\mathfrak{p}_{\gamma,\mathfrak{D}_\ell} := g_s^{\epsilon_s} \cdots g_1^{\epsilon_1}, \tag{2.7}$$

$$\mathfrak{p}_{\gamma,\mathfrak{D}} := \lim_{\ell \to \infty} \mathfrak{p}_{\gamma,\mathfrak{D}_\ell} \in G. \tag{2.8}$$

We call $\mathfrak{p}_{\gamma,\mathfrak{D}}$ the *path-ordered product* (of wall elements in $\mathfrak{D}$) along γ.

Note that $\mathfrak{p}_{\gamma,\mathfrak{D}}$ only depends on the homotopy class of γ in $M_{\mathbb{R}} \setminus \mathrm{Sing}(\mathfrak{D})$.

Definition 2.7 (Equivalence). Two scattering diagrams $\mathfrak{D}$ and $\mathfrak{D}'$ for a common seed $\mathfrak{s}$ are *equivalent* if, for any curve γ that is admissible for both $\mathfrak{D}$ and $\mathfrak{D}'$, the equality $\mathfrak{p}_{\gamma,\mathfrak{D}} = \mathfrak{p}_{\gamma,\mathfrak{D}'}$ holds.

Definition 2.8 (Generality). We say that $z \in M_{\mathbb{R}}$ is *general* if there is at most one rational hyperplane $n^\perp$ $(n \in N,\ n \neq 0)$ such that $z \in n^\perp$.

Lemma 2.9. *(a). The set of all general points is dense in $M_{\mathbb{R}}$.*

(b). For any scattering diagram $\mathfrak{D}$, any point $z \in \mathrm{Sing}(\mathfrak{D})$ is not general.

Proof. (a). This is clear. (b). Suppose that $z \in \partial \mathfrak{d}_\lambda$ for some $\mathfrak{d}_\lambda$. Since $\mathfrak{d}_\lambda$ is a cone of codimension 1, z belongs to a face of $\mathfrak{d}_\lambda$ of codimension 2, which lies in the intersection of two rational hyperplanes. Thus, z is not general. For $z \in \mathfrak{d}_\lambda \cap \mathfrak{d}_{\lambda'}$ with $\dim \mathfrak{d}_\lambda \cap \mathfrak{d}_{\lambda'} = r - 2$, the claim is clear. □

Definition 2.10 (Total wall element at z)**.** For any general $z \in M_{\mathbb{R}}$, we define

$$g_z(\mathfrak{D}) := \prod_{\lambda:\, z \in \mathfrak{d}_\lambda} g_\lambda \in \begin{cases} G^{\parallel}_{n_0} & z \in n_0^{\perp} \text{ for some unique } n_0 \in N^+_{\mathrm{pr}}, \\ \{\mathrm{id}\} & \text{otherwise.} \end{cases} \tag{2.9}$$

The (possibly infinite) product is well-defined due to the finiteness condition of $\mathfrak{D}$. We call it the *total wall element of $\mathfrak{D}$ at z*.

The condition for the equivalence reduces to the following local condition.

Lemma 2.11 ([GHKK18, Lemma 1.9])**.** *Two scattering diagrams $\mathfrak{D}$ and $\mathfrak{D}'$ are equivalent if and only if $g_z(\mathfrak{D}) = g_z(\mathfrak{D}')$ holds for any general $z \in M_{\mathbb{R}}$.*

Proof. Consider the reductions $\mathfrak{D}_\ell$ and $\mathfrak{D}'_\ell$ at ℓ. Note that both have only finitely many walls. Then, thanks to Lemma 2.9, the equivalence condition reduces to the following condition:

- For any general $z \in M_{\mathbb{R}}$ and for any admissible curve γ that intersects walls of $\mathfrak{D}_\ell$ and $\mathfrak{D}'_\ell$ only at z, the following equality holds:

$$\mathfrak{p}_{\gamma, \mathfrak{D}_\ell} \equiv \mathfrak{p}_{\gamma, \mathfrak{D}'_\ell} \mod G^{>\ell}. \tag{2.10}$$

Under the situation, we have $\mathfrak{p}_{\gamma, \mathfrak{D}_\ell} = g_z(\mathfrak{D}_\ell)^{\epsilon}$ with the intersection sign ϵ. Therefore, the above equality is written as $g_z(\mathfrak{D}_\ell) \equiv g_z(\mathfrak{D}'_\ell) \bmod G^{>\ell}$. Since ℓ is arbitrary, we have $g_z(\mathfrak{D}) = g_z(\mathfrak{D}')$. □

Definition 2.12 (Trivial scattering diagram)**.** A scattering diagram $\mathfrak{D}$ is *trivial* if $g_z(\mathfrak{D}) = \mathrm{id}$ for any general $z \in M_{\mathbb{R}}$.

By Lemma 2.11, a scattering diagram is trivial if and only if it is equivalent to the empty scattering diagram $\emptyset$.

Finally, we introduce the most crucial notion for scattering diagrams.

Definition 2.13 (Consistency)**.** A scattering diagram $\mathfrak{D}$ is *consistent* if for any admissible curve γ for $\mathfrak{D}$, the associated path-ordered product $\mathfrak{p}_{\gamma, \mathfrak{D}}$ depends only on the end points of γ.

The following restatement is useful.

Lemma 2.14. *A scattering diagram $\mathfrak{D}$ is consistent if and only if*

$$\mathfrak{p}_{\gamma,\mathfrak{D}} = \mathrm{id} \tag{2.11}$$

holds for any admissible loop (i.e., closed curve) γ for $\mathfrak{D}$.

Proof. This is proved by the standard argument for fundamental groups. Assume that $\mathfrak{D}$ is consistent. Let γ be any admissible loop for $\mathfrak{D}$. Split γ as $\gamma = (\gamma_2)^{-1} \circ \gamma_1$, where γ_1 and γ_2 are admissible curves having the common ending points. Then, by the consistency, we have $\mathfrak{p}_{\gamma_1,\mathfrak{D}} = \mathfrak{p}_{\gamma_2,\mathfrak{D}}$. Therefore, $\mathfrak{p}_{\gamma,\mathfrak{D}} = (\mathfrak{p}_{\gamma_2,\mathfrak{D}})^{-1} \circ \mathfrak{p}_{\gamma_1,\mathfrak{D}} = \mathrm{id}$ holds. The converse is similar. □

2.2 Rank 2 examples: finite type

Based on the pentagon relation in Proposition 1.14, we construct some prototypical examples of consistent scattering diagrams of rank 2. Let $e_1^*, e_2^* \in M$ be the dual basis of $e_1, e_2 \in N$. Accordingly, let $f_1 = e_1^*/\delta_1, f_2 = e_2^*/\delta_2 \in M^\circ$ be the dual basis of $\delta_1 e_1, \delta_2 e_2 \in N^\circ$. Let $\mathbf{e}_1, \mathbf{e}_2 \in \mathbb{Z}^2$ be the unit vectors. We identify $N \simeq \mathbb{Z}^2$, $e_i \mapsto \mathbf{e}_i$, and $M_\mathbb{R} \simeq \mathbb{R}^2$, $f_i \mapsto \mathbf{e}_i$. Then, the canonical pairing $\langle n, z\rangle : N \times M_\mathbb{R} \to \mathbb{R}$ is given by the corresponding vectors $\mathbf{n}$ and $\mathbf{z}$ as

$$\langle \mathbf{n}, \mathbf{z}\rangle = \mathbf{n}^T \begin{pmatrix} \delta_1^{-1} & 0 \\ 0 & \delta_2^{-1} \end{pmatrix} \mathbf{z}. \tag{2.12}$$

Let $B = (b_{ij})$ be the 2×2 integer matrix introduced in (1.7), namely,

$$B = \begin{pmatrix} 0 & -\delta_1 a \\ \delta_2 a & 0 \end{pmatrix}, \quad a = \{e_2, e_1\}. \tag{2.13}$$

(a). Type $A_1 \times A_1$. We start with the case $a = 0$, so that Γ is degenerate. Let δ_1 and δ_2 be arbitrary. By (1.55), we have

$$\Psi[\mathbf{e}_2]^{\delta_2} \Psi[\mathbf{e}_1]^{\delta_1} = \Psi[\mathbf{e}_1]^{\delta_1} \Psi[\mathbf{e}_2]^{\delta_2}. \tag{2.14}$$

This equality is naturally interpreted as a consistent scattering diagram of rank 2 in Figure 8 (a). Namely, it consists of two walls

$$(\mathbf{e}_1^\perp, \Psi[\mathbf{e}_1]^{\delta_1})_{\mathbf{e}_1}, \quad (\mathbf{e}_2^\perp, \Psi[\mathbf{e}_2]^{\delta_2})_{\mathbf{e}_2}. \tag{2.15}$$

The left hand side of the equality (2.14) is the path-ordered product $\mathfrak{p}_{\gamma_1,\mathfrak{D}}$ along γ_1, while the right hand side is the one $\mathfrak{p}_{\gamma_2,\mathfrak{D}}$ along γ_2.

Now assume $a \neq 0$, so that Γ is nondegenerate. Using the rescaling of fixed data in Section 1.5 and interchanging e_1 and e_2, if necessary, we may assume $\{e_2, e_1\} = 1$ without loosing generality. Thus, we have

$$B = \begin{pmatrix} 0 & -\delta_1 \\ \delta_2 & 0 \end{pmatrix} = \begin{pmatrix} \delta_1 & 0 \\ 0 & \delta_2 \end{pmatrix} \begin{pmatrix} 0 & -1 \\ 1 & 0 \end{pmatrix} = \Delta\Omega. \tag{2.16}$$

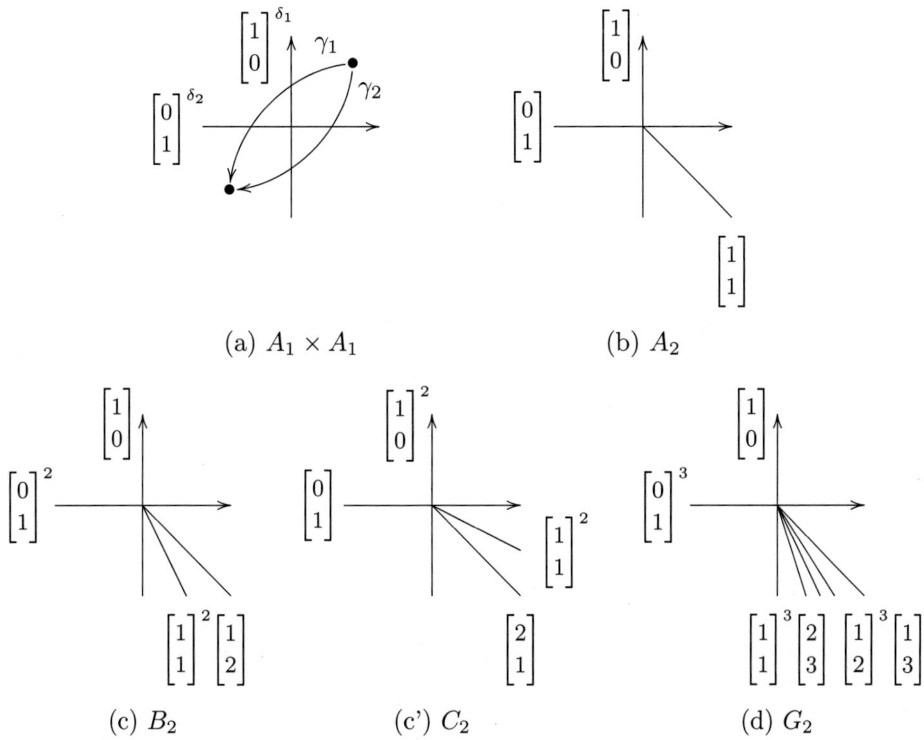

Figure 8: Examples of consistent scattering diagrams of rank 2. Column vectors with exponents represent the wall elements $\Psi[(n_1, n_2)]^c$.

Let $\mathbf{n} = (n_1, n_2) \in \mathbb{Z}^2$ be the corresponding vector to $n \in N^+$. Then, we have $\{\mathbf{n}', \mathbf{n}\} = n'_2 n_1 - n'_1 n_2$.

Here, we concentrate on the case $\delta_1\delta_2 \leq 3$.

(b). Type A_2. Let $\delta_1 = \delta_2 = 1$. Since $\{\mathbf{e}_2, \mathbf{e}_1\} = 1$, we apply the pentagon relation (1.56) with $c = 1$, and we have

$$\Psi[\mathbf{e}_2]\Psi[\mathbf{e}_1] = \Psi[\mathbf{e}_1]\Psi[(1,1)]\Psi[\mathbf{e}_2]. \tag{2.17}$$

This equality is naturally interpreted as a consistent scattering diagram of rank 2 in Figure 8 (b). Namely, it consists of three walls

$$(\mathbf{e}_1^{\perp}, \Psi[\mathbf{e}_1])_{\mathbf{e}_1}, \quad (\mathbf{e}_2^{\perp}, \Psi[\mathbf{e}_2])_{\mathbf{e}_2}, \quad (\mathbb{R}_{\geq 0}(1,-1), \Psi[(1,1)])_{(1,1)}. \tag{2.18}$$

All normal vectors $\mathbf{e}_1$, $\mathbf{e}_2$, $(1,1)$ exhaust the positive roots of the root system of type A_2.

(c). Type B_2 and C_2. Here we consider two cases, $\delta_1 = 1$, $\delta_2 = 2$ (type B_2) and $\delta_1 = 2$, $\delta_2 = 1$ (type C_2). They are essentially the same. Nevertheless, we present both because they are useful to construct the rank 3 scattering diagrams later in Section 5.4. First, we consider the type B_2 case. We consider an element $\Psi[\mathbf{e}_2]^2\Psi[\mathbf{e}_1]$. For simplicity, let us write $\Psi[(n_1, n_2)]$ as $[n_1, n_2]$.

Then, by applying the pentagon relation (1.56) with $c = 1$ repeatedly for adjacent pairs $\mathbf{n}'$, $\mathbf{n}$ with $\{\mathbf{n}', \mathbf{n}\} = 1$, we have

$$\begin{aligned} \begin{bmatrix}0\\1\end{bmatrix}^2 \begin{bmatrix}1\\0\end{bmatrix} &= \begin{bmatrix}0\\1\end{bmatrix}\begin{bmatrix}1\\0\end{bmatrix}\begin{bmatrix}1\\1\end{bmatrix}\begin{bmatrix}0\\1\end{bmatrix} = \begin{bmatrix}1\\0\end{bmatrix}\begin{bmatrix}1\\1\end{bmatrix}\begin{bmatrix}0\\1\end{bmatrix}\begin{bmatrix}1\\1\end{bmatrix}\begin{bmatrix}0\\1\end{bmatrix} \\ &= \begin{bmatrix}1\\0\end{bmatrix}\begin{bmatrix}1\\1\end{bmatrix}^2\begin{bmatrix}1\\2\end{bmatrix}\begin{bmatrix}0\\1\end{bmatrix}^2 . \end{aligned} \tag{2.19}$$

This equality is naturally interpreted as a consistent scattering diagram in Figure 8 (c), which consists of four walls

$$(\mathbf{e}_1^{\perp}, \Psi[\mathbf{e}_1])_{\mathbf{e}_1}, \quad (\mathbf{e}_2^{\perp}, \Psi[\mathbf{e}_2]^2)_{\mathbf{e}_2}, \tag{2.20}$$

$$(\mathbb{R}_{\geq 0}(1,-2), \Psi[(1,1)]^2)_{(1,1)}, \quad (\mathbb{R}_{\geq 0}(1,-1), \Psi[(1,2)])_{(1,2)}. \tag{2.21}$$

Some remarks are in order. Firstly, by (2.12), we have

$$\mathbb{R}_{\geq 0}(1,-2) \subset (1,1)^{\perp}, \quad \mathbb{R}_{\geq 0}(1,-1) \subset (1,2)^{\perp}. \tag{2.22}$$

Secondly, the exponent $c = 1$ or 2 of each wall element $\Psi[\mathbf{n}]^c$ is the normalization factor $\delta(\mathbf{n})$ in Definition 1.18, where N° is identified with the sublattice of $\mathbb{Z}^2$ generated by $(1,0)$ and $(0,2)$. Thirdly, the result is compatible with the formula (1.75). All normal vectors $\mathbf{e}_1$, $\mathbf{e}_2$, $(1,1)$, $(1,2)$ exhaust the positive roots of the root system of type B_2. The type C_2 case is similar, and the corresponding scattering diagram is presented in Figure 8 (c').

(d). Type G_2. Let $\delta_1 = 1$, $\delta_2 = 3$. We consider an element $\Psi[\mathbf{e}_2]^3\Psi[\mathbf{e}_1]$. Again, by applying the pentagon relation (1.56) with $c = 1$ repeatedly, we have

$$\begin{aligned} \begin{bmatrix}0\\1\end{bmatrix}^3 \begin{bmatrix}1\\0\end{bmatrix} &= \begin{bmatrix}0\\1\end{bmatrix}\left(\begin{bmatrix}1\\0\end{bmatrix}\begin{bmatrix}1\\1\end{bmatrix}^2\begin{bmatrix}1\\2\end{bmatrix}\begin{bmatrix}0\\1\end{bmatrix}^2\right) \\ &= \begin{bmatrix}1\\0\end{bmatrix}\begin{bmatrix}1\\1\end{bmatrix}^2\begin{bmatrix}1\\2\end{bmatrix}\begin{bmatrix}1\\1\end{bmatrix}\begin{bmatrix}1\\2\end{bmatrix}^2\begin{bmatrix}1\\3\end{bmatrix}\begin{bmatrix}0\\1\end{bmatrix}^3 \\ &= \begin{bmatrix}1\\0\end{bmatrix}\begin{bmatrix}1\\1\end{bmatrix}^3\begin{bmatrix}2\\3\end{bmatrix}\begin{bmatrix}1\\2\end{bmatrix}^3\begin{bmatrix}1\\3\end{bmatrix}\begin{bmatrix}0\\1\end{bmatrix}^3 . \end{aligned} \tag{2.23}$$

This equality is naturally interpreted as a consistent scattering diagram in Figure 8 (d), which consists of six walls

$$(\mathbf{e}_1^{\perp}, \Psi[\mathbf{e}_1])_{\mathbf{e}_1}, \quad (\mathbf{e}_2^{\perp}, \Psi[\mathbf{e}_2]^3)_{\mathbf{e}_2}, \tag{2.24}$$

$$(\mathbb{R}_{\geq 0}(1,-3), \Psi[(1,1)]^3)_{(1,1)}, \quad (\mathbb{R}_{\geq 0}(1,-2), \Psi[(2,3)])_{(2,3)}, \tag{2.25}$$

$$(\mathbb{R}_{\geq 0}(2,-3), \Psi[(1,2)]^3)_{(1,2)}, \quad (\mathbb{R}_{\geq 0}(1,-1), \Psi[(1,3)])_{(1,3)}. \tag{2.26}$$

The same remarks in the case (c) are applied as before. The case $\delta_1 = 3$, $\delta_2 = 1$ is similar.

These examples are particular cases of *cluster scattering diagrams*, which we are going to study.

2.3 Decompositions of G at general point

Let us begin with a general fact on nilpotent Lie algebras.

Lemma 2.15. *Let $\mathfrak{g}$ be a nilpotent Lie algebra that has a decomposition (as a vector space)*

$$\mathfrak{g} = \mathfrak{g}_1 \oplus \mathfrak{g}_2 \tag{2.27}$$

by its Lie subalgebras $\mathfrak{g}_1$ and $\mathfrak{g}_2$. Let $G = \exp(\mathfrak{g})$ and $G_i = \exp(\mathfrak{g}_i)$ $(i = 1, 2)$ be the corresponding exponential groups whose products are defined by the BCH formula (1.25)*. Then, we have the following decomposition of G:*

$$G = G_1G_2, \quad G_1 \cap G_2 = \{\mathrm{id}\}. \tag{2.28}$$

In other words, any element $g \in G$ is uniquely factorized as $g = g_1g_2$ $(g_i \in G_i)$.

Proof. By the assumption (2.27), any element $g \in G$ is expressed as $g = \exp(X + Y)$, $X \in \mathfrak{g}_1$, $Y \in \mathfrak{g}_2$. Then, apply the Zassenhaus formula (1.32). If the right hand side of (1.32) is already of the form g_1g_2 $(g_i \in G_i)$, we are done. Otherwise, expand the terms $\exp(Z)$ $(Z \notin \mathfrak{g}_1 \cup \mathfrak{g}_2)$ therein by (1.32) again, and transpose the left-most pair $\exp(X_2)\exp(X_1)$ $(X_i \in \mathfrak{g}_i)$ in the opposite order by (1.33). Continue the process until we end up with the expression g_1g_2, $g_i \in G_i$. Due to the nilpotency of $\mathfrak{g}$, the process completes in finitely many steps. Thus, $G = G_1G_2$. If $\exp(X_1) = \exp(X_2)$ $(X_i \in \mathfrak{g}_i)$, we have $X_1 = X_2$. Then, we have $X_1 = X_2 = 0$ by (2.27). Therefore, $G_1 \cap G_2 = \{\mathrm{id}\}$. □

We apply the above lemma to the following situation. For a given seed $\mathfrak{s}$ of a fixed data Γ, let $\mathfrak{g}$, $\widehat{\mathfrak{g}}$, and G be the ones in Section 1.2. Let $z \in M_{\mathbb{R}}$ be any general point. We introduce the *decomposition of $\mathfrak{g}$ at z* as

$$\mathfrak{g} = \mathfrak{g}_+^z \oplus \mathfrak{g}_0^z \oplus \mathfrak{g}_-^z, \tag{2.29}$$

where

$$\mathfrak{g}_+^z = \bigoplus_{\substack{n \in N^+ \\ \langle n, z\rangle > 0}} \mathfrak{g}_n, \quad \mathfrak{g}_0^z = \bigoplus_{\substack{n \in N^+ \\ \langle n, z\rangle = 0}} \mathfrak{g}_n, \quad \mathfrak{g}_-^z = \bigoplus_{\substack{n \in N^+ \\ \langle n, z\rangle < 0}} \mathfrak{g}_n. \tag{2.30}$$

By (1.18), $\mathfrak{g}_+^z$, $\mathfrak{g}_0^z$, $\mathfrak{g}_-^z$ are Lie subagebras of $\mathfrak{g}$. By the generality assumption of z, we have

$$\mathfrak{g}_0^z = \begin{cases} \mathfrak{g}_{n_0}^{\parallel} & z \in n_0^{\perp} \text{ for some unique } n_0 \in N_{\mathrm{pr}}^+, \\ \{0\} & \text{otherwise,} \end{cases} \tag{2.31}$$

where $\mathfrak{g}_n^{\parallel}$ is the one in (1.31).

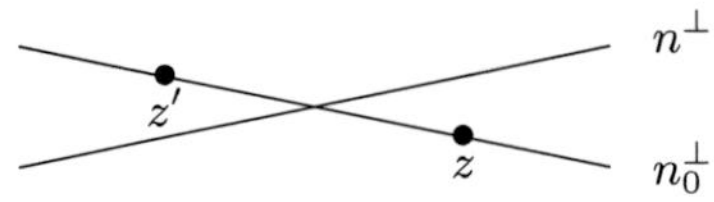

Figure 9: Decomposition at z is not constant on $n_0^\perp$.

Let $\widehat{\mathfrak{g}}^z_+$, $\widehat{\mathfrak{g}}^z_0$, $\widehat{\mathfrak{g}}^z_-$ be their completions, and let G^z_+, G^z_0, G^z_- be the corresponding subgroups of G. In particular,

$$G^z_0 = \begin{cases} G^{\|}_{n_0} & z \in n_0^\perp \text{ for some unique } n_0 \in N^+_{\mathrm{pr}}, \\ \{0\} & \text{otherwise.} \end{cases} \tag{2.32}$$

Correspondingly, we have the *decomposition of G at z* as follows.

Proposition 2.16 ([GHKK18, Proof of Theorem 1.17]). *For any general $z \in M_{\mathbb{R}}$, we have*

$$G = G^z_+ G^z_0 G^z_- \tag{2.33}$$

so that any element g of G is uniquely factorized as

$$g = g^z_+ g^z_0 g^z_- \quad (g^z_+ \in G^z_+,\ g^z_0 \in G^z_0,\ g^z_- \in G^z_-). \tag{2.34}$$

Proof. Apply Lemma 2.15 to the decomposition

$$\mathfrak{g}^{\leq \ell} = (\mathfrak{g}^z_+)^{\leq \ell} \oplus (\mathfrak{g}^z_0)^{\leq \ell} \oplus (\mathfrak{g}^z_-)^{\leq \ell}, \tag{2.35}$$

and we obtain the decomposition of $G^{\leq \ell}$

$$G^{\leq \ell} = (G^z_+)^{\leq \ell} (G^z_0)^{\leq \ell} (G^z_-)^{\leq \ell} \tag{2.36}$$

with the unique factorization property. Moreover, the decomposition is compatible with the projection $\pi_{\ell',\ell}$ in (1.26). Then, taking the inverse limit, we obtain (2.33). The unique factorization property is preserved in the inverse limit. □

By Proposition 2.16 and (2.32), for any $g \in G$ and any general $z \in M_{\mathbb{R}}$, we have a unique g^z_0 such that

$$g^z_0 \in \begin{cases} G^{\|}_{n_0} & z \in n_0^\perp \text{ for some unique } n_0 \in N^+_{\mathrm{pr}}, \\ \{\mathrm{id}\} & \text{otherwise.} \end{cases} \tag{2.37}$$

Compare it with (2.9).

What is important here is that, for a general $z \in n_0^\perp$, g^z_0 varies discontinuously on the whole hyperplane $n_0^\perp$, in general. When z crosses the intersection

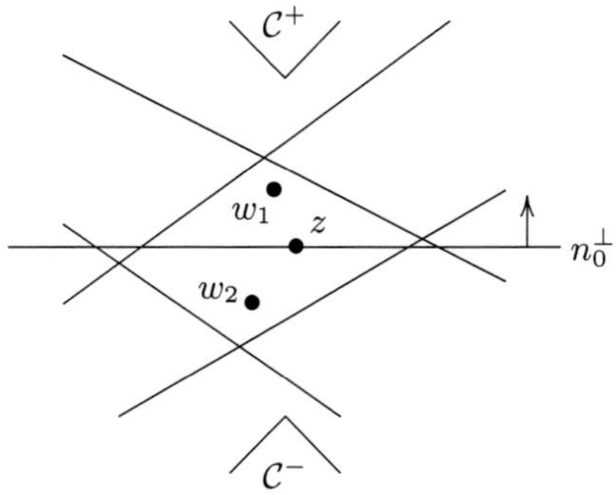

Figure 10: Configuration of z, w_1, w_2 for Lemma 2.17.

$n_0^\perp \cap n^\perp$ with $n \in N_{\mathrm{pr}}^+$ other than n_0, the sign of $\langle n, z\rangle$ changes. See Figure 9. Thus, the decomposition (2.33) changes.

We are going to relate the decomposition of $g \in G$ at z with a consistent scattering diagram. To begin with, let us fix an integer $\ell > 0$. There are only finitely many hyperplanes $n^\perp$ with $n \in (N_{\mathrm{pr}}^+)^{\leq \ell}$. They intersect each other, so that each hyperplane is subdivided into cones. Let z, w_1, and w_2 be general points in $M_{\mathbb{R}}$ as follows: (See Figure 10.)

- z belongs to $n_0^\perp$ for some unique $n_0 \in (N_{\mathrm{pr}}^+)^{\leq \ell}$.
- w_1 and w_2 do not belong to $n^\perp$ for any $n \in (N_{\mathrm{pr}}^+)^{\leq \ell}$.
- $\langle n_0, w_1\rangle > 0$ and $\langle n_0, w_2\rangle < 0$.
- For any $n \in (N_{\mathrm{pr}}^+)^{\leq \ell}$ other than n_0, $\langle n, w_i\rangle > 0$ if and only if $\langle n, z\rangle > 0$.

We have a key lemma.

Lemma 2.17 ([Bri17, Lemma 3.2]). *For any $g \in G$ and the above $z, w_1, w_2 \in M_{\mathbb{R}}$, we have*

$$g_0^z \equiv (g_+^{w_2})^{-1} g_+^{w_1} \mod G^{>\ell}. \tag{2.38}$$

Also, when the configuration of w_1 and w_2 are interchanged, we have

$$(g_0^z)^{-1} \equiv (g_+^{w_2})^{-1} g_+^{w_1} \mod G^{>\ell}. \tag{2.39}$$

Proof. Consider the decomposition $g = g_+^z g_0^z g_-^z$ at z. Under the condition of z, w_1, and w_2, the equality $g \equiv g_+^z g_0^z g_-^z \mod G^{>\ell}$ is interpreted also as the decompositions $g \equiv (g_+^z g_0^z) g_-^z$ at w_1 and $g \equiv g_+^z (g_0^z g_-^z)$ at w_2 mod $G^{>\ell}$ as well, where the middle terms $g_0^{w_1}$ and $g_0^{w_2}$ are the identity mod $G^{>\ell}$. Therefore, $g_+^{w_1} \equiv g_+^z g_0^z$ and $g_+^{w_2} \equiv g_+^z$. Thus, we have (2.38). The formula (2.39) is obtained from (2.38) by taking the inverse. □

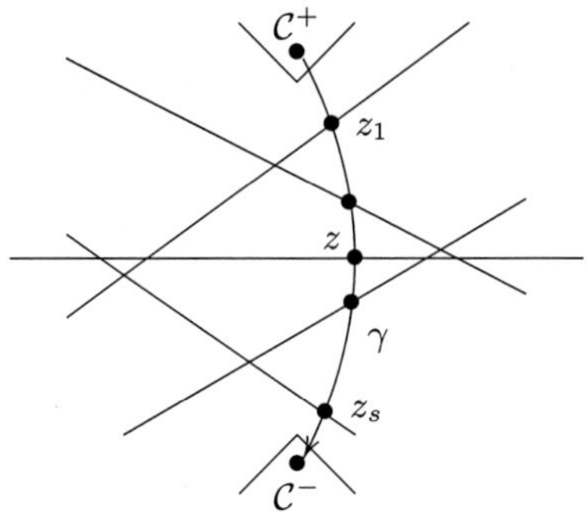

Figure 11: A path γ for the proof of Proposition 2.19.

2.4 Construction of consistent scattering diagrams

For a given seed $\mathfrak{s}$, we define the *positive and negative orthants* $\mathcal{C}^{\pm}$ in $M_{\mathbb{R}}$ by

$$\mathcal{C}^+ = \mathcal{C}^+_{\mathfrak{s}} := \{z \in M_{\mathbb{R}} \mid \langle e_i, z\rangle \geq 0 \ (i = 1, \dots, r)\}, \tag{2.40}$$
$$\mathcal{C}^- = \mathcal{C}^-_{\mathfrak{s}} := \{z \in M_{\mathbb{R}} \mid \langle e_i, z\rangle \leq 0 \ (i = 1, \dots, r)\}. \tag{2.41}$$

Lemma 2.18. *For any wall for $\mathfrak{s}$, its support $\mathfrak{d}$ intersects $\mathcal{C}^{\pm}$ only in the boundary of $\mathcal{C}^{\pm}$.*

Proof. We have $\mathfrak{d} \subset n^{\perp}$ for its normal vector $n \in N^+_{\mathrm{pr}}$. On the other hand, for any $n \in N^+_{\mathrm{pr}}$, $\langle n, z\rangle > 0$ if $z \in \mathrm{Int}(\mathcal{C}^+)$, and $\langle n, z\rangle < 0$ if $z \in \mathrm{Int}(\mathcal{C}^-)$. Thus, $\mathfrak{d} \cap \mathrm{Int}(\mathcal{C}^{\pm}) = \emptyset$. □

Let $\mathfrak{D}$ be a consistent scattering diagram for $\mathfrak{s}$. Let γ_{+-} be any admissible curve for $\mathfrak{D}$ with the initial point in $\mathrm{Int}(\mathcal{C}^+)$ and the final point in $\mathrm{Int}(\mathcal{C}^-)$. By the consistency and Lemma 2.18, the element

$$g(\mathfrak{D}) := \mathfrak{p}_{\gamma_{+-},\mathfrak{D}} \in G \tag{2.42}$$

is independent of the choice of such γ_{+-}.

Surprisingly, the single element $g(\mathfrak{D})$ contains all essential information of the whole consistent scattering diagram $\mathfrak{D}$ up to equivalence.

Proposition 2.19 ([GHKK18, Proof of Theorem 1.17])**.** *Let $\mathfrak{D}$ be a consistent scattering diagram. Let $g = g(\mathfrak{D}) \in G$. Then, for any general $z \in M_{\mathbb{R}}$, we have*

$$g_z(\mathfrak{D}) = g_0^z. \tag{2.43}$$

Proof. Let $z \in M_{\mathbb{R}}$ be a given general point. One can find an admissible curve γ for $\mathfrak{D}$ satisfying the following conditions:

- γ intersects z.
- γ has the initial point in $\mathrm{Int}(\mathcal{C}^+)$ and the final point in $\mathrm{Int}(\mathcal{C}^-)$.
- The velocity vector of γ at any point satisfies $\langle e_i, \gamma' \rangle < 0$ for any i.

For example, take the straight line passing through z from $z + a\sum_{i=1}^{r} e_i^* \in \mathrm{Int}(\mathcal{C}^+)$ to $z - a\sum_{i=1}^{r} e_i^* \in \mathrm{Int}(\mathcal{C}^-)$ for sufficiently large $a > 0$. Then, slightly deform it to avoid intersecting with $\mathrm{Sing}(\mathfrak{D})$.

Consider the reduction $\mathfrak{D}_\ell$ at ℓ. Let $z_1, \dots, z_s$ be the intersections of γ with walls of $\mathfrak{D}_\ell$ in this order such that $z_j = z$. (If z is not on any wall of $\mathfrak{D}_\ell$, we regard it on a wall with the trivial wall element id.) See Figure 11. Thanks to Lemma 2.9, by slightly deforming γ if necessary, we may assume that all $z_1, \dots, z_s$ are general. The intersection signs ϵ_i are all 1 due to the velocity condition. Then, we have

$$g \equiv \mathfrak{p}_{\gamma, \mathfrak{D}_\ell} = g_{z_s}(\mathfrak{D}) \cdots g_{z_{j+1}}(\mathfrak{D}) g_z(\mathfrak{D}) g_{z_{j-1}}(\mathfrak{D}) \cdots g_{z_1}(\mathfrak{D}) \mod G^{>\ell}. \quad (2.44)$$

Let $n_i \in (N_{\mathrm{pr}}^+)^{\leq \ell}$ be the normal vector of the wall crossed by γ at z_i. Then, we have $\langle n_i, z \rangle < 0$ for $i < j$, and $\langle n_i, z \rangle > 0$ for $i > j$. This implies that

$$g_z(\mathfrak{D}) \equiv g_0^z \mod G^{>\ell}. \quad (2.45)$$

Thus, we have $g_z(\mathfrak{D}) = g_0^z$. □

The above result, together with Lemma 2.11, shows that a consistent scattering diagram $\mathfrak{D}$ is recovered from a single element $g(\mathfrak{D})$ up to equivalence. In fact, we have the following stronger result.

Theorem 2.20 ([KS14, Theorem 2.1.6], [GHKK18, Theorem 1.17]). *The assignment $\varphi : \mathfrak{D} \mapsto g(\mathfrak{D})$ gives a one-to-one correspondence between equivalence classes of consistent scattering diagrams and elements in G.*

Proof. **Injectivity of φ.** Let $\mathfrak{D}$ and $\mathfrak{D}'$ be consistent scattering diagrams, and suppose that $g(\mathfrak{D}) = g(\mathfrak{D}') = g$. Then, by (2.43),

$$g_z(\mathfrak{D}) = g_z(\mathfrak{D}') = g_0^z \quad (2.46)$$

holds for any general $z \in M_{\mathbb{R}}$. Then, by Lemma 2.11, $\mathfrak{D}$ and $\mathfrak{D}'$ are equivalent.

Surjectivity of φ. Step 1 (Construction of $\varphi^{-1} : g \mapsto \mathfrak{D}(g)$). Take an arbitrary element $g \in G$. Let g_0^z be the one in Proposition 2.16. First, we construct a scattering diagram $\mathfrak{D} = \mathfrak{D}(g)$ such that

$$g_z(\mathfrak{D}) = g_0^z \quad (2.47)$$

holds for any general $z \in M_{\mathbb{R}}$. Let us fix any $n \in N_{\mathrm{pr}}^+$. For each $\ell > 0$, the hyperplane $n^\perp$ is divided into finitely many cones of codimension 1 by $n'^\perp$

with $n' \in (N_{\mathrm{pr}}^{+})^{\leq \ell}$, $n' \neq n$. Then, for general $z \in n^{\perp}$, $\pi_{\ell}(g_0^z)$ is constant with respect to z on each cone. Let $\mathfrak{d}$ be any of such cones and $z \in \mathfrak{d}$ be any general point. By (2.37), g_0^z is uniquely written as

$$g_0^z = \exp\Bigl(\sum_{j>0} c_{z,jn} X_{jn}\Bigr) \quad (c_{z,jn} \in \Bbbk). \tag{2.48}$$

Then, with $\mathfrak{d}$, we associate a wall $(\mathfrak{d}, g_{\mathfrak{d}})_n$, where

$$g_{\mathfrak{d}} = \begin{cases} \exp(c_{z,jn} X_{jn}) & \ell = j\deg(n) \text{ for some unique integer } j > 0, \\ \mathrm{id} & \text{otherwise.} \end{cases} \tag{2.49}$$

Let $\mathfrak{D}$ be the collection of all such walls for all $n \in N_{\mathrm{pr}}^{+}$ and ℓ. Then, $\mathfrak{D}$ satisfies the finiteness condition by construction. Therefore, $\mathfrak{D}$ is a scattering diagram. Moreover, for any general $z \in M_{\mathbb{R}}$, we have

$$\begin{aligned} g_z(\mathfrak{D}) &= \begin{cases} \prod_{j>0} \exp(c_{z,jn} X_{jn}) & z \in n^{\perp} \text{ for some unique } n \in N_{\mathrm{pr}}^{+}, \\ \mathrm{id} & \text{otherwise} \end{cases} \\ &= g_0^z \end{aligned} \tag{2.50}$$

as desired.

Step 2. Let us show that $\mathfrak{D}$ is consistent. Consider the reduction $\mathfrak{D}_\ell$ at ℓ. Let γ be any admissible curve with the initial point w and and the final point w'. Let $z_1, \dots, z_s$ be the intersections of γ with walls of $\mathfrak{D}_\ell$ in this order. By slightly deforming γ, if necessary, we may assume that all $z_1, \dots, z_s$ are general. Then, by (2.50), we have

$$\mathfrak{p}_{\gamma,\mathfrak{D}_\ell} = (g_0^{z_s})^{\epsilon_s} \cdots (g_0^{z_1})^{\epsilon_1}, \tag{2.51}$$

where ϵ_i is the intersection sign at z_i. Let $w_1 = w$, $w_2, \dots, w_{s+1} = w'$, where w_i $(i = 2, \dots, s)$ is any general points on γ between z_{i-1} and z_i. Note that, when $\epsilon_i = 1$ (resp. $\epsilon_i = -1$), z_i, w_i, w_{i+1} (resp. z_i, w_{i+1}, w_i) are in the same configuration as z, w_1, w_2 in Lemma 2.17. Therefore, by (2.38) and (2.39), we have

$$(g_0^{z_i})^{\epsilon_i} \equiv (g_+^{w_{i+1}})^{-1} g_+^{w_i} \mod G^{>\ell}. \tag{2.52}$$

Putting it into (2.51),

$$\mathfrak{p}_{\gamma,\mathfrak{D}_\ell} \equiv (g_+^{w'})^{-1} g_+^{w} \mod G^{>\ell}. \tag{2.53}$$

Thus, we conclude that

$$\mathfrak{p}_{\gamma,\mathfrak{D}} = (g_+^{w'})^{-1} g_+^{w}, \tag{2.54}$$

which only depends on the end points w and w' of γ.

Step 3. Finally, we show that $g(\mathfrak{D}) = g$. Let γ_{+-} be any admissible curve with the initial point w in $\mathrm{Int}(\mathcal{C}^+)$ and the final point w' in $\mathrm{Int}(\mathcal{C}^-)$. Then, we have the equality (2.54). Meanwhile, for any $n \in N^+$, we have $\langle n, w\rangle > 0$ and $\langle n, w'\rangle < 0$. Thus, the decompositions of g with respect to w and w' yield $g_+^w = g$ and $g_+^{w'} = \mathrm{id}$. Therefore, we have $\mathfrak{p}_{\gamma_{+-},\mathfrak{D}} = g$. Thus, we conclude that φ is surjective. □

Remark 2.21. Theorem 2.20 and its proof only depend on the facts that $\mathfrak{g}$ is N^+-graded and $\mathfrak{g}_n^{\parallel}$ is abelian. Thus, it is applicable to a much wider class of $\mathfrak{g}$.

To summarize the results in this section, all essential information of any consistent scattering diagram $\mathfrak{D}$ is encoded in the corresponding element $g(\mathfrak{D}) = \mathfrak{p}_{\gamma_{+-},\mathfrak{D}}$. Conversely, for any element g of G, one can construct a consistent scattering diagram $\mathfrak{D}(g)$, and admissible curves γ_{+-} therein correspond to various factorizations of g with g_0^z's.

Remark 2.22. The construction of a consistent scattering diagram in Theorem 2.20 also tells that one may assume that, for the support $\mathfrak{d} \subset n^{\perp}$ of any wall, each face of $\mathfrak{d}$ of codimension 2 lies in the intersection with a hyperplane $n'^{\perp}$ with a *positive* normal vector $n' \in N_{\mathrm{pr}}^+$.

2.5 Minimal support

Often it is convenient to consider a consistent scattering diagram whose support is *minimal* among all equivalent diagrams.

Definition 2.23 (Minimal support). Let $\mathfrak{D}$ be a consistent scattering diagram and $g = g(\mathfrak{D})$ be the corresponding element in G. We say that $\mathfrak{D}$ is *with minimal support* if it satisfies the following condition:

- For any general $z \in M_{\mathbb{R}}$ such that $g_0^z = \mathrm{id}$, there is no wall $(\mathfrak{d}, g_{\mathfrak{d}})_n$ of $\mathfrak{D}$ such that $z \in \mathfrak{d}$.

It is clear that the support of such $\mathfrak{D}$ is the minimal set among the supports of all consistent scattering diagrams that are equivalent to $\mathfrak{D}$.

Example 2.24. All examples of rank 2 consistent scattering diagrams in Section 2.2 are with minimal support.

Proposition 2.25. *For any consistent scattering diagram $\mathfrak{D}$, there is a (not unique) consistent scattering diagram with minimal support that is equivalent to $\mathfrak{D}$.*

Proof. By Theorem 2.20, $\mathfrak{D}$ is equivalent to the consistent scattering diagram $\mathfrak{D}(g)$ for some $g \in G$ constructed in the proof of the surjectivity of φ in Theorem 2.20. Let $z \in M_{\mathbb{R}}$ be a general point such that $g_0^z = \mathrm{id}$. Then, for any wall $(\mathfrak{d}, g_{\mathfrak{d}})_n$ in $\mathfrak{D}(g)$ such that $z \in \mathfrak{d}$, we have $g_\sigma = \mathrm{id}$ by (2.49). Therefore, any such wall is removed form $\mathfrak{D}(g)$ up to equivalence. The resulting diagram has the desired property. □

Notes

The contents are taken mostly from [GHKK18, Sect. 1.2], [KS14, Sect. 2] and partly from [Bri17] with added/modified proofs. The consistent scattering diagrams here are a special case of consistent scattering diagrams in [GS11] and *wall-crossing structures* (WCS) in [KS14]. The contents in Section 2.5 is added by us.

3 Cluster scattering diagrams

Here we introduce a special class of consistent scattering diagrams called *cluster scattering diagrams* (CSDs). Then, we show the construction/existence of CSDs. This is the first fundamental result in the application of scattering diagrams to cluster algebras.

3.1 Parametrization of G by parallel subgroups

For a given seed $\mathfrak{s}$ of a (possibly degenerate) fixed data Γ, let $\mathfrak{g}$, $\widehat{\mathfrak{g}}$, and G be the ones in Section 1.2. We introduce another decomposition of G.

For any $n \in N_{\mathrm{pr}}^{+}$, we introduce the *decomposition of* $\mathfrak{g}$ *by* n as

$$\mathfrak{g} = \mathfrak{g}_{+}^{n} \oplus \mathfrak{g}_{0}^{n} \oplus \mathfrak{g}_{-}^{n} \quad (n \in N_{\mathrm{pr}}^{+}), \tag{3.1}$$

where

$$\mathfrak{g}_{+}^{n} = \bigoplus_{\substack{n' \in N^{+} \\ \{n', n\} > 0}} \mathfrak{g}_{n'}, \quad \mathfrak{g}_{0}^{n} = \bigoplus_{\substack{n' \in N^{+} \\ \{n', n\} = 0}} \mathfrak{g}_{n'}, \quad \mathfrak{g}_{-}^{n} = \bigoplus_{\substack{n' \in N^{+} \\ \{n', n\} < 0}} \mathfrak{g}_{n'}. \tag{3.2}$$

By (1.18), $\mathfrak{g}_{+}^{n}$, $\mathfrak{g}_{0}^{n}$, $\mathfrak{g}_{-}^{n}$ are Lie subalgebras of $\mathfrak{g}$. Let $\widehat{\mathfrak{g}}_{+}^{n}$, $\widehat{\mathfrak{g}}_{0}^{n}$, $\widehat{\mathfrak{g}}_{-}^{n}$ be their completions, and let G_{+}^{n}, G_{0}^{n}, G_{-}^{n} be the corresponding subgroups of G.

Example 3.1. In the extreme case of degeneracy $\{\cdot, \cdot\} = 0$, $G = G_{0}^{n}$ for any n.

By Lemma 2.15, we have the *decomposition of* G *by* n.

Lemma 3.2 ([GHKK18, §1.2]). *For any* $n \in N_{\mathrm{pr}}^{+}$*, we have*

$$G = G_{+}^{n} G_{0}^{n} G_{-}^{n} \tag{3.3}$$

so that any element g *of* G *is uniquely factorized as*

$$g = g_{+}^{n} g_{0}^{n} g_{-}^{n} \quad (g_{+}^{n} \in G_{+}^{n},\ g_{0}^{n} \in G_{0}^{n},\ g_{-}^{n} \in G_{-}^{n}). \tag{3.4}$$

In contrast to the decomposition in Proposition 2.16, the above decomposition directly depends on $n \in N_{\mathrm{pr}}^{+}$, not each point z on $n^{\perp}$.

One can further decompose $\mathfrak{g}_{0}^{n}$ as

$$\mathfrak{g}_{0}^{n} = \mathfrak{g}_{n}^{\parallel} \oplus \mathfrak{g}_{n}^{\perp} \quad (n \in N_{\mathrm{pr}}^{+}), \tag{3.5}$$

where $\mathfrak{g}_{n}^{\parallel}$ is the Lie subalgebra of $\mathfrak{g}_{0}^{n}$ in (1.31), and

$$\mathfrak{g}_{n}^{\perp} := \bigoplus_{\substack{n' \in N^{+} \\ \{n', n\} = 0,\, n' \notin \mathbb{Z}n}} \mathfrak{g}_{n'}. \tag{3.6}$$

Lemma 3.3. *$\mathfrak{g}_n^{\perp}$ is a Lie algebra ideal of $\mathfrak{g}_0^n$.*

Proof. It is enough to show that

$$[\mathfrak{g}_0^n, \mathfrak{g}_0^n] \subset \mathfrak{g}_n^{\perp}, \tag{3.7}$$

or equivalently, $[\mathfrak{g}_0^n, \mathfrak{g}_0^n] \cap \mathfrak{g}_n^{\parallel} = \{0\}$. Suppose that $X_{n_1}, X_{n_2} \in \mathfrak{g}_0^n$ and $[X_{n_1}, X_{n_2}] \in \mathfrak{g}_n^{\parallel}$. Then, we have $\{n_i, n\} = 0$, and $n_1 + n_2 = jn$ for some $j > 0$. It follows that $\{n_1, n_2\} = \{n_1, jn\} = 0$. Thus, by (1.18), $[X_{n_1}, X_{n_2}] = \{n_1, n_2\} X_{n_1+n_2} = 0$. □

Let $\widehat{\mathfrak{g}}_n^{\parallel}$ and $\widehat{\mathfrak{g}}_n^{\perp}$ be the completions of $\mathfrak{g}_n^{\parallel}$ and $\mathfrak{g}_n^{\perp}$, and let $G_n^{\parallel}$ and $G_n^{\perp}$ be the corresponding subgroups of G, respectively.

Lemma 3.4 ([GHKK18, §1.2])**.** *For any $n \in N_{\mathrm{pr}}^+$, the following facts hold.*

(a). $G_n^{\perp}$ is a normal subgroup of G_0^n.

(b). We have

$$G_0^n = G_n^{\parallel} G_n^{\perp} \tag{3.8}$$

so that any element g_0^n of G_0^n is uniquely factorized as

$$g_0^n = g_n^{\parallel} g_n^{\perp} \quad (g_n^{\parallel} \in G_n^{\parallel},\ g_n^{\perp} \in G_n^{\perp}). \tag{3.9}$$

(c). The map

$$G_0^n \to G_n^{\parallel}, \quad g_0^n = g_n^{\parallel} g_n^{\perp} \mapsto g_n^{\parallel} \tag{3.10}$$

is a group homomorphism with the kernel $G_n^{\perp}$.

Proof. (a). Let $X \in (\mathfrak{g}_n^{\perp})^{\leq \ell}$ and $Y \in (\mathfrak{g}_0^n)^{\leq \ell}$. Then, by (1.33), the product $\exp(Y)\exp(X)\exp(Y)^{-1}$ is rewritten as a product of $\exp(X)$ and exponentials of the commutators and the higher commutators of X and Y. By Lemma 3.3, it belongs to $(G_n^{\perp})^{\leq \ell}$. (b). This follows from Lemma 2.15. (c). This follows from (a). □

Combining Lemmas 3.2 and 3.4, we obtain a map

$$\psi_n : G \to G_0^n \to G_n^{\parallel}, \quad g \mapsto g_0^n \mapsto g_n^{\parallel} \quad (n \in N_{\mathrm{pr}}^+). \tag{3.11}$$

(The map ψ_n is not a group homomorphism except for $n = e_i$. See Lemma 3.17.) Consider a set (not a group)

$$S^{\parallel} = S_G^{\parallel} := \prod_{n \in N_{\mathrm{pr}}^+} G_n^{\parallel}. \tag{3.12}$$

We then have a map (not a group homomorphism),

$$\psi : G \to S^{\|}, \quad g \mapsto \mathbf{g}^{\|} = (g_n^{\|})_{n \in N_{\mathrm{pr}}^+}. \tag{3.13}$$

We note that

$$(G_n^{\|})^{\leq \ell} = \{\mathrm{id}\} \quad (n \in (N_{\mathrm{pr}}^+)^{>\ell}). \tag{3.14}$$

In particular, we have

$$\pi_\ell(g) = \mathrm{id} \quad (g \in G_n^{\|},\ n \in (N_{\mathrm{pr}}^+)^{>\ell}). \tag{3.15}$$

Theorem 3.5 ([GHKK18, Prop. 1.20]). *The map $\psi : G \to S^{\|}$ is a bijection.*

Our proof is based on the following lemma.

Lemma 3.6 (cf. [GHKK18, Appendix C.1]). *(a). For the projection $\pi_{\ell+1,\ell} : G^{\leq \ell+1} \to G^{\leq \ell}$ in (1.26), $\mathrm{Ker}\, \pi_{\ell+1,\ell}$ is in the center of $G^{\leq \ell+1}$, and it is isomorphic (as a group) to the direct product of groups,*

$$\mathrm{Ker}\, \pi_{\ell+1,\ell} \simeq \prod_{n \in (N^+)^{\ell+1}} G_n, \quad G_n = \exp(\mathfrak{g}_n), \tag{3.16}$$

where the isomorphism is given by $\pi_{\ell+1}(\prod_n g_n) \leftarrow\!\shortmid (g_n)$.

(b). $G^{\leq 1}$ is an abelian group, and it is isomorphic (as a group) to the direct product of groups,

$$G^{\leq 1} \simeq \prod_{n \in (N^+)^{\leq 1}} G_n, \tag{3.17}$$

where the isomorphism is given by $\pi_1(\prod_n g_n) \leftarrow\!\shortmid (g_n)$.

Proof. We work with the corresponding Lie algebras. Let $\pi_{\ell+1,\ell} : \mathfrak{g}^{\leq \ell+1} \to \mathfrak{g}^{\leq \ell}$ be the canonical projection. Then, as a vector space, we have

$$\mathrm{Ker}\, \pi_{\ell+1,\ell} \simeq \bigoplus_{n \in (N^+)^{\ell+1}} \mathfrak{g}_n. \tag{3.18}$$

Moreover, $\mathrm{Ker}\, \pi_{\ell+1,\ell}$ is in the center of $\mathfrak{g}^{\leq \ell+1}$ by (1.18). Therefore, the isomorphism (3.18) holds also as a Lie algebra, where the right hand side is viewed as the direct sum of Lie algebras. The proof of (b) is similar. □

Proof of Theorem 3.5. **Surjectivity of ψ.** Let $\mathbf{h} = (h_n)_{n \in N_{\mathrm{pr}}^+}$ be a given element in $S^{\|}$. Let us construct an element $g \in G$ inductively with respect to degree ℓ such that $\psi(g) = \mathbf{h}$, that is, $g_n^{\|} = h_n$ ($n \in N_{\mathrm{pr}}^+$) holds. To show it, we recursively construct elements $g_\ell \in G$ ($\ell = 1, 2, \dots$) such that

$$\pi_\ell((g_\ell)_n^{\|}) = \pi_\ell(h_n) \quad (n \in (N_{\mathrm{pr}}^+)^{\leq \ell}), \tag{3.19}$$

$$\pi_\ell(g_{\ell+1}) = \pi_\ell(g_\ell). \tag{3.20}$$

By (3.20), the limit $g = \lim_{\ell\to\infty} g_\ell \in G$ is well-defined. Moreover, by (3.19), g has the desired property.

Step 1. We first consider the case $\ell = 1$. We define

$$g_1 = \prod_{n\in(N^+_{\mathrm{pr}})^{\leq 1}} h_n \in G, \tag{3.21}$$

where the order of the (finite) product is arbitrarily chosen, and the element g_1 depends on the order of the product. Then, we have

$$\pi_1(g_1) = \prod_{n\in(N^+_{\mathrm{pr}})^{\leq 1}} \pi_1(h_n), \tag{3.22}$$

where factors $\pi_1(h_n)$ commute with each other, because $G^{\leq 1}$ is abelian by Lemma 3.6 (b). It follows that

$$\pi_1((g_1)^{\|}_n) = \pi_1(h_n) \quad (n \in (N^+_{\mathrm{pr}})^{\leq 1}). \tag{3.23}$$

Step 2. Next, for some $\ell > 0$, suppose that we have $g_\ell \in G$ such that

$$\pi_\ell((g_\ell)^{\|}_n) = \pi_\ell(h_n) \quad (n \in (N^+_{\mathrm{pr}})^{\leq \ell}) \tag{3.24}$$

holds. For any $n \in (N^+_{\mathrm{pr}})^{\leq \ell+1}$, let $h'_n \in G^{\|}_n$ be the one such that

$$(g_\ell)^{\|}_n h'_n = h_n. \tag{3.25}$$

We note that

$$\pi_{\ell+1}(h'_n) \in \operatorname{Ker} \pi_{\ell+1,\ell}. \tag{3.26}$$

This is true for $n \in (N^+_{\mathrm{pr}})^{\leq \ell}$ by (3.24), and for $n \in (N^+_{\mathrm{pr}})^{\ell+1}$ by (3.15). Now, we define

$$g_{\ell+1} = g_\ell \prod_{n\in(N^+_{\mathrm{pr}})^{\leq \ell+1}} h'_n, \tag{3.27}$$

where the order of the (finite) product is arbitrarily chosen, and the element $g_{\ell+1}$ depends on the order of the product. Then, we have

$$\pi_{\ell+1}(g_{\ell+1}) = \pi_{\ell+1}(g_\ell) \prod_{n\in(N^+_{\mathrm{pr}})^{\leq \ell+1}} \pi_{\ell+1}(h'_n). \tag{3.28}$$

By (3.26) and Lemma 3.6 (a), $\pi_{\ell+1}(h'_n)$ commutes with any elements in $G^{\leq \ell+1}$. It follows that

$$\pi_{\ell+1}((g_{\ell+1})^{\|}_n) = \pi_{\ell+1}((g_\ell)^{\|}_n)\pi_{\ell+1}(h'_n) = \pi_{\ell+1}(h_n) \quad (n \in (N^+_{\mathrm{pr}})^{\leq \ell+1}). \tag{3.29}$$

Moreover, by (3.26) and (3.28), we have

$$\pi_\ell(g_{\ell+1}) = \pi_\ell(g_\ell). \tag{3.30}$$

Injectivity of ψ. For given $g, g' \in G$, assume that $\psi(g) = \psi(g') = \mathbf{g}^{\|}$. Let us show that $g = g'$; namely, we have, for any $\ell > 0$,

$$\pi_\ell(g) = \pi_\ell(g'). \tag{3.31}$$

By Lemma 3.6 (b) and the assumption, we have

$$\pi_1(g) = \prod_{n \in (N_{\mathrm{pr}}^+)^{\leq 1}} \pi_1(g_n^{\|}) = \prod_{n \in (N_{\mathrm{pr}}^+)^{\leq 1}} \pi_1(g'^{\|}_n) = \pi_1(g'). \tag{3.32}$$

Thus, the equality (3.31) holds for $\ell = 1$. Suppose that the equality holds for some ℓ. Then, $\pi_{\ell+1}(g)\pi_{\ell+1}(g')^{-1} \in \operatorname{Ker} \pi_{\ell+1,\ell}$. So, by Lemma 3.6 (a), there are some $h_n \in G_n$ such that

$$\pi_{\ell+1}(g) = \pi_{\ell+1}(g') \prod_{n \in (N^+)^{\ell+1}} \pi_{\ell+1}(h_n). \tag{3.33}$$

Since $\pi_{\ell+1}(h_n)$ are central elements in $G^{\leq \ell+1}$, for $n \in (N^+)_{\mathrm{pr}}^{\leq \ell}$ and a positive integer t such that $tn \in (N^+)^{\ell+1}$, we have

$$\pi_{\ell+1}(g_n^{\|}) = \pi_{\ell+1}(g'^{\|}_n)\pi_{\ell+1}(h_{tn}), \tag{3.34}$$

Meanwhile, by the assumption, we have $g_n^{\|} = (g')_n^{\|}$. Therefore, $\pi_{\ell+1}(g_n^{\|}) = \pi_{\ell+1}((g')_n^{\|})$. It follows that $\pi_{\ell+1}(h_n) = \mathrm{id}$ for any $n \in (N^+)^{\ell+1}$. Putting it back in (3.33), we conclude that $\pi_{\ell+1}(g) = \pi_{\ell+1}(g')$. □

3.2 Incoming and outgoing walls

Let $\mathbf{h} = (h_n)_{n \in N_{\mathrm{pr}}^+}$ be an arbitrary element of the set $S^{\|}$ in (3.12). The bijection ψ in Theorem 3.5 uniquely determines the element $g(\mathbf{h}) \in G$ such that $\psi(g(\mathbf{h})) = \mathbf{h}$. In turn, the bijection φ in Theorem 2.20 uniquely determines a consistent scattering diagram $\mathfrak{D}(g(\mathbf{h}))$ up to equivalence such that $\varphi(\mathfrak{D}(g(\mathbf{h}))) = g(\mathbf{h})$. Let us write $\mathfrak{D}(g(\mathbf{h}))$ as $\mathfrak{D}(\mathbf{h})$ for simplicity. Below we will give an alternative description of $\mathfrak{D}(\mathbf{h})$ directly by $\mathbf{h}$.

Let p^* be the map in (1.8). For $n \in N_{\mathrm{pr}}^+$, we have

$$p^*(n) \in n^\perp, \tag{3.35}$$

because $\langle n, p^*(n)\rangle = \{n, n\} = 0$.

Remark 3.7. When Γ is degenerate, $p^*(n)$ is not injective. However, this fact is irrelevant to all results in Section 3.

For any $g \in G$, we have the following relation between g_0^z ($z \in n^\perp$) and $g_n^{\parallel}$ around $p^*(n)$.

Lemma 3.8 ([GHKK18, Theorem 1.21]). *Let $g \in G$ and $n \in N_{\mathrm{pr}}^+$. Then, for any $\ell > 0$, there is an open neighborhood $U_\ell \subset n^\perp$ of $p^*(n)$ such that*

$$g_0^z \equiv g_n^{\parallel} \mod G^{>\ell} \quad (z \in U_\ell,\ z\text{: general}). \tag{3.36}$$

In other words,

$$\lim_{\substack{z \to p^*(n) \\ z \in n^\perp,\, z\text{: general}}} g_0^z = g_n^{\parallel}. \tag{3.37}$$

Proof. *Case 1.* Suppose that $p^*(n)$ is general. This occurs if and only if the fixed data Γ is of rank 2 and nondegenerate. Thus, any hyperplane $n^\perp$ intersects other hyperplane $n'^\perp$ with $n' \in N_{\mathrm{pr}}^+$, $n' \neq n$ only at the origin. Then, $n^\perp$ is divided into two rays, on each of which g_0^z is constant. It follows that, if a general $z \in n^\perp$ and $p^*(n) \neq 0$ belong to the same ray,

$$g_0^z = g_0^{p^*(n)} \tag{3.38}$$

holds. Therefore, it is enough to prove

$$g_0^{p^*(n)} = g_n^{\parallel}. \tag{3.39}$$

By comparing (2.29) and (3.1), we see that two decompositions

$$g = g_+^{p^*(n)} g_0^{p^*(n)} g_-^{p^*(n)}, \tag{3.40}$$

$$g = g_+^n g_0^n g_-^n \tag{3.41}$$

coincide. Thus, we have

$$g_0^n = g_0^{p^*(n)} \in G_n^{\parallel}. \tag{3.42}$$

This further implies that

$$g_0^n = g_n^{\parallel}. \tag{3.43}$$

Thus, we have the equality (3.39).

Case 2. Suppose that $p^*(n)$ is not general. Fix an integer $\ell > 0$. Then, there might be some (possibly no) $n' \in (N_{\mathrm{pr}}^+)^{\leq \ell}$, $n' \neq n$ such that $\langle n', p^*(n) \rangle = \{n', n\} = 0$. We choose an open neighborhood of $U_\ell \subset n^\perp$ of $p^*(n)$ that does not intersect any $n''^\perp$ ($n'' \in (N_{\mathrm{pr}}^+)^{\leq \ell}$) such that $\langle n'', p^*(n) \rangle = \{n'', n\} \neq 0$. See Figure 12. Consider the decomposition of g by n

$$g = g_+^n g_0^n g_-^n. \tag{3.44}$$

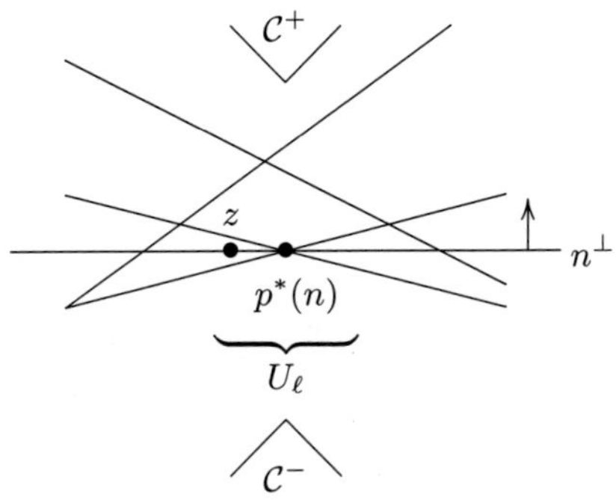

Figure 12: Configuration for Case 2 of the proof of Lemma 3.8.

We further decompose g_0^n at general $z \in U_\ell$,

$$g_0^n = h_+^z h_0^z h_-^z \quad (h_+^z \in G_+^z,\ h_0^z \in G_0^z = G_n^{\parallel},\ h_-^z \in G_-^z). \tag{3.45}$$

Then, $h_\pm^z \in G_n^{\perp}$ and $h_0^z \in G_n^{\parallel}$. Therefore, by applying the group homomorphism in (3.10) to (3.45), we obtain

$$h_0^z = g_n^{\parallel}. \tag{3.46}$$

On the other hand, by the assumption on U_ℓ, we have $\pi_\ell(g_\pm^n) \in (G_\pm^z)^{\leq \ell}$. Thus,

$$g = (g_+^n h_+^z) h_0^z (h_-^z g_-^n) \tag{3.47}$$

is regarded as the decomposition of g at z modulo $G^{>\ell}$. Therefore, we have

$$h_0^z \equiv g_0^z \mod G^{>\ell}. \tag{3.48}$$

Then, by (3.46) and (3.48), we obtain the equality (3.36). □

Definition 3.9. A wall $\mathbf{w} = (\mathfrak{d}, g)_n$ is *incoming* (*resp. outgoing*) if

$$p^*(n) \in \mathfrak{d} \tag{3.49}$$

holds (resp. otherwise) for its normal vector n.

Now we come back to our problem. For any $\mathbf{h} \in S^{\parallel}$, we define a set of incoming walls,

$$\mathrm{In}(\mathbf{h}) := \{(n^\perp, h_n)_n \mid n \in N_{\mathrm{pr}}^+\}. \tag{3.50}$$

For a scattering diagram $\mathfrak{D}$, let $\mathfrak{D}_{\mathrm{in}}$ denote the set of all incoming walls of $\mathfrak{D}$.

The following theorem characterizes a consistent scattering diagram $\mathfrak{D}(\mathbf{h})$ corresponding to $\mathbf{h} \in S^{\parallel}$ up to equivalence.

Theorem 3.10 ([GHKK18, Theorem 1.21]). *For any* $\mathbf{h} \in S^{\|}$, *let* $\mathfrak{D}(\mathbf{h})$ *be a corresponding consistent scattering diagram up to equivalence. Then, the following facts hold.*

(a). There is a consistent scattering diagram $\mathfrak{D}$ *with* $\mathfrak{D}_{\mathrm{in}} = \mathrm{In}(\mathbf{h})$ *such that* $\mathfrak{D}$ *is equivalent to* $\mathfrak{D}(\mathbf{h})$.

(b). Any consistent scattering diagram $\mathfrak{D}$ *with* $\mathfrak{D}_{\mathrm{in}} = \mathrm{In}(\mathbf{h})$ *is equivalent to* $\mathfrak{D}(\mathbf{h})$.

Proof. (a). As a representative of $\mathfrak{D}(\mathbf{h})$, we choose $\mathfrak{D}(g)$ in Step 1 in the proof of the surjectivity of φ in Theorem 2.20 with $g = g(\mathbf{h})$. For each $n \in N^+_{\mathrm{pr}}$, up to the equivalence we rearrange the walls of $\mathfrak{D}(g)$ in $n^\perp$ to have the desired property as follows. For each integer $\ell > 0$, the hyperplane $n^\perp$ is divided into cones, say, $\mathfrak{d}_1, \dots, \mathfrak{d}_p$ such that $\pi_\ell(g_0^z)$ is constant with respect to general z on each cone. Let us write the associated walls therein as $(\mathfrak{d}_1, g_1)_n$, $\dots$, $(\mathfrak{d}_p, g_p)_n$. Let U_ℓ be the open neighborhood of $p^*(n)$ in $n^\perp$ in Lemma 3.8. When $p^*(n)$ is in the interior of some cone, say, $\mathfrak{d}_1$, we may assume that $p^*(n) \in \mathfrak{d}_1$ and $p^*(n) \notin \mathfrak{d}_i$ $(i \neq 1)$. When $p^*(n)$ is on the boundary of some cone, there are several cones that intersect U_ℓ. By the property (3.36), $\pi_\ell(g_0^z)$ is constant on these cones so that

$$\pi_\ell(g_0^z) \equiv g_n^{\|} = h_n \qquad \mathrm{mod}\ G^{>\ell}. \tag{3.51}$$

Therefore, the attached wall element in (2.49) is also constant. Thus, one can join these walls into a single wall. After renumbering cones, we may assume that $p^*(n) \in \mathfrak{d}_1$ and $p^*(n) \notin \mathfrak{d}_i$ $(i \neq 1)$. Then, we do the following:

- Remove the incoming wall $(\mathfrak{d}_1, g_1)_n$.
- Replace the outgoing walls $(\mathfrak{d}_i, g_i)_n$ with $(\mathfrak{d}_i, g_i g_1^{-1})_n$ for $i \neq 1$.

We do this for all $n \in N^+_{\mathrm{pr}}$ and $\ell > 0$, so that all walls that are left are outgoing. Then, finally, we add walls of $\mathrm{In}(\mathbf{h})$. The resulting scattering diagram $\mathfrak{D}$ is equivalent to $\mathfrak{D}(\mathbf{h})$, and it has the desired property.

(b). Suppose that $\mathfrak{D}$ is a consistent scattering diagram with $\mathfrak{D}_{\mathrm{in}} = \mathrm{In}(\mathbf{h})$. Let $g = g(\mathfrak{D})$. By Theorems 2.20 and 3.5, it is enough to prove

$$g_n^{\|} = h_n \quad (n \in N^+_{\mathrm{pr}}). \tag{3.52}$$

Let us fix $n \in N^+_{\mathrm{pr}}$. There is an incoming wall $\mathbf{w}_n = (n^\perp, h_n)_n$, and all other walls with the normal vector n do not contain $p^*(n)$. Then, for each integer $\ell > 0$, there is a neighborhood V_ℓ of $p^*(n)$ in $n^\perp$ that does not intersect any outgoing walls of $\mathfrak{D}_\ell$ with the normal vector n. Thus, for any general $z \in V_\ell$,

$$g_0^z = g_z(\mathfrak{D}) \equiv h_n \quad \mathrm{mod}\ G^{>\ell}. \tag{3.53}$$

It follows that

$$\lim_{\substack{z \to p^*(n) \\ z \in n^\perp,\ z\colon \text{general}}} g_0^z = h_n. \tag{3.54}$$

Meanwhile, by Lemma 3.8, the left hand side coincides with $g_n^{\parallel}$. Thus, the equality (3.52) holds. □

3.3 Cluster scattering diagrams

Now we are ready to introduce a special class of consistent scattering diagrams, which is the main subject of Part III.

Definition 3.11 (Cluster scattering diagram (CSD))**.** For a given seed $\mathfrak{s}$, we define a set of incoming walls

$$\mathrm{In}_{\mathfrak{s}} := \{\mathbf{w}_{e_i} := (e_i^\perp, \Psi[e_i]^{\delta_i})_{e_i} \mid i = 1, \dots, r\}. \tag{3.55}$$

Then, a *cluster scattering diagram* (CSD for short) for $\mathfrak{s}$ is a consistent scattering diagram $\mathfrak{D}_{\mathfrak{s}}$ with $(\mathfrak{D}_{\mathfrak{s}})_{\mathrm{in}} = \mathrm{In}_{\mathfrak{s}}$.

Note that the wall elements in (3.55) is in the normalized form in (1.73) so that this definition is invariant under the rescaling of the fixed data Γ.

The following theorem is the first fundamental result on CSDs.

Theorem 3.12 (Existence/uniqueness of CSD [GHKK18, Theorem 1.21])**.** *For any seed $\mathfrak{s}$, a CSD $\mathfrak{D}_{\mathfrak{s}}$ for $\mathfrak{s}$ uniquely exists up to equivalence.*

Proof. For a given seed $\mathfrak{s}$, let $\mathbf{h}_{\mathfrak{s}} = (h_{\mathfrak{s},n}) \in S^{\parallel}$ be the one with

$$h_{\mathfrak{s},n} = \begin{cases} \Psi[e_i]^{\delta_i} & n = e_i \ (i = 1, \dots, r), \\ \mathrm{id} & \text{other } n \in N_{\mathrm{pr}}^+. \end{cases} \tag{3.56}$$

Then, by Theorem 3.10, there is a unique consistent scattering diagram $\mathfrak{D}$ with $\mathfrak{D}_{\mathrm{in}} = \mathrm{In}(\mathbf{h}_{\mathfrak{s}})$ up to equivalence. Then, by removing all trivial incoming walls $(n^\perp, \mathrm{id})_n$ $(n \neq e_1, \dots, e_r)$, we obtain $\mathfrak{D}_{\mathfrak{s}}$. □

Definition 3.13 (Cluster element)**.** For the above $\mathbf{h}_{\mathfrak{s}}$, the corresponding element $g_{\mathfrak{s}} := g(\mathbf{h}_{\mathfrak{s}}) \in G$ is called the *cluster element* for a seed $\mathfrak{s}$.

Example 3.14. All examples of consistent scattering diagrams in Section 2.2 are CSDs $\mathfrak{D}_{\mathfrak{s}}$ with the cluster element $g_{\mathfrak{s}} = \Psi[\mathbf{e}_2]^{\delta_2}\Psi[\mathbf{e}_1]^{\delta_1}$. By (1.10), we have

$$p^*(e_j) = \sum_{i=1}^{2} b_{ij} f_i. \tag{3.57}$$

Thus, in the identification $N \simeq \mathbb{Z}^2$ and $M_\mathbb{R} \simeq \mathbb{R}^2$ therein, $p^*(\mathbf{n}) = B\mathbf{n} = (-\delta_1 n_1, \delta_2 n_2)$, which is in the second quadrant of $M_\mathbb{R} \simeq \mathbb{R}^2$. It follows that the rays in the fourth quadrant are all outgoing walls. They are of *finite type*, that is, one can choose a representative of $\mathfrak{D}_\mathfrak{s}$ with only finitely many walls. In fact, they are the only rank 2 CSDs of finite type up to equivalence.

3.4 Constancy of total wall-element on $e_i^\perp$

Here we present some property of walls with the normal vector e_i for any consistent scattering diagram. For this purpose, we introduce yet another decomposition of G.

For each e_i $(i = 1, \ldots, r)$, we introduce the *decomposition of* $\mathfrak{g}$ *by* e_i as

$$\mathfrak{g} = \mathfrak{g}_{e_i}^{\parallel} \oplus \mathfrak{g}_{e_i}^{\nparallel}, \tag{3.58}$$

where $\mathfrak{g}_{e_i}^{\nparallel}$ is a Lie subalgebra of $\mathfrak{g}$ defined by

$$\mathfrak{g}_{e_i}^{\nparallel} := \bigoplus_{\substack{n\in N^+ \\ n\notin \mathbb{Z}e_i}} \mathfrak{g}_n = \mathfrak{g}_+^{e_i} \oplus \mathfrak{g}_-^{e_i} \oplus \mathfrak{g}_{e_i}^{\perp}. \tag{3.59}$$

Lemma 3.15. *$\mathfrak{g}_{e_i}^{\nparallel}$ is a Lie algebra ideal of $\mathfrak{g}$.*

Proof. This follows from the following simple fact: For $n, n' \in N^+$, we have $n + n' \in \mathbb{Z}e_i$ if and only if $n, n' \in \mathbb{Z}e_i$. □

Remark 3.16. On the contrary, for $n \in N_{\mathrm{pr}}^+$ other than $e_1, \ldots, e_r$, the Lie subalgebra $\mathfrak{g}_n^{\nparallel}$ defined similarly is not a Lie algebra ideal of $\mathfrak{g}$.

Let $\widehat{\mathfrak{g}}_{e_i}^{\nparallel}$ be the completion of $\mathfrak{g}_{e_i}^{\nparallel}$, and let $G_{e_i}^{\nparallel}$ be the corresponding subgroup of G. We have the *decomposition of* G *by* e_i and a parallel result to Lemma 3.4.

Lemma 3.17. *The following facts hold.*

(a). $G_{e_i}^{\nparallel}$ is a normal subgroup of G.

(b). We have

$$G = G_{e_i}^{\parallel} G_{e_i}^{\nparallel} \tag{3.60}$$

so that any element g of G is uniquely factorized as

$$g = g_{e_i}^{\parallel} g_{e_i}^{\nparallel} \quad (g_{e_i}^{\parallel} \in G_{e_i}^{\parallel},\ g_{e_i}^{\nparallel} \in G_{e_i}^{\nparallel}). \tag{3.61}$$

(c). The map

$$\chi_{e_i} : G \to G_{e_i}^{\parallel}, \quad g = g_{e_i}^{\parallel} g_{e_i}^{\nparallel} \mapsto g_{e_i}^{\parallel} \tag{3.62}$$

is a group homomorphism with the kernel $G_{e_i}^{\nparallel}$.

(d). The map χ_{e_i} in (3.62) coincides with the map ψ_{e_i} in (3.11).

Proof. The proofs of (a)–(c) are the same as Lemma 3.4. Let us prove (d). Let $g = g_+^{e_i} g_{e_i}^{\parallel} g_{e_i}^{\perp} g_-^{e_i}$ be the decomposition in Lemmas 3.2 and 3.4. By applying the map χ_{e_i} to this equality, we obtain $\chi_{e_i}(g) = g_{e_i}^{\parallel} = \psi_{e_i}(g)$. □

We have the following consequence of Lemma 3.17.

Proposition 3.18. *(a). Let g be any element of G. Then, for any general $z \in e_i^{\perp}$, the equality*

$$g_0^z = g_{e_i}^{\parallel} \tag{3.63}$$

holds. In particular, g_0^z is constant with respect to general $z \in e_i^{\perp}$.

(b). Let $\mathfrak{D}$ be any consistent scattering diagram, and let $g = g(\mathfrak{D})$ be the corresponding element in G. Then, for any general $z \in e_i^{\perp}$, the equality

$$g_z(\mathfrak{D}) = g_{e_i}^{\parallel} \tag{3.64}$$

holds. In particular, $g_z(\mathfrak{D})$ is constant with respect to general $z \in e_i^{\perp}$.

Proof. (a). Let g be any element of G. For any general $z \in e_i^{\perp}$, we have the decomposition of g at z in (2.34),

$$g = g_+^z g_0^z g_-^z \quad (g_+^z \in G_+^z,\ g_0^z \in G_{e_i}^{\parallel},\ g_-^z \in G_-^z). \tag{3.65}$$

Then, applying the map χ_{e_i} to this equality, we obtain $\chi_{e_i}(g) = g_0^z$. Therefore, $g_0^z = g_{e_i}^{\parallel}$.

(b). This follows from (a) and Proposition 2.19. □

In particular, we obtain the following property of a CSD, which will be used later.

Proposition 3.19 ([GHKK18, Theorems 1.28, Remark 1.29]). *There is a CSD $\mathfrak{D}_\mathfrak{s}$ such that the normal vector of any outgoing wall is other than e_1, ..., e_r.*

Proof. Take any CSD $\mathfrak{D}_\mathfrak{s}$. The total wall element $g_z(\mathfrak{D}_\mathfrak{s})$ on $e_i^{\perp}$ of $\mathfrak{D}_\mathfrak{s}$ is constant by Proposition 3.18. Since the wall $\mathbf{w}_{e_i}$ in (3.55) is the only incoming wall with the normal vector e_i, $g_z(\mathfrak{D}_\mathfrak{s})$ on $e_i^{\perp}$ coincides with $\Psi[e_i]^{\delta_i}$ by Lemma 3.8. This means that the total contribution to $g_z(\mathfrak{D}_\mathfrak{s})$ on $e_i^{\perp}$ from all outgoing walls is trivial. Therefore, one can remove all outgoing walls whose normal vectors are $e_1, \dots, e_r$ up to equivalence. □

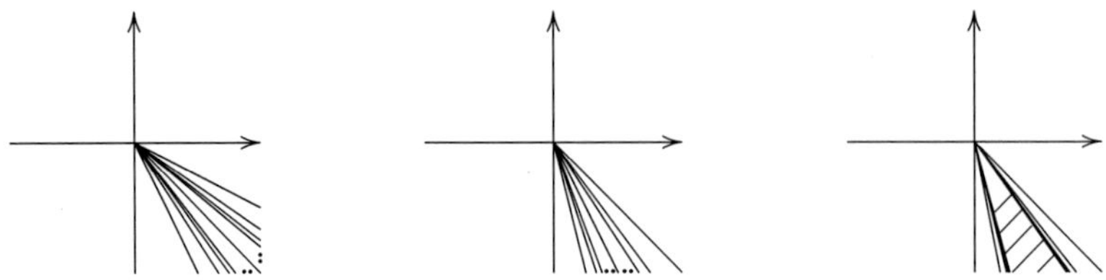

(a) $A_1^{(1)}$ (b) $A_2^{(2)}$ (c) non-affine type

Figure 13: Rank 2 CSD: infinite type

3.5 Rank 2 examples: infinite type

We have already seen all rank 2 CSDs of finite type in Section 2.2. Here we consider rank 2 CSDs of infinite type, where $\delta_1\delta_2 \geq 4$ for B in (2.16).

Affine type. The case $\delta_1\delta_2 = 4$ is exceptional, and it is said to be *of affine type*. There are two cases.

(a). Type $A_1^{(1)}$. Let $\delta_1 = \delta_2 = 2$. Note that, for any primitive $n \in N_{\mathrm{pr}}^+$, $\delta(n) = 2$. The following description of $\mathfrak{D}_\mathfrak{s}$ is well known [Rei09, GPS10, Rea20a]: The walls of $\mathfrak{D}_\mathfrak{s}$ are given by

$$(\mathbf{e}_1^\perp, \Psi[\mathbf{e}_1]^2)_{\mathbf{e}_1}, \quad (\mathbf{e}_2^\perp, \Psi[\mathbf{e}_2]^2)_{\mathbf{e}_2}, \tag{3.66}$$

$$(\mathbb{R}_{\geq 0}(p, -p-1), \Psi[(p+1,p)]^2)_{(p+1,p)} \quad (p \in \mathbb{Z}_{>0}), \tag{3.67}$$

$$(\mathbb{R}_{\geq 0}(p+1, -p), \Psi[(p,p+1)]^2)_{(p,p+1)} \quad (p \in \mathbb{Z}_{>0}), \tag{3.68}$$

$$(\mathbb{R}_{\geq 0}(1,-1), \prod_{j=0}^{\infty} \Psi[2^j \mathbf{n}_0]^{2^{2-j}})_{\mathbf{n}_0}, \tag{3.69}$$

where $\mathbf{n}_0 = (1,1)$. Note that $\delta(2^j\mathbf{n}_0) = 2^{1-j}$ by (1.68). See Figure 13 (a).

Here we demonstrate to derive the above walls, up to some degree, by repeatedly applying the pentagon relation (1.56) in a similar manner for finite type. We use the same notation $[n_1, n_2]^c$ for $\Psi[(n_1,n_2)]^c$ as in Section 2.2. We apply the pentagon relation to the cluster element $g_\mathfrak{s} = \Psi[\mathbf{e}_2]^2\Psi[\mathbf{e}_1]^2$ repeatedly and obtain the equality

$$\begin{bmatrix}0\\1\end{bmatrix}^2 \begin{bmatrix}1\\0\end{bmatrix}^2 = \begin{bmatrix}1\\0\end{bmatrix} \begin{bmatrix}1\\1\end{bmatrix}^2 \begin{bmatrix}1\\2\end{bmatrix} \begin{bmatrix}1\\0\end{bmatrix} \begin{bmatrix}1\\1\end{bmatrix}^2 \begin{bmatrix}1\\2\end{bmatrix} \begin{bmatrix}0\\1\end{bmatrix}^2. \tag{3.70}$$

At this moment, the expression in the right hand side is not yet "ordered" to be presented by a scattering diagram. Here, we say that the above product is *ordered* (resp. *anti-ordered*) if, for any adjacent pair $[\mathbf{n}']^{c'}[\mathbf{n}]^c$, $\{\mathbf{n}', \mathbf{n}\} = n_2'n_1 - n_1'n_2 \leq 0$ (resp. $\{\mathbf{n}', \mathbf{n}\} \geq 0$) holds. Equivalently, if we view $[n_1, n_2]$ as a fraction n_1/n_2, then, the numbers should be aligned in the decreasing order (resp. increasing order) form left to right. The left hand side of (3.70) is anti-ordered. To make the right hand side of (3.70) ordered, we need to interchange $[1,2]$ and $[1,0]$ in the middle, where $\{\mathbf{n}',\mathbf{n}\}=2$ for $\mathbf{n}'=(1,2)$ and $\mathbf{n}=(1,0)$. As the lowest approximation, we consider modulo $G^{>3}$. Then, $[1,2]$ and $[1,0]$ commute, and we have, modulo $G^{>3}$,

$$\begin{bmatrix}0\\1\end{bmatrix}^2\begin{bmatrix}1\\0\end{bmatrix}^2\equiv\begin{bmatrix}1\\0\end{bmatrix}^2\begin{bmatrix}2\\1\end{bmatrix}^2\begin{bmatrix}1\\1\end{bmatrix}^4\begin{bmatrix}1\\2\end{bmatrix}^2\begin{bmatrix}0\\1\end{bmatrix}^2. \tag{3.71}$$

This certainly agrees with the data in (3.66)–(3.69). This is also compatible with the formula (1.75).

To proceed to higher degree, we now apply the pentagon relation (1.56) with $c=2$ to the pair $[1,2]^{1/2}$ and $[1,0]^{1/2}$ in (3.70), which are obtained by factorizing the elements $[1,2]$ and $[1,0]$. To clarify the structure among fractional powers, we use the notation $[n_1,n_2]_2:=[n_1,n_2]^{1/2}$. Then, we have

$$\begin{bmatrix}1\\2\end{bmatrix}\begin{bmatrix}1\\0\end{bmatrix}=\begin{bmatrix}1\\2\end{bmatrix}_2^2\begin{bmatrix}1\\0\end{bmatrix}_2^2=\begin{bmatrix}1\\0\end{bmatrix}_2\begin{bmatrix}2\\2\end{bmatrix}_2^2\begin{bmatrix}3\\4\end{bmatrix}_2\begin{bmatrix}1\\0\end{bmatrix}_2\begin{bmatrix}2\\2\end{bmatrix}_2^2\begin{bmatrix}3\\4\end{bmatrix}_2\begin{bmatrix}1\\2\end{bmatrix}_2^2. \tag{3.72}$$

This is parallel to (3.70). As the next approximation, we consider modulo $G^{>7}$. Then, $[3,4]$ and $[1,0]$ commute, and we have, modulo $G^{>7}$,

$$\begin{bmatrix}1\\2\end{bmatrix}\begin{bmatrix}1\\0\end{bmatrix}\equiv\begin{bmatrix}1\\0\end{bmatrix}_2^2\begin{bmatrix}3\\2\end{bmatrix}_2^2\begin{bmatrix}2\\2\end{bmatrix}_2^4\begin{bmatrix}3\\4\end{bmatrix}_2^2\begin{bmatrix}1\\2\end{bmatrix}_2^2=\begin{bmatrix}1\\0\end{bmatrix}\begin{bmatrix}3\\2\end{bmatrix}\begin{bmatrix}2\\2\end{bmatrix}^2\begin{bmatrix}3\\4\end{bmatrix}\begin{bmatrix}1\\2\end{bmatrix}. \tag{3.73}$$

Then, we plug it into (3.70) and apply the pentagon relation, and we have, modulo $G^{>7}$,

$$\begin{bmatrix}0\\1\end{bmatrix}^2\begin{bmatrix}1\\0\end{bmatrix}^2\equiv\begin{bmatrix}1\\0\end{bmatrix}^2\begin{bmatrix}2\\1\end{bmatrix}^2\begin{bmatrix}3\\2\end{bmatrix}^2\begin{bmatrix}4\\3\end{bmatrix}^2\begin{bmatrix}1\\1\end{bmatrix}^4\begin{bmatrix}2\\2\end{bmatrix}^2\begin{bmatrix}3\\4\end{bmatrix}^2\begin{bmatrix}2\\3\end{bmatrix}^2\begin{bmatrix}1\\2\end{bmatrix}^2\begin{bmatrix}0\\1\end{bmatrix}^2. \tag{3.74}$$

Again, this agrees with the data in (3.66)–(3.69). By repeating this procedure modulo $G^{>2^\ell-1}$, we obtain the equality

$$\begin{bmatrix}0\\1\end{bmatrix}^2\begin{bmatrix}1\\0\end{bmatrix}^2=\begin{bmatrix}1\\0\end{bmatrix}^2\begin{bmatrix}2\\1\end{bmatrix}^2\begin{bmatrix}3\\2\end{bmatrix}^2\cdots\prod_{j=0}^{\infty}\begin{bmatrix}2^j\\2^j\end{bmatrix}^{2^{2-j}}\cdots\begin{bmatrix}2\\3\end{bmatrix}^2\begin{bmatrix}1\\2\end{bmatrix}^2\begin{bmatrix}0\\1\end{bmatrix}^2, \tag{3.75}$$

which is equivalent to the result (3.66)–(3.69). A complete proof of the equality (3.75) based on the pentagon relation is found in [Mat21].

(b). Type $A_2^{(2)}$. Let $\delta_1=1$, $\delta_2=4$. The situation is parallel to $A_1^{(1)}$ though a little more complicated. So, we present results with less explanation. The following description of $\mathfrak{D}_{\mathfrak{s}}$ is known [Rea20a]: The walls of $\mathfrak{D}_{\mathfrak{s}}$ are given by

$$(\mathbf{e}_1^{\perp},\Psi[\mathbf{e}_1])_{\mathbf{e}_1},\quad(\mathbf{e}_2^{\perp},\Psi[\mathbf{e}_2]^4)_{\mathbf{e}_2}, \tag{3.76}$$

$$(\mathbb{R}_{\geq 0}(p,-2p-1),\Psi[(2p+1,4p)])_{(2p+1,4p)}\quad(p\in\mathbb{Z}_{>0}), \tag{3.77}$$

$$(\mathbb{R}_{\geq 0}(2p-1,-4p), \Psi[(p,2p-1)]^4)_{(p,2p-1)} \quad (p \in \mathbb{Z}_{>0}), \tag{3.78}$$

$$(\mathbb{R}_{\geq 0}(p,-2p+1), \Psi[(2p-1,4p)])_{(2p-1,4p)} \quad (p \in \mathbb{Z}_{>0}), \tag{3.79}$$

$$(\mathbb{R}_{\geq 0}(2p+1,-4p), \Psi[(p,2p+1)]^4)_{(p,2p+1)} \quad (p \in \mathbb{Z}_{>0}), \tag{3.80}$$

$$(\mathbb{R}_{\geq 0}(1,-2), \Psi[\mathbf{n}_0]^6 \prod_{j=1}^{\infty} \Psi[2^j \mathbf{n}_0]^{2^{2-j}})_{\mathbf{n}_0}, \tag{3.81}$$

where $\mathbf{n}_0 = (1,2)$. Note that $\delta(2^j \mathbf{n}_0) = 2^{1-j}$. See Figure 13 (b).

We apply the pentagon relation to the cluster element $g_{\mathfrak{s}} = \Psi[\mathbf{e}_2]^4 \Psi[\mathbf{e}_1]$ repeatedly and obtain the equality

$$\begin{bmatrix}0\\1\end{bmatrix}^4 \begin{bmatrix}1\\0\end{bmatrix} = \begin{bmatrix}1\\0\end{bmatrix} \begin{bmatrix}1\\1\end{bmatrix}^3 \begin{bmatrix}2\\3\end{bmatrix} \begin{bmatrix}1\\2\end{bmatrix}^3 \begin{bmatrix}1\\3\end{bmatrix} \begin{bmatrix}1\\1\end{bmatrix} \begin{bmatrix}1\\2\end{bmatrix}^3 \begin{bmatrix}2\\5\end{bmatrix} \begin{bmatrix}1\\3\end{bmatrix}^3 \begin{bmatrix}1\\4\end{bmatrix} \begin{bmatrix}0\\1\end{bmatrix}^4. \tag{3.82}$$

Again, this is not yet ordered, and to order it, we need to interchange $[1,3]$ and $[1,1]$ in the middle.

As the lowest approximation, we consider modulo $G^{>5}$. Then, $[1,3]$ and $[1,1]$ commute, and we have, modulo $G^{>5}$,

$$\begin{bmatrix}0\\1\end{bmatrix}^4 \begin{bmatrix}1\\0\end{bmatrix} \equiv \begin{bmatrix}1\\0\end{bmatrix} \begin{bmatrix}1\\1\end{bmatrix}^4 \begin{bmatrix}2\\3\end{bmatrix}^4 \begin{bmatrix}1\\2\end{bmatrix}^6 \begin{bmatrix}1\\3\end{bmatrix}^4 \begin{bmatrix}1\\4\end{bmatrix} \begin{bmatrix}0\\1\end{bmatrix}^4. \tag{3.83}$$

This certainly agrees with the data in (3.76)–(3.81).

To proceed to higher degree, we do in the same way as before. By apply the pentagon relation (1.56) with $c = 2$, we have

$$\begin{bmatrix}1\\3\end{bmatrix} \begin{bmatrix}1\\1\end{bmatrix} = \begin{bmatrix}1\\3\end{bmatrix}_2^2 \begin{bmatrix}1\\1\end{bmatrix}_2^2 = \begin{bmatrix}1\\1\end{bmatrix}_2 \begin{bmatrix}2\\4\end{bmatrix}_2^2 \begin{bmatrix}3\\7\end{bmatrix}_2 \begin{bmatrix}1\\1\end{bmatrix}_2 \begin{bmatrix}2\\4\end{bmatrix}_2^2 \begin{bmatrix}3\\7\end{bmatrix}_2 \begin{bmatrix}1\\3\end{bmatrix}_2^2, \tag{3.84}$$

which has a common structure with (3.70). As the next approximation, we consider modulo $G^{>11}$. Then, $[3,7]$ and $[1,1]$ commute, and we have, modulo $G^{>11}$,

$$\begin{bmatrix}1\\3\end{bmatrix} \begin{bmatrix}1\\1\end{bmatrix} \equiv \begin{bmatrix}1\\1\end{bmatrix}_2^2 \begin{bmatrix}3\\5\end{bmatrix}_2^2 \begin{bmatrix}2\\4\end{bmatrix}_2^4 \begin{bmatrix}3\\7\end{bmatrix}_2^2 \begin{bmatrix}1\\3\end{bmatrix}_2^2 = \begin{bmatrix}1\\1\end{bmatrix} \begin{bmatrix}3\\5\end{bmatrix} \begin{bmatrix}2\\4\end{bmatrix}^2 \begin{bmatrix}3\\7\end{bmatrix} \begin{bmatrix}1\\3\end{bmatrix}. \tag{3.85}$$

Then, we plug it into (3.82) and apply the pentagon relation, and we have, modulo $G^{>11}$,

$$\begin{aligned} \begin{bmatrix}0\\1\end{bmatrix}^4 \begin{bmatrix}1\\0\end{bmatrix} \equiv{}& \begin{bmatrix}1\\0\end{bmatrix} \begin{bmatrix}1\\1\end{bmatrix}^4 \begin{bmatrix}3\\4\end{bmatrix} \begin{bmatrix}2\\3\end{bmatrix}^4 \begin{bmatrix}3\\5\end{bmatrix}^4 \begin{bmatrix}4\\7\end{bmatrix}^4 \begin{bmatrix}1\\2\end{bmatrix}^6 \begin{bmatrix}2\\4\end{bmatrix}^2 \\ & \times \begin{bmatrix}3\\7\end{bmatrix}^4 \begin{bmatrix}2\\5\end{bmatrix}^4 \begin{bmatrix}3\\8\end{bmatrix} \begin{bmatrix}1\\3\end{bmatrix}^4 \begin{bmatrix}1\\4\end{bmatrix} \begin{bmatrix}0\\1\end{bmatrix}^4. \end{aligned} \tag{3.86}$$

Again, this agrees with the data in (3.76)–(3.81). By repeating this procedure modulo $G^{>3\cdot 2^\ell -1}$, we obtain the equality

$$\begin{aligned}\begin{bmatrix}0\\1\end{bmatrix}^4\begin{bmatrix}1\\0\end{bmatrix} &= \begin{bmatrix}1\\0\end{bmatrix}\begin{bmatrix}1\\1\end{bmatrix}^4\begin{bmatrix}3\\4\end{bmatrix}\begin{bmatrix}2\\3\end{bmatrix}^4\begin{bmatrix}5\\8\end{bmatrix}\begin{bmatrix}3\\5\end{bmatrix}^4\cdots \\ &\quad\times \begin{bmatrix}1\\2\end{bmatrix}^6\prod_{j=1}^{\infty}\begin{bmatrix}2^j\\2^{j+1}\end{bmatrix}^{2^{2-j}}\cdots\begin{bmatrix}5\\12\end{bmatrix}\begin{bmatrix}2\\5\end{bmatrix}^4\begin{bmatrix}3\\8\end{bmatrix}\begin{bmatrix}1\\3\end{bmatrix}^4\begin{bmatrix}1\\4\end{bmatrix}\begin{bmatrix}0\\1\end{bmatrix}^4,\end{aligned} \tag{3.87}$$

which is equivalent to the result (3.66)–(3.69). A complete proof of the equality (3.87) based on the pentagon relation is found in [Mat21]. The case $\delta_1 = 4$, $\delta_2 = 1$ is similar.

Non-affine type. The case $\delta_1\delta_2 > 4$. The entire structure of a CSD is not yet known. Let $\sigma(\mathbf{v}_+, \mathbf{v}_-)$ be the irrational cone spanned by two vectors

$$\mathbf{v}_\pm = \begin{pmatrix}\delta_1\delta_2 \pm \sqrt{\delta_1\delta_2(\delta_1\delta_2 - 4)}\\ -2\delta_2\end{pmatrix}. \tag{3.88}$$

Under the canonical pairing (2.12), an outgoing wall $\mathbf{w}$ is in the region $\sigma(\mathbf{v}_+, \mathbf{v}_-)$ if and only if its normal vector $\mathbf{n} = (n_1, n_2)$ satisfies the condition

$$\frac{2}{\delta_2 + \sqrt{\delta_2(\delta_2 - 4/\delta_1)}} < \frac{n_1}{n_2} < \frac{2}{\delta_2 - \sqrt{\delta_2(\delta_2 - 4/\delta_1)}}. \tag{3.89}$$

See Figure 13 (c), where the cone $\sigma(\mathbf{v}, \mathbf{v}')$ is depicted as a hatched region. The region outside $\sigma(\mathbf{v}_+, \mathbf{v}_-)$ is decomposed into infinitely many chambers by the walls converging to $\mathbf{v}_+$ and $\mathbf{v}_-$ as in the affine case. They are the *G-chambers* in Section 6.6, and its behavior is well-known in cluster algebra theory (e.g., [Rea20a, §3]). On the other hand, in the region $\sigma(\mathbf{v}_+, \mathbf{v}_-)$, which is informally called the *Badlands*, it is expected that every ray with the normal vector $n \in N^+_{\mathrm{pr}}$ in the cone $\sigma(\mathbf{v}, \mathbf{v}')$ appears as a nontrivial wall [GHKK18, Example 1.15]. For the skew-symmetric case $\delta_1 = \delta_2$, this is proved in [DM21, Example 7.10]. However, the wall elements are not known yet. In Section 5.12, we present results calculated by computer, which support the above expectation.

Notes

Most contents are taken from [GHKK18, §1.2] with added/modified proofs. In particular, an alternative and simple proof of Proposition 3.19 is given based on Proposition 3.18. The presentation of rank 2 CSDs of affine type based on the pentagon relation (1.56) is also due to us. The procedure demonstrated above is the implementation of the *perturbation trick* and the *change of lattice trick* described in [GHKK18, Appendix C.3], and it appears later as Algorithm 5.7 in more generality.

4 Principal extension method

In this short section we give a faithful representation of the structure group G for any (possibly degenerate) fixed data based on the principal extension method.

4.1 Principal extension of fixed data

In the previous sections we have successfully constructed a CSD for any (possibly degenerate) fixed data. On the other hand, to study further properties of CSDs, there is a situation where a faithful representation of G is desired even for *degenerate* fixed data. In cluster algebra theory, this corresponds to the situation where the algebraically independence of *$\hat{y}$-variables* in [FZ07] is desirable. To resolve it, the *principal coefficients*, or equivalently, the *principal extensions* of exchange matrices were introduced in [FZ07]. Here we present a parallel method in scattering diagrams by [GHK15, Construction 2.1], which in particular provides a faithful representation of G for any (possibly degenerate) fixed data.

Definition 4.1 (Principal extension of fixed data). Let $\Gamma = \{N, \{\cdot,\cdot\}, N^\circ, (\delta_i), M, M^\circ\}$ be a (possibly degenerate) fixed data of rank r. The *principal extension* Γ_{prin} *of* Γ is a fixed data of rank $2r$ consisting of the following data:

- A lattice $\tilde{N} = N \oplus M^\circ$ of rank $2r$ with a skew-symmetric bilinear form

$$\{\cdot,\cdot\} : \tilde{N} \times \tilde{N} \to \mathbb{Q}, \tag{4.1}$$

$$\{(n,m),(n',m')\} := \{n,n'\} + \langle n',m\rangle - \langle n,m'\rangle. \tag{4.2}$$

- A sublattice $\tilde{N}^\circ = N^\circ \oplus M \subset \tilde{N}$. It is of finite index and

$$\{\tilde{N}^\circ, \tilde{N}\} \subset \mathbb{Z}. \tag{4.3}$$

- A $2r$-tuple of positive integers $(\tilde{\delta}_1, \dots, \tilde{\delta}_{2r}) = (\delta_1, \dots, \delta_r, \delta_1, \dots, \delta_r)$.
- $\tilde{M} = \mathrm{Hom}(\tilde{N}, \mathbb{Z}) \simeq M \oplus N^\circ$, $\tilde{M}^\circ = \mathrm{Hom}(\tilde{N}^\circ, \mathbb{Z}) \simeq M^\circ \oplus N$.

The bilinear form (4.2) is always nondegenerate, so that Γ_{prin} is an nondegenerate fixed data.

The canonical pairing $\langle\cdot,\cdot\rangle : \tilde{N}^\circ \times \tilde{M}^\circ \to \mathbb{Z}$ is given by

$$\langle (n,m),(m',n')\rangle = \langle n,m'\rangle + \langle n',m\rangle. \tag{4.4}$$

Example 4.2. For any seed $\mathfrak{s} = (e_1, \dots, e_r)$ for Γ, let $(f_1, \dots, f_r)$ be the basis of M° in Section 1.1. Then, $\tilde{\mathfrak{s}} = ((e_1,0), \dots, (e_r,0), (0,f_1), \dots, (0,f_r))$ is a seed for Γ_{prin}. Also, $((0,f_1), \dots, (0,f_r), (e_1,0), \dots, (e_r,0))$ is a basis of $\tilde{M}^\circ$.

Remark 4.3. In the above example, by applying the correspondence in (1.7), and using x_i and y_i in (1.7), we have

$$\tilde{\mathbf{x}} = (x_1, \dots, x_r, y_1, \dots, y_r). \tag{4.5}$$

They are regarded as x-variables x_i with *principal coefficients* y_i in [FZ07].

4.2 Principal x-representation of G

To clarify the background, let us briefly describe the scheme of the construction of a CSD in [GHKK18] for degenerate Γ. For a seed $\tilde{\mathfrak{s}}$ for Γ_{prin} in Example 4.2, every result in the previous sections are applicable, where the structure group for scattering diagrams is the group $\tilde{G}$ associated with $\tilde{\mathfrak{s}}$, and the ambient space for their supports is $\tilde{M}_{\mathbb{R}} \simeq \mathbb{R}^{2r}$. Moreover, since Γ_{prin} is nondegenerate, $\tilde{G}$ has the faithful x-representation. Then, one can construct a CSD for $\mathfrak{s}$ by reducing the structure group $\tilde{G}$ to the subgroup G associated with $\mathfrak{s}$. Moreover, G has a faithful representation induced from the faithful x-representation of $\tilde{G}$. The resulting CSD is still in the ambient space $\tilde{M}_{\mathbb{R}} \simeq M_{\mathbb{R}} \oplus \mathbb{R}^r$, and it is invariant along the fiber $\mathbb{R}^r$. Then, we apply the projection $\pi : \tilde{M}_{\mathbb{R}} \to M_{\mathbb{R}}$ to the support, and we obtain the CSD constructed here.

Since we have already constructed a CSD with the ambient space $M_{\mathbb{R}} \simeq \mathbb{R}^r$ in the previous section directly for any (possibly degenerate) fixed data, we do not need to repeat the above construction of a CSD. Thus, our strategy here is just borrowing the faithful representation of G induced from the one of $\tilde{G}$, and applying it to the wall elements when needed.

Let $\mathfrak{s}$ be a seed for a given (possibly degenerate) fixed data Γ. From now on, we also use the internal-sum notation for $\tilde{N} = N \oplus M^\circ$ and $\tilde{M}^\circ = \mathrm{Hom}(\tilde{N}^\circ, \mathbb{Z}) = M^\circ \oplus N$ such as $\tilde{n} = n + m$, $\tilde{m} = m + n$, ($n \in N$, $m \in M^\circ$). This notation is compatible with the notation in (4.2) because $\{(n, 0), (n', 0)\} = \{n, n'\}$. Also, for $n \in N$ and $\tilde{m} = m + n'$ ($m \in M^\circ$, $n' \in N$), we have

$$\langle n, \tilde{m} \rangle = \langle n, m \rangle. \tag{4.6}$$

We introduce a group homomorphism (cf. (1.8)).

$$\begin{array}{cccc} \tilde{p}^* : & N & \to & \tilde{M}^\circ \\ & n & \mapsto & \{\cdot, n\}. \end{array} \tag{4.7}$$

By the definition of $\tilde{p}^*$ and (4.2), we have

$$\langle n' + m', \tilde{p}^*(n) \rangle = \{n', n\} + \langle n, m' \rangle \in \mathbb{Z} \quad (n' \in N^\circ, n \in N, m' \in M). \tag{4.8}$$

Lemma 4.4. *The map $\tilde{p}^*$ is injective. In particular, $\tilde{p}^*(e_1), \cdots, \tilde{p}^*(e_r) \in \tilde{M}^\circ$ are $\mathbb{Z}$-linearly independent.*

Proof. We take a basis $\mathfrak{s} = (e_1, \ldots, e_r)$ of N and a basis $(f_1, \ldots, f_r, e_1, \ldots, e_r)$ of $\tilde{M}^\circ$. Then, as (1.11), we can easily verify that

$$\tilde{p}^*(e_j) = \sum_{i=1}^{r} b_{ij} f_i + e_j. \tag{4.9}$$

Thus, the matrix representation of $\tilde{p}^*$ is given by a $2r \times r$ matrix

$$\tilde{B} = \begin{pmatrix} B \\ I \end{pmatrix}, \tag{4.10}$$

where $B = (b_{ij})_{i,j=1}^r$, $b_{ij} = \{\delta_i e_i, e_j\}$ as before. The matrix $\tilde{B}$ has rank r. □

Remark 4.5. The matrix $\tilde{B}$ is the principal extension of an exchange matrix B in [FZ07].

In place of P in (1.44), we define a monoid $\tilde{P} \subset \tilde{M}^\circ$ as follows:

(i). $\tilde{P} = \sigma \cap \tilde{M}^\circ$, where σ is a $2r$-dimensional strongly convex cone in $\tilde{M}_{\mathbb{R}} = \tilde{M}^\circ \otimes_{\mathbb{R}} \mathbb{R}$.

(ii). $\tilde{p}^*(e_1), \cdots, \tilde{p}^*(e_r) \in \tilde{P}$.

Such $\tilde{P}$ is not unique at all, and we choose one arbitrarily. The results we will present here do not depend on the choice of $\tilde{P}$. For example, we may take the monoid generated by $f_1, \ldots, f_r, \tilde{p}^*(e_1), \ldots, \tilde{p}^*(e_r)$, which are a basis of $\tilde{M}^\circ$ by (4.9).

Lemma 4.6. *For any $n \in N$ with $n \neq 0$, there is some $\tilde{m} \in \tilde{P}$ such that $\langle n, \tilde{m} \rangle \neq 0$.*

Proof. By the condition (i), if $\langle n, \tilde{m} \rangle = 0$ for any $\tilde{m} \in \tilde{P}$, $\langle n, \tilde{m} \rangle = 0$ for any $\tilde{m} \in \tilde{M}^\circ$ by linearity. This is a contradiction. □

Let $\Bbbk[\tilde{P}]$ be the monoid algebra of $\tilde{P}$ over $\Bbbk$. Let J be the maximal ideals of $\Bbbk[\tilde{P}]$ generated by $\tilde{P} \setminus \{0\}$, and let $\Bbbk[[\tilde{P}]]$ be the completion by J. We express an arbitrary element $f \in \Bbbk[[\tilde{P}]]$ as a formal power series in x as

$$f = \sum_{\tilde{m} \in \tilde{P}} c_{\tilde{m}} x^{\tilde{m}}. \tag{4.11}$$

In parallel to (1.46), we define the action of $X_n \in \mathfrak{g}$ $(n \in N^+)$ on $\Bbbk[[\tilde{P}]]$ by

$$\tilde{X}_n(x^{\tilde{m}}) := \langle n, m \rangle x^{\tilde{m} + \tilde{p}^*(n)} \quad (\tilde{m} \in \tilde{P}). \tag{4.12}$$

Here and in the related formulas below, $\tilde{m} = m + n'$ $(m \in M^\circ, n' \in N)$.

Proposition 4.7. *We have an injective group homomorphism*

$$\begin{array}{rccc} \tilde{\rho}_x : & G & \to & \mathrm{Aut}(\Bbbk[[\tilde{P}]]) \\ & \exp(X) & \mapsto & \mathrm{Exp}(\tilde{X}). \end{array} \tag{4.13}$$

Proof. We only need to prove the injectivity, since the rest is the same with Propositions 1.8 and 1.9. To show the injectivity, it is enough to prove the linear independence of $\tilde{X}_n$ ($n \in N^+$) in (4.12). This follows from the injectivity of $\tilde{p}^*$ and Lemma 4.6. □

The representation $\tilde{\rho}_x$ is closely related with the mutations of x-variables with principal coefficients in (4.5). Thus, we call it the *principal x-representation* of G.

4.3 Faithful representation of wall elements

Let $\mathfrak{s}$ be any seed for a given (possibly degenerate) fixed data.

Lemma 4.8 ([GHKK18, Lemma 1.3]). *For any $n \in N^+_{\mathrm{pr}}$, let*

$$g = \exp\left(\sum_{j>0} c_j X_{jn} \right) \in G_n^{\parallel} \tag{4.14}$$

be an arbitrary element in $G_n^{\parallel}$. Correspondingly, let

$$f_g := \exp\left(\sum_{j>0} j c_j x^{j\tilde{p}^*(n)} \right) \in \Bbbk[[x^{\tilde{p}^*(n)}]] \subset \Bbbk[[\tilde{P}]]. \tag{4.15}$$

Then, under the principal x-representation $\tilde{\rho}_x$, g acts on $\Bbbk[[\tilde{P}]]$ by

$$g(x^{\tilde{m}}) = x^{\tilde{m}} f_g^{\langle n,m \rangle} \quad (\tilde{m} \in \tilde{P}). \tag{4.16}$$

Proof. We have

$$\begin{aligned} g(x^{\tilde{m}}) &= x^{\tilde{m}} \exp\left(\sum_{j>0} c_j x^{j\tilde{p}^*(n)} \langle jn, m \rangle \right) \\ &= x^{\tilde{m}} \exp\left(\sum_{j>0} j c_j x^{j\tilde{p}^*(n)} \right)^{\langle n,m \rangle} = x^{\tilde{m}} f_g^{\langle n,m \rangle}. \end{aligned} \tag{4.17}$$

□

Thanks to the injectivity of $\tilde{p}^*$ and $\tilde{\rho}_x$, $g \in G$ is uniquely specified by f_g and its action (4.16). Also, if we consider the normalized form of g in (1.71), then f_{g_0} corresponding to $g_0 = g^{1/\delta(n)}$ is also invariant under the rescaling of Γ. In view of this, we introduce the following notion.

Definition 4.9 (Normalized automorphism $\mathfrak{p}_{n,f}$). For any $n \in N^+_{\mathrm{pr}}$ and

$$f = 1 + \sum_{j>0} c_j x^{j\tilde{p}^*(n)} \in \Bbbk[[x^{\tilde{p}^*(n)}]] \subset \Bbbk[[\tilde{P}]], \tag{4.18}$$

we define an algebra automorphism $\mathfrak{p}_{n,f}$ of $\Bbbk[[\tilde{P}]]$ by

$$\mathfrak{p}_{n,f}(x^{\tilde{m}}) := x^{\tilde{m}} f^{\langle \delta(n)n, m \rangle} \quad (\tilde{m} \in \tilde{P}) \tag{4.19}$$

We call $\mathfrak{p}_{n,f}$ the *normalized automorphism* by (n, f).

Definition 4.10 (Wall function). Let $\mathbf{w} = (\mathfrak{d}, g)_n$ ($n \in N_{\mathrm{pr}}^+$) be any wall for $\mathfrak{s}$. Let $g_0 := g^{1/\delta(n)}$, and let $f = f_{g_0}$ be the formal power series in $x^{\tilde{p}^*(n)}$ in Lemma 4.8 corresponding to g_0. Then, $g_{\mathfrak{d}}$ acts on $\Bbbk[[\tilde{P}]]$ by $\mathfrak{p}_{n,f}$. Thus, we can specify the wall $\mathbf{w}$ also by $\mathbf{w} = [\mathfrak{d}, f]_n$ without ambiguity, where we use the square bracket to avoid confusion with the usual notation. We call f the *wall function* of a wall $\mathbf{w}$.

The above specification of a wall is used in [GHKK18].

The following example is particularly important.

Example 4.11 (Cf. Proposition 1.12). For any $n \in N_{\mathrm{pr}}^+$ and $t \in \mathbb{Z}_{>0}$, let $g = \Psi[tn]^{\delta(tn)} = \Psi[tn]^{\delta(n)/t}$, where $\Psi[tn]$ is the dilogarithm element in (1.50). Let

$$g_0 := g^{1/\delta(n)} = \Psi[tn]^{1/t} = \exp\left(\frac{1}{t} \sum_{j>0} \frac{(-1)^{j+1}}{j^2} X_{jtn} \right). \tag{4.20}$$

Then, the corresponding wall function $f = f_{g_0}$ is given by $f[tn]$, where

$$f[tn] := \exp\left(\sum_{j>0} \frac{(-1)^{j+1}}{j} x^{j\tilde{p}^*(tn)} \right) = 1 + x^{\tilde{p}^*(tn)}. \tag{4.21}$$

Note that the factor t^{-1} in (4.20) cancels the factor t from X_{jtn} in (4.21). Therefore, we have

$$\Psi[tn]^{\delta(tn)}(x^{\tilde{m}}) = \mathfrak{p}_{n,f[tn]} = x^{\tilde{m}}(1 + x^{\tilde{p}^*(tn)})^{\langle \delta(n)n, m \rangle}. \tag{4.22}$$

This automorphism is identified with the automorphism part of the Fock-Goncharov decomposition of mutations of x-variables with principal coefficients in cluster algebras [FG09a, §2.1]. In terms of $f[tn]$, the set of incoming walls in (3.55) is represented as

$$\mathrm{In}_{\mathfrak{s}} = \{ \mathbf{w}_{e_i} = [e_i^{\perp}, 1 + x^{\tilde{p}^*(e_i)}]_{e_i} \mid i = 1, \ldots, r \}. \tag{4.23}$$

This is the description of a CSD given in [GHKK18].

Notes

All results in this section are taken from [GHKK18, §1, Appendix B] and [GHK15, §2], and adapted for the present context.

5 Positive realization of CSD and pentagon relation

In this section we present an alternative construction of CSDs following [GHKK18]. The exponents of wall elements for the resulting CSDs are all positive; thus we call them *positive realizations* of CSDs. This realization is a key to prove the Laurent positivity of the corresponding cluster algebra. The construction also reveals the fundamental role of the pentagon relation for CSDs.

5.1 Positive realization of CSD

We start with a general fact on a consistent scattering diagram, which easily follows from Proposition 1.13 and the proof of Theorem 2.20. Recall the correspondence between $\Psi[tn]^{\delta(tn)}$ and $f[tn]$ in Example 4.11.

Proposition 5.1. *Let $\mathfrak{D}$ be any consistent scattering diagram for a seed $\mathfrak{s}$. Then, there is a consistent scattering diagram $\mathfrak{D}'$ that is equivalent to $\mathfrak{D}$ such that the wall element/function of any wall $\mathbf{w} = (\mathfrak{d}, g)_n = [\mathfrak{d}, f]_n$ of $\mathfrak{D}'$ has the following form*

$$g = \Psi[tn]^{s\delta(tn)}, \quad f = (1 + x^{\tilde{p}^*(tn)})^s \quad (n \in N^+_{\mathrm{pr}},\ t \in \mathbb{Z}_{>0},\ s \in \Bbbk). \tag{5.1}$$

Proof. Let $\mathfrak{D} = \mathfrak{D}(g)$ be a consistent scattering diagram corresponding to $g \in G$. We construct $\mathfrak{D}'$ by slightly modifying the construction of $\mathfrak{D}(g)$ in the proof of the surjectivity of φ in Theorem 2.20. As in the proof of Proposition 1.13, we uniquely expand g^z_0 in (2.48) as

$$g^z_0 = \prod_{j>0} \Psi[jn]^{c'_{z,jn}} \quad (c'_{z,jn} \in \Bbbk). \tag{5.2}$$

Accordingly, we place $g_{\mathfrak{d}}$ in (2.49) with

$$g_{\mathfrak{d}} = \begin{cases} \Psi[jn]^{c'_{z,jn}} & \ell = j\deg(n) \text{ for some unique integer } j > 0, \\ \mathrm{id} & \text{otherwise.} \end{cases} \tag{5.3}$$

Then, the resulting consistent scattering diagram has the property (5.1). □

The goal of this section is to prove the following positivity result on a CSD, where the fact $s \in \mathbb{Z}_{>0}$ is crucial.

Theorem 5.2 ([GHKK18, Theorem 1.13]). *Let $\mathfrak{D}_{\mathfrak{s}}$ be a CSD for a seed $\mathfrak{s}$. Then, there is a consistent scattering diagram $\mathfrak{D}$ that is equivalent to $\mathfrak{D}_{\mathfrak{s}}$ such that the wall element/function of any wall $\mathbf{w} = (\mathfrak{d}, g)_n = [\mathfrak{d}, f]_n$ of $\mathfrak{D}$ has the following form*

$$g = \Psi[tn]^{s\delta(tn)}, \quad f = (1 + x^{\tilde{p}^*(tn)})^s \quad (n \in N^+_{\mathrm{pr}},\ s, t \in \mathbb{Z}_{>0}). \tag{5.4}$$

In particular, the wall function f is a polynomial in x with positive coefficients.

We call the above CSD a *positive realization* of $\mathfrak{D}_{\mathfrak{s}}$. By decomposing the wall in (5.4) into s walls with $g = \Psi[tn]^{\delta(tn)}$, we may assume that $s = 1$ for any wall whenever necessary.

Example 5.3. One can directly confirm the property (5.4) for finite and affine types of rank 2 by the results presented in Sections 2.2 and 3.5, respectively. For walls other than (3.69) and (3.81), we have $s = 1$. For the wall (3.69), one can split it into infinitely many walls so that the wall elements are given by

$$g = \Psi[2^j \mathbf{n}_0]^{2\delta(2^j \mathbf{n}_0)} \quad (j = 0, 1, 2, \dots). \tag{5.5}$$

Thus, only the case $s = 2$ in (5.4) appears. Similarly, for the wall (3.81), we have

$$g = \Psi[\mathbf{n}_0]^{3\delta(\mathbf{n}_0)},\ \Psi[2^j \mathbf{n}_0]^{2\delta(2^j \mathbf{n}_0)} \quad (j = 1, 2, \dots). \tag{5.6}$$

Thus, the cases $s = 2$ and 3 in (5.4) appear.

5.2 Ordering Algorithm

Here we concentrate on a fixed data Γ of rank 2. The following result on dilogarithm elements is a key to prove Theorem 5.2. Let us recall the notion of ordered and anti-ordered products in Section 3.5. Namely, a (possibly infinite) product of $\Psi[n]^c$'s ($n \in N^+$, $c \in \mathbb{Q}_{>0}$) is *ordered* (resp. *anti-ordered*), if, for any adjacent pair $\Psi[n']^{c'}\Psi[n]^c$, $\{n', n\} \leq 0$ (resp. $\{n', n\} \geq 0$) holds.

Proposition 5.4 (Ordering Lemma). *Let $\mathfrak{s}$ be a seed for a fixed data Γ of rank 2. Let*

$$C^{\mathrm{in}} = \Psi[t'_j n'_j]^{s'_j \delta(t'_j n'_j)} \cdots \Psi[t'_1 n'_1]^{s'_1 \delta(t'_1 n'_1)} \quad (n'_i \in N^+_{\mathrm{pr}},\ s'_i, t'_i \in \mathbb{Z}_{>0}) \tag{5.7}$$

be any finite anti-ordered product. Then, C^{in} equals to a (possibly infinite) ordered product C^{out} of factors of the same form

$$\Psi[t_i n_i]^{s_i \delta(t_i n_i)} \quad (n_i \in N^+_{\mathrm{pr}},\ s_i, t_i \in \mathbb{Z}_{>0}). \tag{5.8}$$

Moreover, after gathering all powers of a common $\Psi[t_i n_i]$, the above product is unique up to the reordering of the commuting adjacent pairs with $n_i = n_{i+1}$.

When Γ is degenerate, the claim holds trivially by Proposition 1.14 (a). Thus, we may concentrate on the case where Γ is nondegenerate.

The uniqueness is easily shown by a routine argument with Lemma 3.6. Namely, let $\pi_\ell(C^{\mathrm{out}}) \in G^{\leq \ell}$, which is also regarded as an ordered product. The uniqueness of the ordered product $\pi_1(C^{\mathrm{out}})$ holds by Lemma 3.6 (b). Suppose that the uniqueness holds for the ordered product $\pi_\ell(C^{\mathrm{out}})$. Then, the uniqueness of the ordered product $\pi_{\ell+1}(C^{\mathrm{out}})$ follows from the assumption and Lemma 3.6 (a).

To prove the existance, we introduce an algorithm based on the pentagon relation (1.56), which is a systematic generalization of the one used in Section 3.5. In the algorithm, without loss of generality, we may assume that $\{e_2, e_1\} = 1$ as before. Let us write $\Psi[n]$ as $[n]$, for simplicity, as in Section 3.5.

Definition 5.5 (p-exchangeable pair)**.** We say that, for any integer $p > 0$, an anti-ordered adjacent pair $[n']^{c'}[n]^c$ is *p-exchangeable* if $\{n', n\} = p$ and $c' = j'/p$ and $c = j/p$ for some $j', j \in \mathbb{Z}_{>0}$. Note that we can apply the pentagon relation (1.56) for such a pair by decomposing them into factors $[n']^{1/p}$ and $[n]^{1/p}$.

Example 5.6. Let $[t_1n_1]^{s_1/t_1}[t_2n_2]^{s_2/t_2}$ ($n_i \in N^+_{\mathrm{pr}}$, $s_i, t_i \in \mathbb{Z}_{>0}$) be an anti-ordered adjacent pair with $\{t_1n_1, t_2n_2\} = p$. Then, $p = t_1t_2\{n_1, n_2\}$ is a multiple of both t_1 and t_2. Thus, it is p-exchangable.

Here is the algorithm with annotation.

Algorithm 5.7 (Ordering Algorithm)**.** Fix any degree $\ell > 0$. For any finite anti-ordered product C^{in} with factors of the form $[tn]^{s/t}$ ($n \in N^+_{\mathrm{pr}}$, $s, t \in \mathbb{Z}_{>0}$), obtain a finite ordered product $C^{\mathrm{out}}(\ell)$ of factors of the form $[tn]^{s/t}$ ($n \in N^+_{\mathrm{pr}}$, $s, t \in \mathbb{Z}_{>0}$) by the following procedure. (The fact $s \in \mathbb{Q}_{>0}$ is obvious from the algorithm, but the fact $s \in \mathbb{Z}_{>0}$ is not so, and it will be shown in the proof of Lemma 5.10.)

(p_0). This is the main routine. Below $p_0 := 1$.

1. Set $C_0 = C^{\mathrm{in}}$. (initial data)
2. Decompose every factor $[n]^c$ ($c \in \mathbb{Z}_{>0}$) in C_0 into a product of a single factor $[n]$. (Here we distinguish a single factor $[n]^c$ and a product $[n] \cdots [n]$ of a single factor $[n]$.)
3. Repeat the following operation to C_0 in this order until we do nothing.
    i. Pick up any adjacent pair $[n']^{c'}[n]^c$ in C_0. (c' and c are 1 or non-integers.)
    ii. If $\{n', n\} < 0$ (ordered), we do nothing.
    iii. If $\{n', n\} = 0$ (parallel),
        a. If $\deg(n') > \deg(n)$, we exchange it as $[n]^c[n']^{c'}$. (Align parallel vectors from left to right in the increasing order of degree.)
        b. Otherwise, we do nothing.
    iv. If $\{n', n\} > 0$ (anti-ordered),

a. If $\deg(n'+n) > \ell$, we exchange it as $[n]^c[n']^{c'}$. (They are commutative modulo $G^{>\ell}$.)
b. If $\{n', n\} = p_0$ and $c' = c = 1$ (p_0-exchangeable), replace it with $[n][n'+n][n']$. (Apply the pentagon relation (1.56) with $c = p_0$.)
c. Otherwise, we do nothing.

4. If the resulting C_0 is ordered, we join all factors $[n]^{c_1}, \dots, [n]^{c_j}$ with common n (which are now adjacent to each other) into a single factor $[n]^{c_1+\cdots+c_j}$. Then, we set $C^{\mathrm{out}}(\ell) := C_0$, and the process completes.
5. Otherwise, pick up any p_1-exchangeable adjacent pair in C_0,

$$C_1^{\mathrm{in}} = [n_1']^{c'}[n_1]^c, \quad \{n_1', n_1\} = p_1 \quad (c', c \in (1/p_1)\mathbb{Z}_{>0}) \tag{5.9}$$

for some integer $p_1 > p_0$, and go to the subroutine (p_1).

(p_1). 1. Set $C_1 = C_1^{\mathrm{in}}$.
2. Decompose every factor $[n]^c$ ($c \in (1/p_1)\mathbb{Z}_{>0}$) in C_1 into a product of a single factor $[n]_{p_1} := [n]^{1/p_1}$.
3. Repeat the same process (3) of (p_0) for C_1, where

   * $[n]$ is replaced with $[n]_{p_1}$.
   * p_0 is replaced with p_1.

   (In particular, the pentagon relation (1.56) with $c = p_1$ is applied to p_1-exchangeable anti-ordered adjacent pairs in C_1.)
4. If the resulting C_1 is ordered, we join all factors $[n]_{p_1}^{c_1}, \dots, [n]_{p_1}^{c_j}$ with common n into a single factor $[n]^{(c_1+\cdots+c_j)/p_1}$. (The case $c_i \neq 1$ may happen by the result of the forthcoming subroutine (p_2).) Then, we set $C_1^{\mathrm{out}} := C_1$, replace C_1^{in} in C_0 with C_1^{out}, and we come back to the process (2) of (p_0). (The "compatibility" of the exponent $(c_1 + \cdots + c_j)/p_1$ in the parental routine (p_0) is the issue we are going to examine.)
5. Otherwise, pick up any p_2-exchangeable adjacent pair

$$C_2^{\mathrm{in}} = [n_2']_{p_1}^{c'}[n_2]_{p_1}^{c} \quad \{n_2', n_2\} = p_2 \quad (c', c \in (p_1/p_2)\mathbb{Z}_{>0}) \tag{5.10}$$

in C_1 for some integer $p_2 \neq p_1$, and go to the subroutine (p_2). (Since n_2' and n_2 are $\mathbb{Z}$-linear combinations of n_1' and n_1, $p_2 > p_1$ and p_2 is a multiple of p_1.)

(p_2). 1–3. Repeat the same process (1)–(3) of (p_1) by replacing p_1 with p_2.
4. If the resulting C_2 is ordered, join all factors $[n]_{p_2}^{c_1}, \dots, [n]_{p_2}^{c_j}$ with common n into a factor $[n]_{p_1}^{(c_1+\cdots+c_j)p_1/p_2}$. Then, we set $C_2^{\mathrm{out}} := C_2$, replace C_2^{in} in C_1 with C_2^{out}, and we come back to the process (2) of

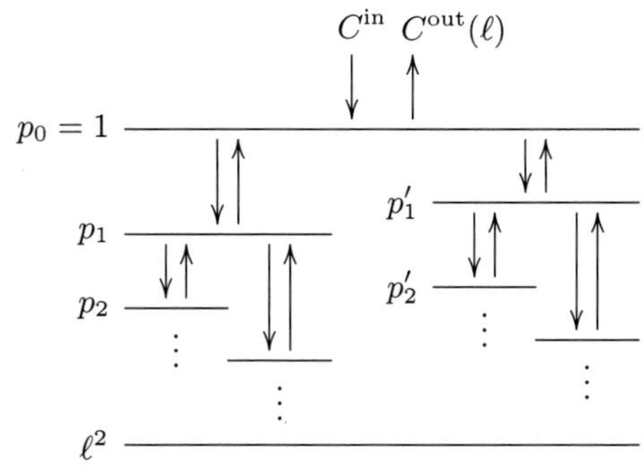

Figure 14: Flow of Ordering Algorithm.

(p_1). (Again, the "compatibility" of the exponent $(c_1+\cdots+c_j)p_1/p_2$ in the parental routine (p_1) is the issue.)

5. Otherwise, go to the next subroutine (p_3) for some $p_3 > p_2$.

See Figure 14 for a schematic diagram of the flow of the algorithm. Note that each sequence of subroutines $p_0 = 1 < p_1 < p_2 < \cdots$ has a finite depth, because $p_i \leq \ell^2$.

We say that the above algorithm *fails* if, at some stage, C_i is not ordered and all remaining anti-ordered adjacent pair therein is not p-exchangeable for any integer $p > 0$. As mentioned in Algorithm 5.7, some fractional power created by a subroutine may cause the failure of the process. For example, in (3.73), if the power of $[3, 2]$ is $1/2$, instead of 1 therein, one cannot carry out the remaining ordering in (3.74), because the pentagon relation with $c = 1$ is not applicable. We are going to show in Lemma 5.10 that such a situation never occurs.

If the algorithm does not fail, then the algorithm completes (successfully) in finite steps by the following reason:

- Due to the structure of the pentagon relation (1.56), all factors created in the subroutine (p_i) are of higher degree than the factors of the initial p_i-exchangeable adjacent pair.
- In particular, for the lowest degree ℓ_0 of the factors in C^{in}, there is no new factor of degree ℓ_0 that is created during the ordering.
- Then, by the induction on ℓ' with $\ell_0 \leq \ell' \leq \ell$, we can show that the number of the factors of degree ℓ' that are created during the ordering has some upper bound.

Let us go back to the situation in Proposition 5.4. For a given C^{in} therein, suppose that the algorithm does not fail for any ℓ. Then, we have

$$C^{\text{in}} \equiv C^{\text{out}}(\ell) \mod G^{>\ell}, \quad \pi_{\ell',\ell}(C^{\text{out}}(\ell')) = C^{\text{out}}(\ell). \tag{5.11}$$

Let $C^{\mathrm{out}} := \lim_{\ell\to\infty} C^{\mathrm{out}}(\ell)$. Then, we have

$$C^{\mathrm{in}} = C^{\mathrm{out}}, \tag{5.12}$$

and C^{out} is a (possibly infinite) ordered product of factors

$$\Psi[t_i n_i]^{s_i \delta(t_i n_i)} \quad (n_i \in N^+_{\mathrm{pr}},\ t_i \in \mathbb{Z}_{>0},\ s_i \in \mathbb{Q}_{>0}), \tag{5.13}$$

where we may assume that all $\Psi[t_i n_i]$ are mutually distinct by joining the factors with common $\Psi[t_i n_i]$. The positivity of s_i follows from the fact that the pentagon relation does not involve negative factors.

To prove that s_i is an integer, we use the following property of the principal x-representation $\tilde{\rho}_x$.

Lemma 5.8. *Under the principal x-representation $\tilde{\rho}_x$, $\Psi[tn]^{s\delta(tn)}$ acts on $\mathbb{Z}[[\tilde{P}]]$ if and only if s is an integer.*

Proof. Recall (see Example 4.11) that the action of $\Psi[tn]^{s\delta(tn)}$ ($n \in N^+_{\mathrm{pr}}$, $s \in \mathbb{Q}$, $t \in \mathbb{Z}_{>0}$) on $\Bbbk[[\tilde{P}]]$ is given by

$$\Psi[tn]^{s\delta(tn)}(x^{\tilde{m}}) = x^{\tilde{m}}(1 + x^{\tilde{p}^*(tn)})^{s\langle \delta(n)n, m\rangle} \quad (\tilde{m} \in \tilde{P}). \tag{5.14}$$

Thus, the if-part is clear. Suppose that s is not an integer. Note that the coefficient of the term $x^{\tilde{m}}x^{\tilde{p}^*(tn)}$ in (5.14) is $c = s\langle \delta(n)n, m\rangle$. Then, it is enough to show that c is not an integer for some $\tilde{m} \in \tilde{P}$. Let $a_0 > 0$ be the smallest positive integer such that $\langle \delta(n)n, m\rangle \in a_0\mathbb{Z}$ for any $\tilde{m} \in \tilde{P}$. By the condition (i) of the definition of $\tilde{P}$ in Section 4.2, $\tilde{P}$ contains a $\mathbb{Z}$-basis of $\tilde{M}^\circ$, for example $\tilde{m} + f_i, \tilde{m} + e_i$ ($i = 1, \dots, r$) for some $\tilde{m} \in \tilde{P}$. Then, by $\mathbb{Z}$-linearity, we have $\langle \delta(n)n, m\rangle \in a_0\mathbb{Z}$ for any $\tilde{m} \in \tilde{M}^\circ$. If $a_0 \neq 1$, we have $\delta(n)n/a_0 \in N^\circ$, which contradicts the definition of $\delta(n)$. Thus, $a_0 = 1$. Then, $c = s$ is not an integer for $\tilde{m} \in \tilde{P}$ attaining the minimum $a_0 = 1$. □

Lemma 5.9. *The factor s_i in (5.13) is an integer.*

Proof. By Lemma 5.8, the product C^{in} acts on $\mathbb{Z}[[\tilde{P}]]$. Then, again by Lemma 5.8, we can prove that all s_i in (5.13) are integers by the induction on $\deg(t_i n_i)$. □

It remains to show that Algorithm 5.7 never fails. This is proved by the same method for Lemma 5.9.

Lemma 5.10. *For any C^{in}, Algorithm 5.7 never fails, so that it completes in finitely many steps.*

Proof. We first note a key observation to prove the lemma.

For the first subroutine (p_1) entered from the main routine (p_0), the following facts hold:

(**i**). The initial product C_1^{in} in (5.9) has the form

$$C_1^{\mathrm{in}} = [t_1 n_1]^{s_1/t_1} [t_2 n_2]^{s_2/t_2} \quad (n_i \in N_{\mathrm{pr}}^+,\ s_i, t_i \in \mathbb{Z}_{>0}). \tag{5.15}$$

(**o**). If the algorithm does not fail, any factor in the final product C_1^{out} also has the form $[tn]^{s/t}$ with the same condition in (5.15).

The property (**o**) follows from the property (**i**) by applying the same proof of Lemma 5.9, where we set $N^\circ = N$, $M^\circ = M$ and consider the action on $\mathbb{Z}[[\tilde{P}]]$.

Similarly, but slightly differently, suppose that a subroutine (p_i) $(i \geq 1)$ starts with the initial product

$$C_i^{\mathrm{in}} = [\tilde{n}_1]_{p_i}^{s_1} [\tilde{n}_2]_{p_i}^{s_2}, \quad \{\tilde{n}_1, \tilde{n}_2\} = p_i \quad (\tilde{n}_i \in N^+,\ s_i \in \mathbb{Z}_{>0}). \tag{5.16}$$

Let

$$N_{p_i}^+ := \{a_1 \tilde{n}_1 + a_2 \tilde{n}_2 \in N^+ \mid a_i \in \mathbb{Z}_{\geq 0},\ a_1 + a_2 > 0\}. \tag{5.17}$$

Let $(N_{p_i}^+)_{\mathrm{pr}}$ be the set of all primitive vectors in $N_{p_i}^+$ with respect to $N_{p_i}^+$ (not with respect to N^+). Note that, for any $n_1, n_2 \in N_{p_i}^+$, $\{n_1, n_2\}$ is a multiple of p_i. Then, for a subroutine (p_{i+1}) $(i \geq 1)$ entered from a subroutine (p_i), the following facts hold:

(**i**). The initial product C_{i+1}^{in} has the form

$$C_{i+1}^{\mathrm{in}} = [t_1 n_1]_{p_i}^{s_1/t_1} [t_2 n_2]_{p_i}^{s_2/t_2} \quad (n_i \in (N_{p_i}^+)_{\mathrm{pr}},\ s_i, t_i \in \mathbb{Z}_{>0}). \tag{5.18}$$

(**o**). If the algorithm does not fail, any factor in the final product C_{i+1}^{out} also has the form $[tn]_{p_i}^{s/t}$ with the same condition in (5.18).

Again, the property (**o**) follows from the property (**i**) by the same argument as above, where we consider the action on $(1/p_i)\mathbb{Z}[[\tilde{P}]]$, and N and M° for $\tilde{P}$ are replaced with the sublattice N_{p_i} generated by $\tilde{n}_1$ and $\tilde{n}_2$ and its dual lattice M_{p_i}, respectively.

Let us prove the lemma together with the property (**i**) in both cases along the flow of the algorithm.

1. The algorithm starts from the main routine (p_0), where the initial product C^{in} certainly has factors of the form in (5.15) by assumption. Also, any factor created by the pentagon relation from a 1-exchangeable pair has the form $[tn]^s$ with $s \in \mathbb{Z}_{>0}$; in particular, it has the form in (5.15). Suppose that we are in the process (5) of the routine (p_0). Take any remaining anti-ordered pair $[t_1 n_1]^{s_1/t_1} [t_2 n_2]^{s_2/t_2}$, and set $p_1 := \{t_1 n_1, t_2 n_2\}$. Then, by Example 5.6, it is p_1-exchangeable. Thus, we go to the subroutine (p_1) with C_1^{in} satisfying the property (**i**).

2. In the subroutine (p_1) we decompose $C_1^{\rm in}$ into integer powers of $[n]_{p_1}$, and after ordering all p_1-exchangeable pairs by the pentagon relation, we only obtain integer powers of $[n]_{p_1}$. Thus, if we have an anti-ordered pair in the process (5) of the routine (p_1), it has the form in (5.18) with $\{t_1n_1, t_2n_2\} = p_2 > p_1$ and s_i/t_i are integers. In particular, the pair is p_2-exchangeable, and we go to the next subroutine (p_2) with $C_2^{\rm in}$ satisfying the property (**i**). (So far, we do not encounter a serious problem.)
3. The process continues until we get to the deepest subroutine (p_k). Then, by definition, (p_k) returns the ordered product $C_k^{\rm out}$ satisfying the property (**o**) to the subroutine (p_{k-1}). We claim that the subroutine (p_{k-1}) continues to run. (The following is the key argument.) In fact, suppose that we still have any anti-ordered pair in C_{k-1}. Then, it has the form
$$[t_1n_1]_{p_{k-1}}^{s_1/t_1}[t_2n_2]_{p_{k-1}}^{s_2/t_2} \quad (n_i \in (N^+_{p_{k-1}})_{\rm pr},\ s_i, t_i \in \mathbb{Z}_{>0}) \tag{5.19}$$
thanks to the property (**o**) just we had. Let $p'_k = \{t_1n_1, t_2n_2\} > 0$. Since $\{n_1, n_2\}$ is a multiple of p_{k-1} as already mentioned, p'_k is a multiple of both t_1p_{k-1} and t_2p_{k-1}. Thus, the pair is p'_k-exchangeable. If $p'_k = p_{k-1}$, we stay in the subroutine (p_{k-1}). If $p'_k > p_{k-1}$, we go to another subroutine (p'_k) again with $C_k^{\rm in}$ satisfying the property (**i**). The same story continues.

Thus, the algorithm never fails. □

By Lemmas 5.9 and 5.10, we obtain Proposition 5.4. Moreover, as a special case of Proposition 5.4, we obtain Theorem 5.2 in the rank 2 case.

Corollary 5.11. *Theorem 5.2 holds for* rank $N = 2$.

Proof. Apply Proposition 5.4 to the product $C^{\rm in} = \Psi[e_2]^{\delta_2}\Psi[e_1]^{\delta_1}$. □

Example 5.12. It is not difficult to translate Algorithm 5.7 into a computer program. The source code is given in Appendix A for the reader's convenience. Let us exhibit some examples of non-affine infinite type calculated by computer. Under the same assumption in Section 3.5, we have the following results modulo $G^{>7}$:

$$\begin{aligned}\begin{bmatrix}0\\1\end{bmatrix}^5\begin{bmatrix}1\\0\end{bmatrix}^1 &\equiv \begin{bmatrix}1\\0\end{bmatrix}^1\begin{bmatrix}1\\1\end{bmatrix}^5\begin{bmatrix}3\\4\end{bmatrix}^5\left(\begin{bmatrix}2\\3\end{bmatrix}^{10}\begin{bmatrix}1\\2\end{bmatrix}^{10}\begin{bmatrix}2\\4\end{bmatrix}^{10}\begin{bmatrix}2\\5\end{bmatrix}^{27}\begin{bmatrix}1\\3\end{bmatrix}^{10}\right)\\ &\quad\times\begin{bmatrix}1\\4\end{bmatrix}^5\begin{bmatrix}1\\5\end{bmatrix}^1\begin{bmatrix}0\\1\end{bmatrix}^5,\end{aligned} \tag{5.20}$$

$$\begin{aligned}\begin{bmatrix}0\\1\end{bmatrix}^6\begin{bmatrix}1\\0\end{bmatrix}^1 \equiv{}& \begin{bmatrix}1\\0\end{bmatrix}^1\begin{bmatrix}1\\1\end{bmatrix}^6\left(\begin{bmatrix}3\\4\end{bmatrix}^{15}\begin{bmatrix}2\\3\end{bmatrix}^{20}\begin{bmatrix}1\\2\end{bmatrix}^{15}\begin{bmatrix}2\\4\end{bmatrix}^{30}\begin{bmatrix}2\\5\end{bmatrix}^{102}\begin{bmatrix}1\\3\end{bmatrix}^{20}\begin{bmatrix}1\\4\end{bmatrix}^{15}\right)\\ &\times\begin{bmatrix}1\\5\end{bmatrix}^6\begin{bmatrix}1\\6\end{bmatrix}^1\begin{bmatrix}0\\1\end{bmatrix}^6,\end{aligned} \tag{5.21}$$

$$\begin{aligned}\begin{bmatrix}0\\1\end{bmatrix}^3\begin{bmatrix}1\\0\end{bmatrix}^2 \equiv{}& \begin{bmatrix}1\\0\end{bmatrix}^2\begin{bmatrix}2\\1\end{bmatrix}^3\left(\begin{bmatrix}3\\2\end{bmatrix}^{6}\begin{bmatrix}4\\3\end{bmatrix}^{14}\begin{bmatrix}1\\1\end{bmatrix}^{6}\begin{bmatrix}2\\2\end{bmatrix}^{6}\begin{bmatrix}3\\3\end{bmatrix}^{6}\right.\\ &\times\left.\begin{bmatrix}3\\4\end{bmatrix}^{36}\begin{bmatrix}2\\3\end{bmatrix}^{14}\begin{bmatrix}1\\2\end{bmatrix}^{6}\begin{bmatrix}2\\4\end{bmatrix}^{6}\right)\begin{bmatrix}2\\5\end{bmatrix}^3\begin{bmatrix}1\\3\end{bmatrix}^2\begin{bmatrix}0\\1\end{bmatrix}^3,\end{aligned} \tag{5.22}$$

$$\begin{aligned}\begin{bmatrix}0\\1\end{bmatrix}^4\begin{bmatrix}1\\0\end{bmatrix}^2 \equiv{}& \begin{bmatrix}1\\0\end{bmatrix}^2\begin{bmatrix}2\\1\end{bmatrix}^4\left(\begin{bmatrix}3\\2\end{bmatrix}^{12}\begin{bmatrix}4\\3\end{bmatrix}^{44}\begin{bmatrix}1\\1\end{bmatrix}^{8}\begin{bmatrix}2\\2\end{bmatrix}^{12}\begin{bmatrix}3\\3\end{bmatrix}^{24}\right.\\ &\times\left.\begin{bmatrix}3\\4\end{bmatrix}^{182}\begin{bmatrix}2\\3\end{bmatrix}^{44}\begin{bmatrix}1\\2\end{bmatrix}^{12}\begin{bmatrix}2\\4\end{bmatrix}^{40}\begin{bmatrix}2\\5\end{bmatrix}^{44}\begin{bmatrix}1\\3\end{bmatrix}^{8}\right)\begin{bmatrix}1\\4\end{bmatrix}^2\begin{bmatrix}0\\1\end{bmatrix}^4,\end{aligned} \tag{5.23}$$

$$\begin{aligned}\begin{bmatrix}0\\1\end{bmatrix}^3\begin{bmatrix}1\\0\end{bmatrix}^3 \equiv{}& \begin{bmatrix}1\\0\end{bmatrix}^3\begin{bmatrix}3\\1\end{bmatrix}^3\left(\begin{bmatrix}5\\2\end{bmatrix}^{9}\begin{bmatrix}2\\1\end{bmatrix}^{9}\begin{bmatrix}4\\2\end{bmatrix}^{18}\begin{bmatrix}3\\2\end{bmatrix}^{39}\begin{bmatrix}4\\3\end{bmatrix}^{204}\begin{bmatrix}1\\1\end{bmatrix}^{9}\begin{bmatrix}2\\2\end{bmatrix}^{18}\right.\\ &\times\left.\begin{bmatrix}3\\3\end{bmatrix}^{54}\begin{bmatrix}3\\4\end{bmatrix}^{204}\begin{bmatrix}2\\3\end{bmatrix}^{39}\begin{bmatrix}1\\2\end{bmatrix}^{9}\begin{bmatrix}2\\4\end{bmatrix}^{18}\begin{bmatrix}2\\5\end{bmatrix}^{9}\right)\begin{bmatrix}1\\3\end{bmatrix}^3\begin{bmatrix}0\\1\end{bmatrix}^3.\end{aligned} \tag{5.24}$$

Let us examine the behavior in the Badlands $\sigma(\mathbf{v}_+, \mathbf{v}_-)$ defined in (3.88). The condition (3.89) for a normal vector $\mathbf{n} = (n_1, n_2)$ is given as follows:

$$(\delta_1, \delta_2) = (1,5), \quad \frac{2}{5+\sqrt{5}} = 0.276\ldots \le \frac{n_1}{n_2} \le \frac{2}{5-\sqrt{5}} = 0.723\ldots, \tag{5.25}$$

$$(\delta_1, \delta_2) = (1,6), \quad \frac{1}{3+\sqrt{3}} = 0.211\ldots \le \frac{n_1}{n_2} \le \frac{1}{3-\sqrt{3}} = 0.788\ldots, \tag{5.26}$$

$$(\delta_1, \delta_2) = (2,3), \quad \frac{2}{3+\sqrt{3}} = 0.422\ldots \le \frac{n_1}{n_2} \le \frac{2}{3-\sqrt{3}} = 1.577\ldots, \tag{5.27}$$

$$(\delta_1, \delta_2) = (2,4), \quad \frac{1}{2+\sqrt{2}} = 0.292\ldots \le \frac{n_1}{n_2} \le \frac{1}{2-\sqrt{2}} = 1.707\ldots, \tag{5.28}$$

$$(\delta_1, \delta_2) = (3,3), \quad \frac{2}{3+\sqrt{5}} = 0.381\ldots \le \frac{n_1}{n_2} \le \frac{2}{3-\sqrt{5}} = 2.618\ldots. \tag{5.29}$$

The corresponding elements $\Psi[n]$ are inside the parentheses in (5.20)–(5.24). We observe that, in each case, *every* $n \in N^+$ with $\deg(n) \le 7$ satisfying the above condition appears in the product. This agrees with (and even stronger than) the expectation of [GHKK18] explained in Section 3.5. We expect this is true up to any degree, and we do not have any counterexample so far.

Let us also present an example of the ordering of a more general anti-ordered product in the left hand side. We have, modulo $G^{>9}$,

$$
\begin{aligned}
&\begin{bmatrix}0\\1\end{bmatrix}^2\begin{bmatrix}2\\4\end{bmatrix}^{1/2}\begin{bmatrix}3\\3\end{bmatrix}^{2/3}\begin{bmatrix}1\\0\end{bmatrix}^2\\
\equiv&\begin{bmatrix}1\\0\end{bmatrix}^2\begin{bmatrix}2\\1\end{bmatrix}^2\begin{bmatrix}6\\3\end{bmatrix}^{40/3}\begin{bmatrix}5\\3\end{bmatrix}^{10}\begin{bmatrix}3\\2\end{bmatrix}^2\begin{bmatrix}4\\3\end{bmatrix}^6\begin{bmatrix}5\\4\end{bmatrix}^{146}\begin{bmatrix}1\\1\end{bmatrix}^4\begin{bmatrix}2\\2\end{bmatrix}^2\begin{bmatrix}3\\3\end{bmatrix}^{2/3}\begin{bmatrix}4\\4\end{bmatrix}^{47}\begin{bmatrix}4\\5\end{bmatrix}^{230}\\
&\times\begin{bmatrix}3\\4\end{bmatrix}^{10}\begin{bmatrix}2\\3\end{bmatrix}^2\begin{bmatrix}3\\5\end{bmatrix}^{34}\begin{bmatrix}1\\2\end{bmatrix}^2\begin{bmatrix}2\\4\end{bmatrix}^{1/2}\begin{bmatrix}3\\6\end{bmatrix}^{220/3}\begin{bmatrix}2\\5\end{bmatrix}^2\begin{bmatrix}2\\6\end{bmatrix}^3\begin{bmatrix}2\\7\end{bmatrix}^2\begin{bmatrix}0\\1\end{bmatrix}^2.
\end{aligned}
\tag{5.30}
$$

As we see in these examples, some exponents of $\Psi[n]$ quickly increases when $\deg(n)$ increases. In fact, they easily exceed 1,000 when the degree is around 10 even for these tiny examples. This puts a limitation of the explicit computation of higher degrees even with computer, unfortunately.

5.3 Alternative construction of CSDs

Here we present an alternative construction of CSDs based on Proposition 5.4 (Ordering lemma). The resulting CSD is a positive realization in Theorem 5.2.

Let us introduce the following notion following [GHKK18].

Definition 5.13 (Joint). Let $\mathfrak{D}$ be a scattering diagram.

(a). For any pair of walls $\mathbf{w}_i = (\mathfrak{d}_i, g_i)_{n_i}$ $(i = 1, 2)$, the intersection of their supports $\mathfrak{j} = \mathfrak{d}_1 \cap \mathfrak{d}_2$ is called a *joint* of $\mathfrak{D}$ if $\mathfrak{j}$ is a cone of codimension 2.

(b) For any joint $\mathfrak{j}$, let

$$N_{\mathfrak{j}} := \{n \in N \mid \langle n, z\rangle = 0 \ (z \in \mathfrak{j})\}. \tag{5.31}$$

Then, a joint $\mathfrak{j}$ is *parallel* (resp. *perpendicular*) if the skew-symmetric form $\{\cdot,\cdot\}$ restricted on $N_{\mathfrak{j}}$ vanishes (resp. otherwise). Thus, the joint $\mathfrak{j}$ in (a) is parallel if and only if $\{n_1, n_2\} = 0$.

By the topological reason, any consistency relation of $\mathfrak{D}$ reduces to the consistency relations for admissible loops around joints.

For a given seed $\mathfrak{s}$, we will inductively construct scattering diagrams $\mathfrak{D}_\ell$ $(\ell = 1, 2, \dots)$ satisfying the following properties:

(a). $\mathfrak{D}_1 = \mathrm{In}_{\mathfrak{s}}$, where $\mathrm{In}_{\mathfrak{s}}$ is the set of incoming walls in (3.55).

(b). $\mathfrak{D}_1 \subset \mathfrak{D}_2 \subset \cdots$, and all walls of $\mathfrak{D}_{\ell+1} \setminus \mathfrak{D}_\ell$ are outgoing walls and have the form in (5.4) with $\deg(tn) = \ell + 1$.

(c). Let $\mathfrak{j}$ be any perpendicular joint of $\mathfrak{D}_\ell$, and let $\gamma_{\mathfrak{j}}$ be any sufficiently small loop around $\mathfrak{j}$ such that it only crosses the walls of $\mathfrak{D}_\ell$ containing

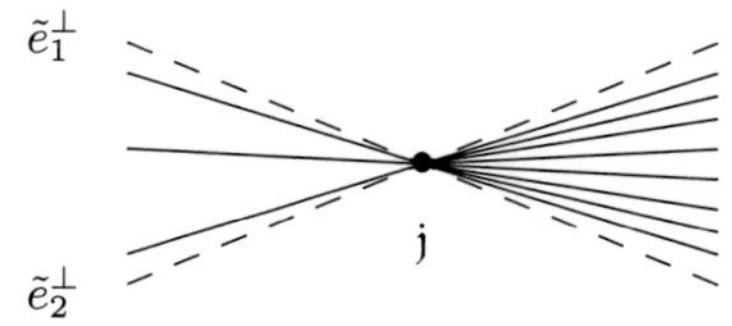

Figure 15: Walls containing a perpendicular joint $\mathfrak{j}$.

$\mathfrak{j}$. Then, we have

$$\mathfrak{p}_{\gamma_\mathfrak{j},\mathfrak{D}_\ell} \equiv \mathrm{id} \mod G^{>\ell}. \tag{5.32}$$

Note that we do not directly impose the condition (5.32) for parallel joints.

Construction 5.14 (Cf. [GHKK18, Construction C.1, Appendix C.3]).

(1). We start with $\mathfrak{D}_1 = \mathrm{In}_\mathfrak{s}$.

(2). Next, we describe the construction of $\mathfrak{D}_2$ from $\mathfrak{D}_1$ to illustrate the idea for the general case (3). For any *perpendicular* joint $\mathfrak{j} = e_i^\perp \cap e_j^\perp$ of $\mathfrak{D}_1$, where $\{e_i, e_j\} \neq 0$, let $N_\mathfrak{j}$ be the one in (5.31), which is the rank 2 sublattice of N generated by e_i and e_j. By exchanging i and j if necessary, we may assume that that $c = \{e_j, e_i\} > 0$. Then, for any $n = n_i e_i + n_j e_j \in N_\mathfrak{j}^+ := N^+ \cap N_\mathfrak{j}$, we have

$$\langle e_i, p^*(n)\rangle = \{e_i, n\} \leq 0, \quad \langle e_j, p^*(n)\rangle = \{e_j, n\} \geq 0. \tag{5.33}$$

Thus, $p^*(n)$ is in the "second quadrant" in $M_\mathbb{R}$ with respect to $e_i^\perp$ and $e_j^\perp$. Now, we apply Proposition 5.4 to the anti-ordered product $C^{\mathrm{in}} = \Psi[e_j]^{\delta_j}\Psi[e_i]^{\delta_i}$ modulo $G^{>2}$, and obtain the ordered product

$$\Psi[e_i]^{\delta_i}\Psi[e_i+e_j]^{c\delta_i\delta_j}\Psi[e_j]^{\delta_j}. \tag{5.34}$$

Then, we add a new wall

$$(\sigma(\mathfrak{j}, -p^*(e_i+e_j)), \Psi[e_i+e_j]^{c\delta_i\delta_j})_{e_i+e_j} \tag{5.35}$$

to $\mathfrak{D}_1$, where $\sigma(\mathfrak{j}, -p^*(e_i+e_j))$ is the cone in $M_\mathbb{R}$ spanned by $\mathfrak{j}$ and $-p^*(e_i+e_j)$. The perpendicular condition implies that $p^*(e_i+e_j) \notin \mathfrak{j}$. Thus, the cone $\sigma(\mathfrak{j}, -p^*(e_i+e_j))$ is of codimension 1. Also, it is outgoing because it is in the "fourth quadrant" in $M_\mathbb{R}$ with respect to $e_i^\perp$ and $e_j^\perp$. We do the procedure for all perpendicular joints of $\mathfrak{D}_1$ to obtain $\mathfrak{D}_2$.

(3). Generalizing the procedure in (2), we construct $\mathfrak{D}_{\ell+1}$ from $\mathfrak{D}_\ell$ as follows. By decomposing the supports of walls of $\mathfrak{D}_\ell$ if necessary, we may assume that any pair of joints $\mathfrak{j}_1 \neq \mathfrak{j}_2$ may intersect each other only in their boundaries. For any *perpendicular* joint $\mathfrak{j}$ of $\mathfrak{D}_\ell$, let $N_\mathfrak{j}$ be the one in (5.31). Let $N_\mathfrak{j}^+ := N^+ \cap N_\mathfrak{j}$ and $N_{\mathfrak{j},\mathrm{pr}}^+ := N_{\mathrm{pr}}^+ \cap N_\mathfrak{j}$. Then, there exist a unique pair $\tilde{e}_1, \tilde{e}_2 \in N_{\mathfrak{j},\mathrm{pr}}^+$ such that

$$N_\mathfrak{j}^+ \subset \mathbb{Q}_{\geq 0}\tilde{e}_1 + \mathbb{Q}_{\geq 0}\tilde{e}_2, \quad \{\tilde{e}_2, \tilde{e}_1\} > 0. \tag{5.36}$$

Then, for any $n \in N_{\mathfrak{j}}^{+}$,

$$\langle \tilde{e}_1, p^*(n) \rangle = \{\tilde{e}_1, n\} \leq 0, \quad \langle \tilde{e}_2, p^*(n) \rangle = \{\tilde{e}_2, n\} \geq 0. \tag{5.37}$$

See Figure 15. Thus, $p^*(n)$ is in the "second quadrant" in $M_{\mathbb{R}}$ with respect to $\tilde{e}_1^{\perp}$ and $\tilde{e}_2^{\perp}$. Then, by the construction of walls as described below, all wall elements in $\mathfrak{D}_\ell$ has the form in (5.4). We take the wall elements of all walls that contain $\mathfrak{j}$ and intersect the second quadrant with respect to $\tilde{e}_1^{\perp}$ and $\tilde{e}_2^{\perp}$. Then, we consider the anti-ordered product of them

$$C^{\mathrm{in}} = \Psi[t'_{j'} n'_{j'}]^{s'_{j'} \delta(t'_{j'} n'_{j'})} \cdots \Psi[t'_1 n'_1]^{s'_1 \delta(t'_1 n'_1)} \quad (\deg(t'_i n'_i) \leq \ell), \tag{5.38}$$

where $n'_i \in N_{\mathfrak{j},\mathrm{pr}}^{+}$ and $s'_i, t'_i \in \mathbb{Z}_{>0}$. Then, we apply Proposition 5.4 to C^{in} modulo $G^{>\ell+1}$, and obtain a finite ordered product

$$C^{\mathrm{out}}(\ell+1) = \Psi[t_j n_j]^{s_j \delta(t_j n_j)} \cdots \Psi[t_1 n_1]^{s_1 \delta(t_1 n_1)} \quad (\deg(t_i n_i) \leq \ell+1), \tag{5.39}$$

such that $C^{\mathrm{in}} \equiv C^{\mathrm{out}}(\ell+1) \bmod G^{>\ell+1}$. By the construction of walls which we describe below, all factors in (5.39) with $\deg(t_i n_i) \leq \ell$ already appear as wall elements in $\mathfrak{D}_\ell$. Now, we add new walls

$$(\sigma(\mathfrak{j}, -p^*(n_i)), \Psi[t_i n_i]^{s_i \delta(t_i n_i)})_{n_i} \tag{5.40}$$

to $\mathfrak{D}_\ell$ for all factors in (5.39) with $\deg(t_i n_i) = \ell + 1$. By the perpendicular condition, the cone $\sigma(\mathfrak{j}, -p^*(n_i))$ is of codimension 1. Also, it is outgoing because it is in the "fourth quadrant" with respect to $\tilde{e}_1^{\perp}$ and $\tilde{e}_2^{\perp}$. We call such a wall in (5.40) *a wall attached to* $\mathfrak{j}$. We do the procedure for all perpendicular joints of $\mathfrak{D}_\ell$ to obtain $\mathfrak{D}_{\ell+1}$.

Next we prove the condition (c) for $\mathfrak{D}_{\ell+1}$ (not by the induction on ℓ). It is useful to note the following property.

Lemma 5.15. *Let* $\mathbf{w}$ *be any wall of* $\mathfrak{D}_{\ell+1} \setminus \mathfrak{D}_\ell$ *attached to a joint* $\mathfrak{j}$ *in* $\mathfrak{D}_\ell$ *in* (5.40). *Let* $\mathbf{w}'$ *be any wall of* $\mathfrak{D}_{\ell+1}$ *other than* $\mathbf{w}$. *Let* $\mathfrak{d}$ *and* $\mathfrak{d}'$ *be the supports of* $\mathbf{w}$ *and* $\mathbf{w}'$, *respectively. Suppose that* $\mathfrak{j}' = \mathfrak{d} \cap \mathfrak{d}'$ *is a joint of* $\mathfrak{D}_{\ell+1}$ *such that* $\mathfrak{j}' \not\subset \mathfrak{j}$.

(a). If $\mathfrak{j}' \subset \partial\mathfrak{d}$, *then* $\mathfrak{j}'$ *is parallel.*

(b). If $\mathfrak{j}'$ *is perpendicular, the wall* $\mathbf{w}$ *does not contribute to* $\mathfrak{p}_{\gamma_{\mathfrak{j}'}, \mathfrak{D}_{\ell+1}}$ *modulo* $G^{>\ell+1}$

Proof. (a). Let $\mathfrak{d} = \sigma(\mathfrak{j}, -p^*(n))$ for some $n \in N_{\mathfrak{j},\mathrm{pr}}^{+}$. Then, by the assumptions that $\mathfrak{j}' \not\subset \mathfrak{j}$ and $\mathfrak{j}' \subset \partial\mathfrak{d}$, the joint $\mathfrak{j}' = \mathfrak{d} \cap \mathfrak{d}'$ is contained in a vector space of codimension 2 that contains $p^*(n)$. Thus, $p^*(n) \in (n')^{\perp}$, where $n' \in N_{\mathrm{pr}}^{+}$ is the normal vector for $\mathbf{w}'$. It follows that $0 = \langle n', p^*(n) \rangle = \{n', n\}$. Therefore, $\mathfrak{j}'$ is parallel.

(b). By (a), we have $\mathfrak{j}' \not\subset \partial\mathfrak{d}$. Then, $\gamma_{\mathfrak{j}'}$ crosses $\mathfrak{d}$ twice with opposite intersection signs. Since the wall element of $\mathbf{w}$ is in the center of $G^{\leq \ell+1}$, two contributions of the wall $\mathbf{w}$ to $\mathfrak{p}_{\gamma_{\mathfrak{j}'},\mathfrak{D}_{\ell+1}}$ cancel modulo $G^{>\ell+1}$. □

Let $\mathfrak{j}'$ be any perpendicular joint of $\mathfrak{D}_{\ell+1}$. There are two cases:

- Suppose that $\mathfrak{j}' \not\subset \mathfrak{j}$ for any perpendicular joint $\mathfrak{j}$ of $\mathfrak{D}_\ell$. Then, $\mathfrak{j}'$ is a new joint created by some added wall to $\mathfrak{D}_\ell$. Then, we have $\mathfrak{p}_{\gamma_{\mathfrak{j}'},\mathfrak{D}_{\ell+1}} \equiv \mathrm{id} \mod G^{>\ell+1}$ by Lemma 5.15 (b).
- Suppose that $\mathfrak{j}' \subset \mathfrak{j}$ for some perpendicular joint $\mathfrak{j}$ of $\mathfrak{D}_\ell$. Again by Lemma 5.15 (b), one can ignore the contribution to $\mathfrak{p}_{\gamma_{\mathfrak{j}'},\mathfrak{D}_{\ell+1}}$ modulo $G^{>\ell+1}$ from the added walls other than the ones attached to $\mathfrak{j}$. Then, by the construction of the added walls attached to $\mathfrak{j}$ we have $\mathfrak{p}_{\gamma_{\mathfrak{j}'},\mathfrak{D}_{\ell+1}} \equiv \mathfrak{p}_{\gamma_{\mathfrak{j}},\mathfrak{D}_{\ell+1}} \equiv \mathrm{id} \mod G^{>\ell+1}$.

Thus, the condition (c) holds for $\mathfrak{D}_{\ell+1}$.

Using the diagrams $\mathfrak{D}_1 \subset \mathfrak{D}_2 \subset \cdots$ constructed above, we define

$$\mathfrak{D} = \bigcup_{\ell=1}^{\infty} \mathfrak{D}_\ell, \tag{5.41}$$

whose reduction at ℓ is $\mathfrak{D}_\ell$. The scattering diagram $\mathfrak{D}$ has the following properties:

(a). $\mathfrak{D}_{\mathrm{in}} = \mathrm{In}_{\mathfrak{s}}$.

(b). All outgoing walls of $\mathfrak{D}$ have the form in (5.35) with $\delta(tn) > 1$.

(c). For any perpendicular joint $\mathfrak{j}$ of $\mathfrak{D}_\ell$, $\mathfrak{p}_{\gamma_{\mathfrak{j}},\mathfrak{D}_\ell} \equiv \mathrm{id} \mod G^{>\ell}$ holds.

However, we have not yet shown that $\mathfrak{D}$ is consistent, because the consistency around parallel joints is not examined.

The proof of Theorem 5.2 completes by showing the following lemma.

Lemma 5.16 ([GHKK18, Lemma C.7]). *Let $\mathfrak{D}$ be the one constructed above. Let $\mathfrak{D}_{\mathfrak{s}}$ be a CSD in Theorem 3.12. Then, $\mathfrak{D}$ and $\mathfrak{D}_{\mathfrak{s}}$ are equivalent. In particular, $\mathfrak{D}$ is consistent and it is a CSD for $\mathfrak{s}$.*

Proof. We note that any CSD $\mathfrak{D}_{\mathfrak{s}}$ also satisfies the property (a) by definition and the property (c) by the consistency. Let $\mathfrak{C} = \mathfrak{D}_{\mathfrak{s}} - \mathfrak{D}$ be a scattering diagram obtained as follows. Take all walls of $\mathfrak{D}$, invert all wall elements, add all walls of $\mathfrak{D}_{\mathfrak{s}}$, and eliminate all incoming walls because they cancel each other. In particular, $\mathfrak{D}_{\mathfrak{s}}$ is equivalent to $\mathfrak{D} \sqcup \mathfrak{C}$ and $\mathfrak{C}$ has no incoming wall.

We show that $\mathfrak{C}$ is trivial in the sense of Definition 2.12. Let $\mathfrak{C}_\ell$ be the reduction of $\mathfrak{C}$ at ℓ. Then, we prove the following equivalent claim:

Claim. For any general $z \in M_{\mathbb{R}}$, we have

$$g_z(\mathfrak{C}_\ell) \equiv \mathrm{id} \mod G^{>\ell}. \tag{5.42}$$

We show the claim by the induction on ℓ. The claim is true for $\ell = 1$, because $(\mathfrak{D}_{\mathfrak{s}})_1 = \mathfrak{D}_1 = \mathrm{In}(\mathfrak{s})$, so that $\mathfrak{C}_1 = \emptyset$. Now assume that Claim holds for ℓ. Take any perpendicular joint $\mathfrak{j}$ of $\mathfrak{D}_{\ell+1}$. By the induction assumption, for any general $z \in M_{\mathbb{R}}$, $g_z(\mathfrak{C}_{\ell+1})$ is central in $G^{\leq \ell+1}$. Thus, we have

$$\mathfrak{p}_{\gamma_{\mathfrak{j}},\mathfrak{C}_{\ell+1}} \equiv \mathfrak{p}_{\gamma_{\mathfrak{j}},(\mathfrak{D}_{\mathfrak{s}})_{\ell+1}} \mathfrak{p}^{-1}_{\gamma_{\mathfrak{j}},\mathfrak{D}_{\ell+1}} \mod G^{>\ell+1}. \tag{5.43}$$

On the other hand, by the consistency of $\mathfrak{D}_{\mathfrak{s}}$, we have $\mathfrak{p}_{\gamma_{\mathfrak{j}},(\mathfrak{D}_{\mathfrak{s}})_{\ell+1}} \equiv \mathrm{id} \mod G^{>\ell+1}$. Also, by the property (c) of $\mathfrak{D}$, we have $\mathfrak{p}_{\gamma_{\mathfrak{j}},\mathfrak{D}_{\ell+1}} \equiv \mathrm{id} \mod G^{>\ell+1}$. It follows that

$$\mathfrak{p}_{\gamma_{\mathfrak{j}},\mathfrak{C}_{\ell+1}} \equiv \mathrm{id} \mod G^{>\ell+1}. \tag{5.44}$$

Take any $n \in N^+_{\mathrm{pr}}$ such that $\mathfrak{j} \subset n^{\perp}$. Suppose that $\gamma_{\mathfrak{j}}$ intersects $n^{\perp}$ at only two general z_1 and z_2, and that they locate in the opposite side of $\mathfrak{j}$ on $n^{\perp}$ each other. Then, the congruence (5.44) implies that $g_{z_1}(\mathfrak{C}) \equiv g_{z_2}(\mathfrak{C}) \mod G^{>\ell+1}$. Thus, $n^{\perp}$ is decomposed into walls only by parallel joints so that $g_z(\mathfrak{C})$ is constant mod $G^{>\ell+1}$ on each wall. Then, each wall should contain the line $\mathbb{R}p^*(n)$. Thus, if $g_z(\mathfrak{C}_\ell) \not\equiv \mathrm{id} \mod G^{>\ell+1}$ on some wall, the wall is incoming. This is a contradiction. Therefore, Claim holds for $\ell + 1$. □

The above construction of a CSD also reveals the following fundamental result on the structure of a CSD, which is implicit in [GHKK18].

Theorem 5.17 (cf. [GHKK18, Appendix C.3]). *For any admissible loop γ for $\mathfrak{D}_{\mathfrak{s}}$, the consistency relation $\mathfrak{p}_{\gamma,\mathfrak{D}_{\mathfrak{s}}} = \mathrm{id}$ is reduced to a trivial one (i.e., $a = a$) by applying the relations in Proposition 1.14 possibly infinitely many times.*

Proof. Consider the reduction $\mathfrak{D}_{\mathfrak{s},\ell+1}$ at $\ell + 1$. We may concentrate on the consistency relation for an admissible loop around a joint in $\mathfrak{D}_{\mathfrak{s},\ell+1}$. For a parallel joint γ, the consistency relation around γ has the form

$$\begin{aligned} &\Psi[t'_{j'}n'_{j'}]^{s'_{j'}\delta(t'_{j'}n'_{j'})} \cdots \Psi[t'_1n'_1]^{s'_1\delta(t'_1n'_1)} \\ &= \Psi[t'_1n'_1]^{s'_1\delta(t'_1n'_1)} \cdots \Psi[t'_{j'}n'_{j'}]^{s'_{j'}\delta(t'_{j'}n'_{j'})}, \end{aligned} \tag{5.45}$$

where the left hand side is an anti-ordered product. By applying (1.55), it reduces to a trivial relation. For a perpendicular joint γ, the consistency relation around γ has the form

$$\begin{aligned} &\Psi[t'_{j'}n'_{j'}]^{s'_{j'}\delta(t'_{j'}n'_{j'})} \cdots \Psi[t'_1n'_1]^{s'_1\delta(t'_1n'_1)} \\ &\equiv \Psi[t_jn_j]^{s_j\delta(t_jn_j)} \cdots \Psi[t_1n_1]^{s_1\delta(t_1n_1)} \mod G^{>\ell+1}, \end{aligned} \tag{5.46}$$

where the left and right hand sides are the ones in (5.38) and (5.39), respectively. The right hand side was constructed from the left hand side by

Algorithm 5.7, which only depends on the relations (1.55) and (1.56). Thus, by applying the relations in the reverse way, the relation (5.46) is reduced to a trivial one. □

In summary, a CSD consists only of the dilogarithm elements of the form $\Psi[tn]^{s\delta(tn)}$ $(s \in \mathbb{Z}_{>0})$, or $\Psi[tn]^{\delta(tn)}$ by decomposing them, and its consistency relations are generated by the pentagon relation.

We also record the following properties of the CSD constructed here.

Proposition 5.18. *The CSD $\mathfrak{D}_{\mathfrak{s}}$ constructed in Construction 5.14 is with minimal support. Also, the normal vector of any outgoing wall is not e_i $(i = 1, \dots, r)$.*

Proof. The wall elements in the form (5.4) never cancel each other due to the positivity of s. Thus, for any general z on a wall of $\mathfrak{D}_{\mathfrak{s}}$, $g_z(\mathfrak{D}_{\mathfrak{s}}) \neq \mathrm{id}$ holds. Also, any outgoing wall of $\mathfrak{D}_{\mathfrak{s}}$ created in (5.39) has the normal vector other than e_i due to Algorithm 5.7. (This is true also by Proposition 3.19.) □

5.4 Rank 3 examples: finite type

Let us demonstrate how Construction 5.14 really works for rank 3 CSDs. To represent rank 3 CSDs, we use the projection of $\mathrm{Supp}(\mathfrak{D}_{\mathfrak{s}})$ to the unit 2-sphere S^2 in $M_{\mathbb{R}} \simeq \mathbb{R}^2$, then apply the standard stereographic projection of S^2 to the plane $\mathbb{R}^2$ centered at the point $(1/\sqrt{3})(1,1,1) \in S^2$. Then, the supports of incoming walls (resp. outgoing walls) in $\mathfrak{D}_{\mathfrak{s}}$ are represented by circles (resp. arcs), respectively. Below, figures are only topological, that is, without caring the precise metrics.

Here we concentrate on rank 3 CSDs of *finite type*.

(a). Type A_3. Let B be the matrix in (1.4). Consider the case

$$B = \begin{pmatrix} 0 & -1 & 0 \\ 1 & 0 & -1 \\ 0 & 1 & 0 \end{pmatrix}, \quad \delta_1 = \delta_2 = \delta_3 = 1. \tag{5.47}$$

This is the simplest nontrivial example of a rank 3 CSD, where all perpendicular joints are of type A_2 in Section 2.2. We follow the procedure of Construction 5.14 step by step. See Figure 16, where only perpendicular joints are marked by $\bullet$ in figures.

At degree 1, there are three incoming walls $\mathbf{w}_{e_i}$ $(i = 1, 2, 3)$ as the initial input. Correspondingly, we have three circles representing their supports $\mathbf{e}_i^{\perp}$. Note that the inside and the outside of each circle $\mathbf{e}_i^{\perp}$ corresponds to the half spaces

$$\mathcal{H}_i^{>0} := \{z \in M_{\mathbb{R}} \mid \langle e_i, z\rangle > 0\}, \quad \mathcal{H}_i^{<0} := \{z \in M_{\mathbb{R}} \mid \langle e_i, z\rangle < 0\}, \tag{5.48}$$

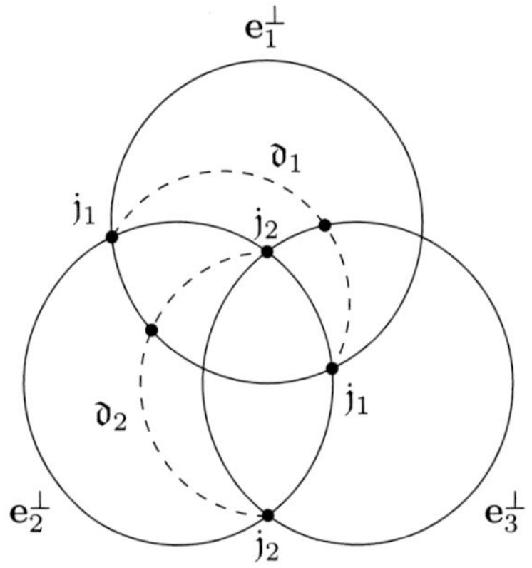

(a) walls of degree 2

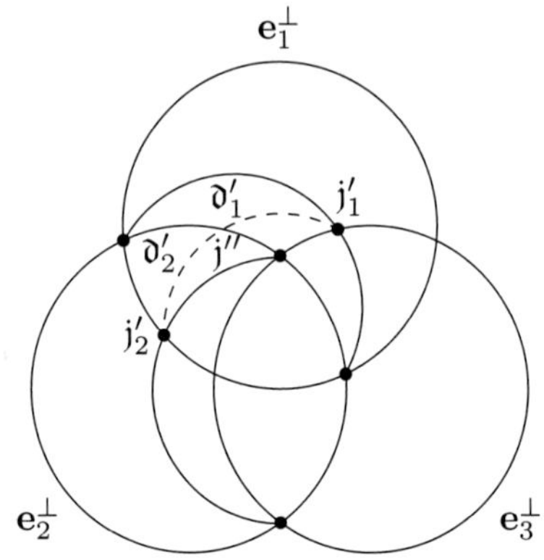

(b) walls of degree 3

Figure 16: Construction of CSD of type A_3. A filled circle • represents a perpendicular joint. A dashed arc represents an attached wall of a given degree.

respectively.

Now, we construct walls of degree 2. We have two perpendicular joints $\mathfrak{j}_1 = \mathbf{e}_1^\perp \cap \mathbf{e}_2^\perp = \mathbb{R}\mathbf{e}_3$, $\mathfrak{j}_2 = \mathbf{e}_2^\perp \cap \mathbf{e}_3^\perp = \mathbb{R}\mathbf{e}_1$ to take care of, and the other joint $\mathfrak{j}_3 = \mathbf{e}_1^\perp \cap \mathbf{e}_3^\perp = \mathbb{R}\mathbf{e}_2$ is parallel. Note that (only) these joints intersect the unit circle S^2 twice in $M_\mathbb{R}$. Therefore, there are two points representing each joint in the figure. First, consider the joint $\mathfrak{j}_1$. Since $\{\mathbf{e}_2, \mathbf{e}_1\} = 1$, the attached wall is

$$\mathbf{w}_1 = (\mathfrak{d}_1, \Psi[(1,1,0)])_{(1,1,0)}, \tag{5.49}$$

$$\mathfrak{d}_1 := \sigma(\mathfrak{j}_1, -p^*((1,1,0))) = \sigma((0,0,1),(0,0,-1),(1,-1,-1)). \tag{5.50}$$

where we note that $-p^*(\mathbf{n}) = -B\mathbf{n}$. Since $\mathfrak{d}_1$ is in the "fourth quadrant", it is inside the circle $\mathbf{e}_1^\perp$ and outside the circle $\mathbf{e}_2^\perp$ except for the boundary. It is depicted in the dashed arc in Figure 16 (a). Similarly, for the joint $\mathfrak{j}_2$, we have $\{\mathbf{e}_3, \mathbf{e}_2\} = 1$, and the attached wall is

$$\mathbf{w}_1 = (\mathfrak{d}_2, \Psi[(0,1,1)])_{(0,1,1)}, \tag{5.51}$$

$$\mathfrak{d}_2 := \sigma(\mathfrak{j}_2, -p^*((0,1,1))) = \sigma((1,0,0),(-1,0,0),(1,1,-1)). \tag{5.52}$$

This finishes the degree 2 walls.

Next, we construct walls of degree 3. All joints that appeared above are already consistent. Thus, we concentrate on the newly created perpendicular joints $\mathfrak{j}'_1 = \mathfrak{d}_1 \cap \mathbf{e}_3^\perp = \mathbb{R}_{\geq 0}(1,-1,0)$ and $\mathfrak{j}'_2 = \mathfrak{d}_2 \cap \mathbf{e}_1^\perp = \mathbb{R}_{\geq 0}(0,1,-1)$ in Figure 16 (b). First, consider the joint $\mathfrak{j}'_1$. Since $\{\mathbf{e}_3, (1,1,0)\} = 1$, the attached wall is

$$\mathbf{w}'_1 = (\mathfrak{d}'_1, \Psi[(1,1,1)])_{(1,1,1)}, \tag{5.53}$$

$$\mathfrak{d}'_1 := \sigma(\mathfrak{j}'_1, -p^*((1,1,1))) = \sigma((1,-1,0),(1,0,-1)). \tag{5.54}$$

Similarly, for the joint $\mathfrak{j}'_2$, we have $\{(0,1,1), \mathbf{e}_1\} = 1$, and the attached wall is

$$\mathbf{w}'_2 = (\mathfrak{d}'_2, \Psi[(1,1,1)])_{(1,1,1)}, \tag{5.55}$$

$$\mathfrak{d}'_2 := \sigma(\mathfrak{j}'_2, -p^*((1,1,1))) = \sigma((0,1,-1),(1,0,-1)). \tag{5.56}$$

The supports $\mathfrak{d}'_1$ and $\mathfrak{d}'_2$ are in the same hyperplane $(1,1,1)^{\perp}$. Moreover, they satisfy

$$\mathfrak{j}'' := \mathfrak{d}'_1 \cap \mathfrak{d}'_2 = \mathbb{R}_{\geq 0}(1,0,-1), \quad \mathfrak{d}'_1 \cup \mathfrak{d}'_2 = \sigma(\mathfrak{j}'_1, \mathfrak{j}'_2). \tag{5.57}$$

Thus, $\mathfrak{d}'_1$ and $\mathfrak{d}'_2$ "shake hands" at $\mathfrak{j}''$ to make a cone $\sigma(\mathfrak{j}'_1, \mathfrak{j}'_2)$. (This shake-hands phenomenon is guaranteed to occur due to the consistency proved in Lemma 5.16.) Note that the algorithm does not request us to joint $\mathbf{w}'_1$ and $\mathbf{w}'_2$ into one wall. The cone $\sigma(\mathfrak{j}'_1, \mathfrak{j}'_2)$ is depicted in the dashed arc in Figure 16 (b). Moreover, we have $\sigma(\mathfrak{j}'_1, \mathfrak{j}'_2) \cap e_2^{\perp} = \mathfrak{j}''$, and $\{(1,1,1), \mathbf{e}_2\} = 0$. Thus, the joint $\mathfrak{j}''$ is parallel (as it should be because $-p^*(1,1,1) \in \mathfrak{j}''$ by construction). Therefore, the construction completes. Observe that all normal vectors

$$\mathbf{e}_1,\ \mathbf{e}_2,\ \mathbf{e}_3,\ (1,1,0),\ (0,1,1),\ (1,1,1) \tag{5.58}$$

exhaust the *positive roots* of the root system of type A_3.

Remark 5.19. As this is of finite type, $\mathrm{Supp}(\mathfrak{D}_{\mathfrak{s}})$ in Figure 16 (b) coincides with the *cluster complex* of a cluster pattern of type A_3 [FZ03a]. We give more accounts in Section 6.6. It also coincides with the *picture* for the *picture group* of type A_3 in [IOTW, ITW, IT21], where the picture groups are defined based on the representation theory of (valued) quivers. Moreover, it also coincides with the *mutation fan* in [Rea14, Rea20b]. The same remark applies to the following examples (b) and (c). The appearance of the positive roots is explained by these theories. Nevertheless, the phenomenon that the pentagon relation reproduces root systems is remarkable and important, and it deserves to be investigated further from this new perspective.

(b). Type B_3. Consider the case

$$B = \begin{pmatrix} 0 & -1 & 0 \\ 1 & 0 & -1 \\ 0 & 2 & 0 \end{pmatrix}, \quad \delta_1 = \delta_2 = 1,\ \delta_3 = 2. \tag{5.59}$$

This is again of finite type. The construction is similar to Type A_3, but it involves joints of type B_2 in Section 2.2 as well. For example, to the perpendicular joint $\mathfrak{j} = \mathbf{e}_2^{\perp} \cap \mathbf{e}_3^{\perp}$, a wall $\mathbf{w} = (\mathfrak{d}, \Psi[(0,1,1)]^2)_{(0,1,1)}$ of degree 2 and a wall $\mathbf{w}' = (\mathfrak{d}', \Psi[(0,1,2)])_{(0,1,2)}$ of degree 3 are attached. The details are

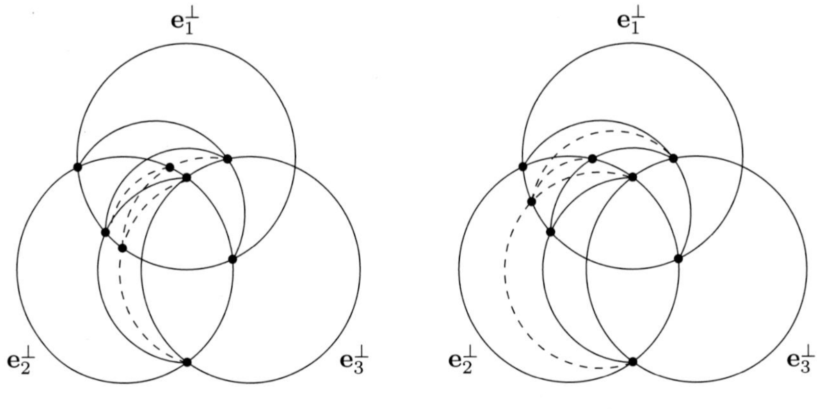

(a) B_3 (b) C_3

Figure 17: Construction of CSDs of types B_3 and C_3. The solid arcs are the supports of type A_3, while the dashed arcs are the additional supports.

left to the reader as an exercise. The resulting CSD is depicted in Figure 17 (a). As observed therein, all supports of type A_3 in the case (a) appear again, and they are indicated by solid arcs. Newly added supports are indicated by dashed arcs, where they have normal vectors $(0,1,2)$, $(1,1,2)$, $(0,2,2)$ from longer to shorter arcs. All normal vectors

$$\mathbf{e}_1,\ \mathbf{e}_2,\ \mathbf{e}_3,\ (1,1,0),\ (0,1,1),\ (1,1,1),\ (0,1,2),\ (1,1,2),\ (1,2,2) \tag{5.60}$$

exhaust the positive roots of the root system of type B_3 as expected.

(c). Type C_3. Consider the case

$$B = \begin{pmatrix} 0 & -1 & 0 \\ 1 & 0 & -2 \\ 0 & 1 & 0 \end{pmatrix}, \quad \delta_1 = \delta_2 = 2,\ \delta_3 = 1. \tag{5.61}$$

This is again of finite type. The construction is similar to type B_3. However, it involves joints of type C_2, instead of type B_2, in Section 2.2. Note that this difference affects the configuration of attached walls. Also, for joints of type A_2 we apply the pentagon relation (1.56) with $c = 1/2$. The resulting CSD is depicted in Figure 17 (b). Once again, all supports of type A_3 appear, and they are indicated by solid arcs. Newly added supports are indicated by dashed arcs, where they have normal vectors $(0,2,1)$, $(2,2,1)$, $(1,2,1)$ from longer to shorter arcs. All normal vectors

$$\mathbf{e}_1,\ \mathbf{e}_2,\ \mathbf{e}_3,\ (1,1,0),\ (0,1,1),\ (1,1,1),\ (0,2,1),\ (1,2,1),\ (2,2,1) \tag{5.62}$$

exhaust the positive roots of the root system of type C_3. Moreover, we see in Figure 17 (a) and (b) that the CSDs of type B_3 and C_3 are topologically identical even though the process of the construction and the assignment of normal vectors are different. This agrees with the well-known *duality* between the cluster patterns of types B_r and C_r [FZ03a, FZ03c], and a more general result called the *synchronicity* on cluster patterns in [Nak21]; namely, any pair of cluster patterns with the initial exchange matrices B and $-B^T$ share the same structure.

The examples (a)–(c) exhaust all rank 3 CSD of finite type with indecomposable B up to mutation-equivalence in [FZ03a]. We will continue to study examples of rank 3 CSDs of infinite type in Section 6.7 after clarifying the *chamber structure* of CSDs.

Notes

The construction of a CSD in Construction 5.14 is a hybrid of the construction by [KS06, §10] and [GS11, §3] presented in [GHKK18, Appendix C.1] and the proof of the positivity in Theorem 5.2 in [GHKK18, Appendix C.3]. Algorithm 5.7, which is based on the pentagon relation in Proposition 1.14, substitutes for the *perturbation trick* and the *change of lattice trick* in [GHKK18, Appendix C.3]. Theorem 5.17 is implicit in [GHKK18, Appendix C.3].

6 Mutation invariance of CSDs

In this section we study the mutation invariance of CSDs, which is important to make the connection with cluster patterns.

6.1 Mutations of seed and CSD

Throughout the section, we fix $k \in \{1, \dots, r\}$ arbitrarily.

For any integer a, we define

$$[a]_+ = \begin{cases} a & a \geq 0, \\ 0 & a < 0. \end{cases} \tag{6.1}$$

We have a useful formula

$$[a]_+ - [-a]_+ = a. \tag{6.2}$$

The following notion originates in [FG09a, §1.2].

Definition 6.1 (Seed mutation). Let $\mathfrak{s} = (e_1, \dots, e_r)$ be a seed for a given fixed data Γ, and let $k \in \{1, \dots, r\}$. We define a new seed $\mu_k(\mathfrak{s}) = \mathfrak{s}' = (e'_1, \dots, e'_r)$ for Γ by

$$e'_i = \begin{cases} -e_k & i = k, \\ e_i + [\{\delta_k e_k, e_i\}]_+ e_k & i \neq k, \end{cases} \tag{6.3}$$

which is called the *mutation of* $\mathfrak{s}$ *in direction* k.

Lemma 6.2. *The above* $\mathfrak{s}'$ *is indeed a seed for* Γ.

Proof. The transformation (6.3) is written in the following matrix form,

$$(e'_1, \dots, e'_r) = (e_1, \dots, e_r) \begin{pmatrix} 1 & & & & \\ & \ddots & & & \\ [b_{k1}]_+ & \cdots & -1 & \cdots & [b_{kr}]_+ \\ & & & \ddots & \\ & & & & 1 \end{pmatrix}, \tag{6.4}$$

where we recall that $b_{ki} = \{\delta_k e_k, e_i\}$. The matrix is unimodular. Thus, $\mathfrak{s}'$ is a basis of N. Also, the basis $(\delta_1 e_1, \dots, \delta_r e_r)$ of N° transforms as

$$\delta_i e'_i = \begin{cases} -\delta_k e_k & i = k, \\ \delta_i e_i + [\{e_k, \delta_i e_i\}]_+ \delta_k e_k & i \neq k. \end{cases} \tag{6.5}$$

Then, $(\delta_1 e'_1, \dots, \delta_r e'_r)$ is a basis of N° by the same reason as above. □

Remark 6.3. The mutation μ_k above is not involutive. Namely, $\mu_k(\mathfrak{s}') \neq \mathfrak{s}$ in general. Indeed, the inverse of (6.3) is given by

$$e_i = \begin{cases} -e'_k & i = k, \\ e'_i + [\{-\delta_k e'_k, e'_i\}]_+ e'_k & i \neq k. \end{cases} \tag{6.6}$$

Even though μ_k is not involutive, it is "involutive" in the following sense.

Lemma 6.4. *Let $\mathfrak{s}'' = \mu_k(\mathfrak{s}')$. Let $B_{\Gamma,\mathfrak{s}}$ be the matrix in (1.4). Then, we have $B_{\Gamma,\mathfrak{s}''} = B_{\Gamma,\mathfrak{s}}$. Thus, the associated groups $G_{\Gamma,\mathfrak{s}''}$ and $G_{\Gamma,\mathfrak{s}}$ are isomorphic.*

Proof. By (6.3), we have

$$e''_i = \begin{cases} e_k & i = k, \\ e_i + \{\delta_k e_k, e_i\} e_k & i \neq k. \end{cases} \tag{6.7}$$

Then, it is easy to check the equality $B_{\Gamma,\mathfrak{s}''} = B_{\Gamma,\mathfrak{s}}$. The last claim follows from Proposition 1.23. □

Under the change of a basis of N from $\mathfrak{s}$ to $\mathfrak{s}'$, we have the following coordinate transformation.

Lemma 6.5. *For $n = \sum_{i=1}^r a_i e_i \in N$, let $n = \sum_{i=1}^r a'_i e'_i$. Then, we have*

$$a'_i = \begin{cases} -a_k + \displaystyle\sum_{j=1}^r [\{\delta_k e_k, e_j\}]_+ a_j & i = k, \\ a_i & i \neq k. \end{cases} \tag{6.8}$$

Proof. Let A be the matrix in (6.4). Then, $A^2 = I$ holds. Thus, the column vector $\mathbf{a} = (a_1, \dots, a_r)$ transforms by $\mathbf{a}' = A\mathbf{a}$. □

Below we use the principal x-representation in Section 4 and its variation. Thus, a wall is expressed by $\mathbf{w} = (\mathfrak{d}, g)_n = [\mathfrak{d}, f]_n$ interchangeably as specified in Definition 4.10. As a general rule to remember, we apply p^* for $\mathfrak{d}$ and n, while we apply $\tilde{p}^*$ for wall functions f.

For any seed $\mathfrak{s} = (e_1, \dots, e_r)$ and $k \in \{1, \dots, r\}$, we define linear maps

$$\begin{aligned} S_k = S_{k,\mathfrak{s}} : \ M_{\mathbb{R}} &\to M_{\mathbb{R}} \\ z &\mapsto z + \langle \delta_k e_k, z\rangle p^*(e_k), \end{aligned} \tag{6.9}$$

$$\begin{aligned} \tilde{S}_k = \tilde{S}_{k,\mathfrak{s}} : \ \tilde{M}^\circ &\to \tilde{M}^\circ \\ \tilde{m} &\mapsto \tilde{m} + \langle \delta_k e_k, m\rangle \tilde{p}^*(e_k), \end{aligned} \tag{6.10}$$

$$\begin{aligned} S^*_k = S^*_{k,\mathfrak{s}} : \ N &\to N \\ n &\mapsto n + \{\delta_k e_k, n\} e_k, \end{aligned} \tag{6.11}$$

where we continue to use the notation $\tilde{m} = m + n'$ ($m \in M^\circ$, $n' \in N$). They are dual to each other in the following sense.

Lemma 6.6. *(a). The following equalities hold.*

$$\langle S_k^*(n), S_k(z)\rangle = \langle n, z\rangle, \tag{6.12}$$

$$\langle S_k^*(n), \tilde{S}_k(\tilde{m})\rangle = \langle n, \tilde{m}\rangle = \langle n, m\rangle, \tag{6.13}$$

$$S_k(p^*(n)) = p^*(S_k^*(n)), \tag{6.14}$$

$$\tilde{S}_k(\tilde{p}^*(n)) = \tilde{p}^*(S_k^*(n)). \tag{6.15}$$

(b). We have

$$S_k^*(\delta_i e_i) = \begin{cases} \delta_k e_k & i = k, \\ \delta_i e_i + \{e_k, \delta_i e_i\}\delta_k e_k & i \neq k. \end{cases} \tag{6.16}$$

In particular, S_k^* preserves N°.

(c). For any $n \in N$, we have $n \in N^\circ$ if and only if $S_k^*(n) \in N^\circ$.

(d). Let $\delta(n)$ ($n \in N$) be the normalization factor in Definition 1.18. Then, we have

$$\delta(S_k^*(n)) = \delta(n). \tag{6.17}$$

Proof. (a). We have

$$\begin{aligned} \langle S_k^*(n), S_k(z)\rangle &= \langle n + \{\delta_k e_k, n\}e_k, z + \langle \delta_k e_k, z\rangle p^*(e_k)\rangle \\ &= \langle n, z\rangle + \langle \delta_k e_k, z\rangle\{n, e_k\} + \{\delta_k e_k, n\}\langle e_k, z\rangle \\ &= \langle n, z\rangle. \end{aligned} \tag{6.18}$$

$$\begin{aligned} \langle S_k^*(n), \tilde{S}_k(\tilde{m})\rangle &= \langle n + \{\delta_k e_k, n\}e_k, \tilde{m} + \langle \delta_k e_k, m\rangle \tilde{p}^*(e_k)\rangle \\ &= \langle n, \tilde{m}\rangle + \langle \delta_k e_k, m\rangle\{n, e_k\} + \{\delta_k e_k, n\}\langle e_k, \tilde{m}\rangle \\ &= \langle n, \tilde{m}\rangle = \langle n, m\rangle. \end{aligned} \tag{6.19}$$

$$\begin{aligned} S_k(p^*(n)) &= p^*(n) + \langle \delta_k e_k, p^*(n)\rangle p^*(e_k) \\ &= p^*(n + \{\delta_k e_k, n\}e_k) = p^*(S_k^*(n)). \end{aligned} \tag{6.20}$$

$$\begin{aligned} \tilde{S}_k(\tilde{p}^*(n)) &= \tilde{p}^*(n) + \langle \delta_k e_k, \tilde{p}^*(n)\rangle \tilde{p}^*(e_k) \\ &= \tilde{p}^*(n + \{\delta_k e_k, n\}e_k) = \tilde{p}^*(S_k^*(n)). \end{aligned} \tag{6.21}$$

(b). This is clear by (6.11).

(c). The only-if-part follows from (b). The inverse of S_k^* is given by

$$S_k^{*-1} : N \to N, \quad n \mapsto n - \{\delta_k e_k, n\}e_k, \tag{6.22}$$

which also preserves N°. Thus, the if-part also holds.

(d). This follows from (c). □

We define the half spaces of $M_{\mathbb{R}}$

$$\mathcal{H}_k^+ = \mathcal{H}_{k,\mathfrak{s}}^+ := \{z \in M_{\mathbb{R}} \mid \langle e_k, z\rangle \geq 0\}, \tag{6.23}$$

$$\mathcal{H}_k^- = \mathcal{H}_{k,\mathfrak{s}}^- := \{z \in M_{\mathbb{R}} \mid \langle e_k, z \rangle \leq 0\}. \tag{6.24}$$

Then, we define a piecewise-linear transformation

$$\begin{aligned} T_k = T_{k,\mathfrak{s}} : \quad M_{\mathbb{R}} \quad &\to \qquad\qquad M_{\mathbb{R}} \\ z \quad &\mapsto \quad z + [\langle \delta_k e_k, z \rangle]_+ p^*(e_k) = \begin{cases} S_k(z) & z \in \mathcal{H}_k^+, \\ z & z \in \mathcal{H}_k^-. \end{cases} \end{aligned} \tag{6.25}$$

Definition 6.7 (Mutation of CSD). Let $\mathfrak{D}_{\mathfrak{s}}$ be the CSD for a seed $\mathfrak{s}$ in Construction 5.14, thus, satisfying the properties in Proposition 5.18. Any wall of $\mathfrak{D}_{\mathfrak{s}}$ has the form

$$(\mathfrak{d}, g)_n = \ (\mathfrak{d}, \Psi[tn]^{s\delta(tn)})_n = [\mathfrak{d}, (1 + x^{\tilde{p}^*(tn)})^s]_n. \tag{6.26}$$

Then, we define

$$\begin{aligned} S_k(\mathfrak{d}, g)_n &:= (S_k(\mathfrak{d}), \Psi[S_k^*(tn)]^{s\delta(S_k^*(tn))})_{S_k^*(n)} && (6.27) \\ &= [S_k(\mathfrak{d}), (1 + x^{\tilde{S}_k(\tilde{p}^*(tn))})^s]_{S_k^*(n)}. && (6.28) \end{aligned}$$

Note that for $n \in N^+$, $S_k^*(n)$ does not necessarily belong to N^+. However, we temporarily ignore this fact. For each $k \in \{1, \dots, r\}$, let $T_k(\mathfrak{D}_{\mathfrak{s}})$ be a collection of "walls" obtained from $\mathfrak{D}_{\mathfrak{s}}$ as follows:

(a). For each wall $\mathbf{w} = (\mathfrak{d}, g)_n$ of $\mathfrak{D}_{\mathfrak{s}}$ other than $\mathbf{w}_{e_k} = (e_k^{\perp}, \Psi[e_k]^{\delta_k})_{e_k}$, let $\mathfrak{d}_+ = \mathfrak{d} \cap \mathcal{H}_k^+$, and $\mathfrak{d}_- = \mathfrak{d} \cap \mathcal{H}_k^-$. We split the wall $\mathbf{w}$ into two walls

$$\mathbf{w}_+ = (\mathfrak{d}_+, g)_n, \quad \mathbf{w}_- = (\mathfrak{d}_-, g)_n, \tag{6.29}$$

where, if one of $\mathfrak{d}_+$ and $\mathfrak{d}_-$ is codimension more than 1, we throw that part away. Then, we replace $\mathbf{w}_+$ with a new wall $S_k(\mathbf{w}_+)$ in (6.27), while we leave $\mathbf{w}_-$ as it is.

(b). Replace $\mathbf{w}_{e_k}$ with $\mathbf{w}_{e'_k} = (e_k'^{\perp}, \Psi[e'_k]^{\delta_k})_{e'_k}$. (Clearly, $e'_k = -e_k \notin N^+$.)

We call $T_k(\mathfrak{D}_{\mathfrak{s}})$ the *mutation of $\mathfrak{D}_{\mathfrak{s}}$ in direction k*.

By construction, we have

$$\mathrm{Supp}(T_k(\mathfrak{D}_{\mathfrak{s}})) = T_k(\mathrm{Supp}(\mathfrak{D}_{\mathfrak{s}})). \tag{6.30}$$

Let $n = \sum_{i=1}^r a_i e_i \in N_{\mathfrak{s}}^+$ be the normal vector of any wall in $\mathfrak{D}_{\mathfrak{s}}$. Then, by Lemma 6.6, the normal vector n' of the corresponding wall in $T_k(\mathfrak{D}_{\mathfrak{s}})$ defined above is given by

$$n' = \sum_{i=1}^r b_i e_i = \begin{cases} -e_k & n = e_k \\ n + \{\delta_k e_k, n\} e_k. & n \text{ is the normal vector of } \mathfrak{d}_+, \\ n & n \text{ is the normal vector of } \mathfrak{d}_-. \end{cases} \tag{6.31}$$

Lemma 6.8. *Let n' be as above. Then, under the change of the coordinate transformation $n' = \sum_{i=1}^{r} b_i e_i = \sum_{i=1}^{r} b'_i e'_i$, we have $b'_i = a_i$ for $i \neq k$, while*

$$b'_k = \begin{cases} 1 & n = e_k, \\ -a_k + \displaystyle\sum_{j=1}^{r} [\mp\{\delta_k e_k, e_j\}]_+ a_j & n \text{ is the normal vector of } \mathfrak{d}_\pm. \end{cases} \tag{6.32}$$

Proof. This follows from (6.31) and Lemma 6.5, where we use (6.2) for the second case. □

The goal of this section to prove the following theorem, which is the second fundamental result on cluster scattering diagrams.

Theorem 6.9 (Mutation invariance of CSD [GHKK18, Theorem 1.24]). *The collection of walls $T_k(\mathfrak{D}_\mathfrak{s})$ is a consistent scattering diagram for the seed $\mathfrak{s}' = \mu_k(\mathfrak{s})$. Moreover, $T_k(\mathfrak{D}_\mathfrak{s})$ is equivalent to a CSD $\mathfrak{D}_{\mathfrak{s}'}$ for $\mathfrak{s}'$.*

We see in (6.32) that $b'_k < 0$ may occur. Therefore, for $T_k(\mathfrak{D}_\mathfrak{s})$ to be a scattering diagram for $\mathfrak{s}'$, $g_z(\mathfrak{D}_\mathfrak{s}) = \mathrm{id}$ should hold if z is on a wall of $\mathfrak{D}_\mathfrak{s}$ such that $b'_k < 0$. This property is nontrivial, and so far we have no direct proof. (In the rank 2 case, however, we can directly prove it. See Example 6.22.) Below we follow the strategy by [GHKK18] to avoid this difficulty. The main idea of [GHKK18] is that we temporarily extend the notion of scattering diagrams by accepting the walls with $b'_k < 0$, then prove the uniqueness property of a CSD $\mathfrak{D}_{\mathfrak{s}'}$ *up to equivalence among such extended scattering diagrams.*

6.2 Extended structure group

As mentioned above, following the strategy by [GHKK18], we temporarily loosen the condition $n \in N^+_{\mathrm{pr}}$ for the normal vector n of a wall. We define

$$\overline{N}^+ = \overline{N}^+_{k,\mathfrak{s}} := \left\{ \sum_{i=1}^{r} a_i e_i \mid a_i \in \mathbb{Z}_{\geq 0} \text{ for } i \neq k,\ a_k \in \mathbb{Z},\ \sum_{\substack{i=1 \\ i \neq k}}^{n} a_i > 0 \right\}. \tag{6.33}$$

In other words, we allow negative coefficients of e_k in n, while we exclude $n = je_k$ ($j \in \mathbb{Z}$). Let $\overline{N}^+_{\mathrm{pr}}$ be the set of all primitive elements in $\overline{N}^+$.

Lemma 6.10. *The following facts hold:*

(a). $e_k \notin \overline{N}^+_{\mathrm{pr}}$.

(b). For $n \in N^+_{\mathrm{pr}}$ with $n \neq e_k$, we have $n, S^*_k(n) \in \overline{N}^+_{\mathrm{pr}}$.

Proof. The only nontrivial claim is that $S^*_k(n)$ is primitive in (b). If it is not, $S^*_k(n) = jn'$ for some $j \in \mathbb{Z}_{>1}$ and $n' \in N$. Then, applying S_k^{*-1} in (6.22), we see that $n = S_k^{*-1}(jn') = jS_k^{*-1}(n')$ is not primitive. □

We define a degree function

$$\begin{array}{rccc} \overline{\deg}: & \overline{N}^+ & \to & \mathbb{Z}_{>0} \\ & \sum_{i=1}^{r} a_i e_i & \mapsto & \sum_{\substack{i=1 \\ i\neq k}}^{r} a_i . \end{array} \tag{6.34}$$

Define an $\overline{N}^+$-graded Lie algebra $\overline{\mathfrak{g}}$,

$$\overline{\mathfrak{g}} = \bigoplus_{n \in \overline{N}^+} \mathfrak{g}_n, \tag{6.35}$$

which is parallel to $\mathfrak{g}$ in (1.17). Then, we repeat the construction of $\widehat{g}$ and G in Section 1, where N^+ and $\deg(n)$ therein are replaced with $\overline{N}^+$ and $\overline{\deg}(n)$. Note that the corresponding Lie algebras $\overline{\mathfrak{g}}^{\leq \ell}$ are still nilpotent. Therefore, the construction works, and we obtain the completion $\widehat{\overline{\mathfrak{g}}}$ and the corresponding exponential group $\overline{G}$. The reason to exclude e_k from $\overline{N}^+$ is now clear, because it is incompatible with the completion of $\overline{\mathfrak{g}}$. We call $\overline{G}$ the *extended structure group for* $\mathfrak{D}_{\mathfrak{s}}$ *and* k.

This, however, is not yet enough for our purpose. Let $G(\Psi[e_k])$ be the group freely generated by $\Psi[e_k]$. Since $\Psi[e_k] \notin \overline{G}$, the product of $\Psi[e_k]$ and elements of $\overline{G}$ is not defined in $\overline{G}$. Instead of further extending $\overline{G}$, we extend the algebra $\Bbbk[[\tilde{P}]]$ so that both $G(\Psi[e_k])$ and $\overline{G}$ act faithfully on it.

First, we define a monoid $\overline{P} \subset \tilde{M}^\circ$ as follows:

(i). $\overline{P} = \overline{\sigma} \cap \tilde{M}^\circ$, where $\overline{\sigma}$ is a $2r$-dimensional (not strongly convex) cone in $\tilde{M}_{\mathbb{R}}$ such that $\overline{\sigma} \cap (-\overline{\sigma}) = \mathbb{R}\tilde{p}^*(e_k)$.

(ii). $\tilde{p}^*(e_i), \cdots, \tilde{p}^*(e_r) \in \overline{P}$.

We may take $\tilde{P}$ in Section 4.2 as a submonoid of $\overline{P}$ in the positive side of $\tilde{p}^*(e_k)$. Let $\Bbbk[\overline{P}]$ be the monoid ring of $\overline{P}$ over $\Bbbk$. Then, the group of units of $\overline{P}$ is given by $\overline{P}^\times = \overline{P} \cap \mathbb{R}\tilde{p}^*(e_k)$. Let $\overline{J}$ be the maximal ideal of $\Bbbk[\overline{P}]$ generated by $\overline{P} \setminus \overline{P}^\times$, and let $\Bbbk[[\overline{P}]]$ be the completion with respect to $\overline{J}$. Furthermore, we consider the localization $R_k := \Bbbk[[\overline{P}]]_{f[e_k]}$ of $\Bbbk[[\overline{P}]]$ by $f[e_k] = 1 + x^{\tilde{p}^*(e_k)}$. Namely, any element $f \in R_k$ is written as

$$f = \sum_{\tilde{m} \in \overline{P}} b_{\tilde{m}} (1 + x^{\tilde{p}^*(e_k)})^{-a_{\tilde{m}}} x^{\tilde{m}} \quad (a_{\tilde{m}} \in \mathbb{Z}_{\geq 0},\ b_{\tilde{m}} \in \Bbbk), \tag{6.36}$$

where in the infinite sum there are only finitely many $j \in \mathbb{Z}$ such that $b_{\tilde{m}+j\tilde{p}^*(e_k)} \neq 0$ for each $\tilde{m} \in \overline{P}$. We define the action of $\Psi[n] \in \overline{G}$ ($n \in \overline{N}^+$)

and $\Psi[e_k] \in G(\Psi[e_k])$ on R_k in the usual way as

$$\begin{aligned}\Psi[n](f) &= \sum_{\tilde{m}\in\overline{P}} b_{\tilde{m}}(1+x^{\tilde{p}^*(e_k)}(1+x^{\tilde{p}^*(n)})^{\langle n,\tilde{p}^*(e_k)\rangle})^{-a_{\tilde{m}}} x^{\tilde{m}}(1+x^{\tilde{p}^*(n)})^{\langle n,m\rangle}\\ &= \sum_{\tilde{m}\in\overline{P}} b_{\tilde{m}}(1+x^{\tilde{p}^*(e_k)})^{-a_{\tilde{m}}}\left(1+x^{\tilde{p}^*(e_k)}\frac{(1+x^{\tilde{p}^*(n)})^{\langle n,\tilde{p}^*(e_k)\rangle}-1}{1+x^{\tilde{p}^*(e_k)}}\right)^{-a_{\tilde{m}}}\\ &\qquad \times x^{\tilde{m}}(1+x^{\tilde{p}^*(n)})^{\langle n,m\rangle},\end{aligned} \tag{6.37}$$

$$\Psi[e_k](f) = \sum_{\tilde{m}\in\overline{P}} b_{\tilde{m}}(1+x^{\tilde{p}^*(e_k)})^{-a_{\tilde{m}}} x^{\tilde{m}}(1+x^{\tilde{p}^*(e_k)})^{\langle e_k,m\rangle}. \tag{6.38}$$

The right hand sides are in R_k. Moreover, the actions are faithful as Proposition 1.8. Thus, we have the following result.

Lemma 6.11. *Both groups $G(\Psi[e_k])$ and $\overline{G}$ act faithfully as automorphisms of R_k. In particular, the product of $\Psi[e_k]$ and $g \in \overline{G}$ is defined in* $\mathrm{Aut}(R_k)$.

The wall elements $\Psi[e_k]^{\pm 1}$ appear only for the consistency relation around a joint $\mathfrak{j} \in e_k^{\perp}$. Moreover, the consistency relation has the form

$$g_1\Psi[e_k]^{\pm\delta_k} g_2 \Psi[e_k]^{\mp\delta_k} = \mathrm{id} \quad (g_1, g_2 \in \overline{G}_{\mathfrak{j}}), \tag{6.39}$$

where $G_{\mathfrak{j}}$ is the one in the following lemma. Then, the lemma tells that the consistency relation essentially reduces to the one in $\overline{G}$.

Lemma 6.12 ([GHKK18, Lemma C.8]). *Let $n \in \overline{N}^+_{\mathrm{pr}}$ and $\mathfrak{j} = e_k^{\perp} \cap n^{\perp}$. Let $N_{\mathfrak{j}}$ be the one in* (5.31)*, and let $\overline{N}^+_{\mathfrak{j},\mathrm{pr}} := N_{\mathfrak{j}} \cap \overline{N}^+_{\mathrm{pr}}$. Let $\overline{G}_{\mathfrak{j}}$ be the subgroup of $\overline{G}$ generated by the parallel subgroups $\overline{G}^{\parallel}_n$ ($n \in \overline{N}^+_{\mathfrak{j},\mathrm{pr}}$). Then, we have*

$$\Psi[e_k]^{\delta_k}\overline{G}_{\mathfrak{j}}\Psi[e_k]^{-\delta_k} \subset \overline{G}_{\mathfrak{j}}, \quad \text{if } \{e_k, n\} \geq 0, \tag{6.40}$$

$$\Psi[e_k]^{-\delta_k}\overline{G}_{\mathfrak{j}}\Psi[e_k]^{\delta_k} \subset \overline{G}_{\mathfrak{j}}, \quad \text{if } \{e_k, n\} \leq 0. \tag{6.41}$$

Proof. This is clear if $\{e_k, n\} = 0$. Suppose that $\{e_k, n\} \neq 0$. Then, we obtain the same formula in Proposition 1.24 (a) by working with the action on R_k. The claim immediately follows from it. □

For the seed $\mathfrak{s}' = \mu_k(\mathfrak{s}) = (e'_1, \dots, e'_r)$, we repeat the same construction as above. Then, despite the asymmetry of the mutation in Remark 6.3, the following result holds, where $'$ indicates the corresponding ones for $\mathfrak{s}'$:

Lemma 6.13.

$$\overline{N}^+ = \overline{N}'^+, \quad \bar{\mathfrak{g}} = \bar{\mathfrak{g}}', \quad \overline{G} = \overline{G}'. \tag{6.42}$$

Moreover, $\overline{P}$ serves as $\overline{P}'$ for $\mathfrak{s}'$ so that

$$\Bbbk[[\overline{P}]] = \Bbbk[[\overline{P}']], \quad R_k = R'_k. \tag{6.43}$$

Proof. For example, the equality $\overline{N}^+ = \overline{N}'^+$ follows from (6.3) and (6.33). The last equality in (6.43) follows from the identity

$$1 + x^{\tilde{p}^*(e_k)} = x^{-\tilde{p}^*(e'_k)}(1 + x^{\tilde{p}^*(e'_k)}). \tag{6.44}$$

□

6.3 Extended scattering diagrams

Definition 6.14 (Extended scattering diagram). For a seed $\mathfrak{s}$ and $k \in \{1, \dots, r\}$, an *extended scattering diagram* $\overline{\mathfrak{D}}$ for $(\mathfrak{s}, k)$ is a collection of walls satisfying the following properties:

- A wall is either $\mathbf{w}_{e_k} = (e_k^\perp, \Psi[e_k]^{\delta_k})_{e_k}$, or $\mathbf{w} = (\mathfrak{d}, g)_n$ with $n \in \overline{N}^+_{\mathrm{pr}}$, $\mathfrak{d} \subset n^\perp$, $g \in \overline{G}^{\parallel}_n$, where $\mathfrak{d}$ is a cone of codimension 1 as usual. (In particular, $n \neq e_k$.)
- It satisfies the finiteness condition with respect to $\overline{G}$ and $\overline{\deg}$ for walls other than $\mathbf{w}_{e_k}$.
- There is at most one $\mathbf{w}_{e_k}$.

Thanks to Lemma 6.11, for any admissible curve γ in $\overline{\mathfrak{D}}$, the path-ordered product $\mathfrak{p}_{\gamma,\overline{\mathfrak{D}}}$ is defined in $\mathrm{Aut}(R_k)$ in the same way as before. Thus, the equivalence and the consistency of extended scattering diagrams and other related notions are defined in the same way as before. (In particular, the consistency condition around a joint $\mathfrak{j} \subset e_k^\perp$ makes sense in $\overline{G}$ thanks to Lemma 6.12.)

Remark 6.15. In [GHKK18] the distinguished wall $\mathbf{w}_{e_k}$ is excluded from walls, and called the *slab*.

Lemma 6.16. *(a). Let $\mathfrak{D}_{\mathfrak{s}}$ be the CSD constructed in Construction 5.14. Then, $\mathfrak{D}_{\mathfrak{s}}$ is a consistent extended scattering diagram for $(\mathfrak{s}, k)$.*

(b). For the same CSD $\mathfrak{D}_{\mathfrak{s}}$ as above, the mutation $T_k(\mathfrak{D}_{\mathfrak{s}})$ is an extended scattering diagram for $(\mathfrak{s}', k)$, where $\mathfrak{s}' = \mu_k(\mathfrak{s})$.

Proof. (a). For any wall of $\mathfrak{D}_{\mathfrak{s}}$ other than $\mathbf{w}_{e_k}$, its normal vector $n \in N^+_{\mathrm{pr}}$ is not e_k by Proposition 5.18. Thus, n belongs to $n \in \overline{N}^+_{\mathrm{pr}}$ by Lemma 6.10 (b).

Next, we show that $\mathfrak{D}_{\mathfrak{s}}$ satisfies the finiteness condition with respect to $\overline{G}$. This is not trivial, and it is equivalent to the following claim:

Claim. For each $n \in N^+$ with $n - e_k \notin N^+$, there are only finitely many $j \in \mathbb{Z}_{>0}$ such that $\Psi[n + je_k]^{s\delta(n+je_k)}$ is a wall element of $\mathfrak{D}_{\mathfrak{s}}$.

Let us prove Claim by the induction on $\ell = \overline{\deg}(n)$. Let $\ell = 1$, namely, $n = e_i$ $(i \neq k)$. Then, by Construction 5.14, such walls are created only from the perpendicular joint $\mathfrak{j} \subset e_k^\perp \cap e_i^\perp$ with an anti-ordered product of $\Psi[e_i]^{\delta_i}$ and $\Psi[e_k]^{\delta_k}$. By Proposition 1.24 (b), j is bounded as

$$j \leq |\{\delta_k e_k, e_i\}|. \tag{6.45}$$

Suppose that Claim holds up to ℓ. Let $\overline{\deg}(n) = \ell + 1$. Then, a wall element of the form $\Psi[n + je_k]^{s\delta(n+je_k)}$ is created in the following two ways:

(a). It is created from a perpendicular joint $\mathfrak{j} \not\subset e_k^\perp$ with an anti-ordered product of $\Psi[n' + j'e_k]$'s with $\overline{\deg}(n') \leq \ell$.

(b). It is created from a perpendicular joint $\mathfrak{j} \subset e_k^\perp \cap n^\perp$ with an anti-ordered product of $\Psi[e_k]$ and $\Psi[n + j'e_k]$, where the latter is created in (a).

There are only finitely many such walls by the induction assumption and Proposition 1.24 (b). Thus, Claim holds for $\ell + 1$. Therefore, $\mathfrak{D}_\mathfrak{s}$ is an extended scattering diagram for $(\mathfrak{s}, k)$.

Finally, we show the consistency of $\mathfrak{D}_\mathfrak{s}$ as an extended scattering diagram for $(\mathfrak{s}, k)$. Recall that any wall element $\Psi[n]^{s\delta(n)}$ of $\mathfrak{D}_\mathfrak{s}$ also acts on R_k. Thus, for any admissible loop γ, the consistency relation $\mathfrak{p}_{\gamma,\mathfrak{D}_\mathfrak{s}} = \mathrm{id}$ along γ holds in $\mathrm{Aut}(R_k)$, where the path-ordered product is defined in the limit with respect to deg. Let us temporally write the path-ordered products along γ defined in the limit with respect to deg and $\overline{\deg}$ as $\mathfrak{p}_\gamma$ and $\overline{\mathfrak{p}}_\gamma$, respectively. Then, it is enough to prove the equality

$$\mathfrak{p}_\gamma = \overline{\mathfrak{p}}_\gamma. \tag{6.46}$$

For any integer $\ell > 0$, let C_ℓ be the largest j for n with $\overline{\deg}(n) = \ell$ in Claim above. Then, the following inequalities hold for any $n \in N^+$ (the first inequality is trivial):

$$\overline{\deg}(n) \leq \deg(n) \leq \overline{\deg}(n) + C_{\overline{\deg}(n)}. \tag{6.47}$$

Thus, the limit $\ell = \deg(n) \to \infty$ implies the limit $\overline{\ell} = \overline{\deg}(n) \to \infty$, and *vice versa*. Thus, we have the equality (6.46). Therefore, $\mathfrak{D}_\mathfrak{s}$ is consistent as an extended scattering diagram for $(\mathfrak{s}, k)$.

(b). For any wall of $T_k(\mathfrak{D}_\mathfrak{s})$ other than $\mathbf{w}_{e_k}$, its normal vector $S_k^*(n)$ belongs to $\overline{N}_{\mathrm{pr}}^+$ by Lemma 6.10 (b). The finiteness condition with respect to $\overline{G}$ follows from the finiteness condition of $\mathfrak{D}_\mathfrak{s}$, because $\overline{\deg}(S_k^*(n)) = \overline{\deg}(n)$. Thus, $T_k(\mathfrak{D}_\mathfrak{s})$ is an extended scattering diagram for $(\mathfrak{s}', k)$. □

We will soon show that $T_k(\mathfrak{D}_\mathfrak{s})$ is also consistent as an extended scattering diagram for $(\mathfrak{s}', k)$. We first prove the following uniqueness theorem, which is a key of the strategy by [GHKK18]. The claim is parallel to Theorem 3.10 (b), but it will be proved by a different argument.

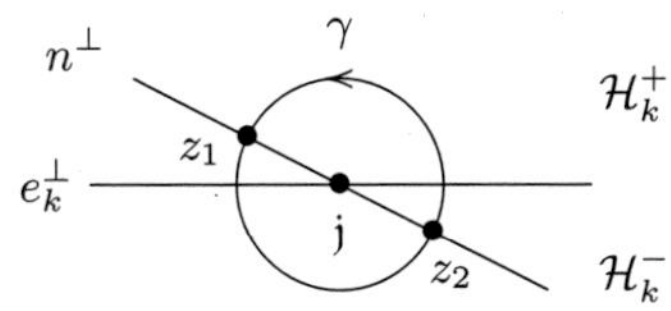

Figure 18: Admissible loop γ in the proof of Proposition 6.17.

Proposition 6.17 (Cf. [GHKK18, Theorem 1.28]). *Let* $\mathrm{In}_{\mathfrak{s}}$ *be the one in* (3.55). *Then, any consistent extended scattering diagram* $\overline{\mathfrak{D}}$ *for* $(\mathfrak{s}, k)$ *with*

$$\overline{\mathfrak{D}}_{\mathrm{in}} = \mathrm{In}_{\mathfrak{s}} \tag{6.48}$$

is equivalent to a CSD $\mathfrak{D}_{\mathfrak{s}}$ *as an extended scattering diagram.*

Proof. Any CSD $\mathfrak{D}_{\mathfrak{s}}$ satisfies the condition (6.48) by Lemma 6.16 and Definition 3.11. Thus, we prove the following more general claim.

Claim. Two consistent extended scattering diagrams $\overline{\mathfrak{D}}$, $\overline{\mathfrak{D}}'$ satisfying the condition (6.48) are equivalent.

Suppose that both $\overline{\mathfrak{D}}$ and $\overline{\mathfrak{D}}'$ satisfy the condition (6.48). Let $\overline{\mathfrak{C}} = \overline{\mathfrak{D}}' - \overline{\mathfrak{D}}$. be an extended scattering diagram defined in the same way as in the proof of Lemma 5.16. Namely, take all walls of $\overline{\mathfrak{D}}$, invert all wall elements, add all walls of $\overline{\mathfrak{D}}'$, and eliminate all incoming walls because they cancel each other. Then, we have the following properties:

- $\overline{\mathfrak{D}} \sqcup \overline{\mathfrak{C}}$ and $\overline{\mathfrak{D}}'$ are equivalent.
- All walls of $\overline{\mathfrak{C}}$ are outgoing.
- All walls of $\overline{\mathfrak{C}}$ have normal vectors other than e_k.

Now we prove that, for any $n \in \overline{N}^+_{\mathrm{pr}}$,

$$\pi_\ell(g_z(\overline{\mathfrak{C}})) = \mathrm{id} \tag{6.49}$$

holds for any general $z \in n^\perp$. We prove it by the induction on ℓ. Let $\overline{\mathfrak{C}}_\ell$ be the reduction of $\overline{\mathfrak{C}}$ at ℓ. We note the following:

- A parallel result to Lemma 3.6 holds for $\overline{G}$.

First, consider the case $\ell = 1$. Let us fix $n \in (\overline{N}^+_{\mathrm{pr}})^{\leq 1}$, namely, $n = e_i + j e_k$ $(i \neq k)$ for some $j \in \mathbb{Z}$. We consider the case $\{e_k, n\} \geq 0$. (We work with all such n simultaneously.) We first note that $p^*(n) \in \mathcal{H}_k^+$ and the claim (6.49) holds for $\ell = 1$ in some neighborhood of $p^*(n)$ due to the absence of any incoming wall. Let z_1 and z_2 be any general points in $n^\perp$. We take an admissible loop γ intersecting $n^\perp$ transversally only at z_1 and z_2. Let $\mathfrak{j} = n^\perp \cap e_k^\perp$. First, suppose that $\mathfrak{j}$ is not inside the loop γ. Then, every nontrivial element appearing in the products $\mathfrak{p}_{\gamma,\overline{\mathfrak{D}}_1}$ and $\mathfrak{p}_{\gamma,\overline{\mathfrak{D}}'_1}$ belong to $\overline{G}$. Since $\overline{G}^{\leq 1}$ is abelian and $\overline{\mathfrak{D}}$ and $\overline{\mathfrak{D}}'$ are consistent, we have

$$\pi_1(\mathfrak{p}_{\gamma,\overline{\mathfrak{C}}_1}) = \pi_1(\mathfrak{p}_{\gamma,\overline{\mathfrak{D}}'_1})\pi_1(\mathfrak{p}^{-1}_{\gamma,\overline{\mathfrak{D}}_1}) = \mathrm{id}. \tag{6.50}$$

Then, by the decomposition of $\overline{G}^{\leq 1}$, which is parallel to Lemma 3.6, we have

$$\pi_1(g_{z_1}(\overline{\mathfrak{C}})) = \pi_1(g_{z_2}(\overline{\mathfrak{C}})). \tag{6.51}$$

It follows that $\pi_1(g_z(\overline{\mathfrak{C}})) = \mathrm{id}$ holds in $\mathrm{Int}(\mathcal{H}_k^+)$, while $\pi_1(g_z(\overline{\mathfrak{C}}))$ is constant in $\mathrm{Int}(\mathcal{H}_k^-)$.

Next, suppose that $\mathfrak{j}$ is inside the loop γ. One can choose $z_1 \in n^\perp \cap \mathrm{Int}(\mathcal{H}_k^+)$ and $z_2 \in n^\perp \cap \mathrm{Int}(\mathcal{H}_k^-)$ sufficiently close to each other so that γ intersects only walls whose normal vectors are in $\overline{N}^+_{\mathfrak{j},\mathrm{pr}}$ of Lemma 6.12. See Figure 18. It follows from the result just obtained above that $\pi_1(g_{z_1}(\overline{\mathfrak{C}})) = \mathrm{id}$ on $n^\perp \cap \mathrm{Int}(\mathcal{H}_k^+)$. Also, any other $n'^\perp$ ($n' \in \overline{N}^+_{\mathfrak{j},\mathrm{pr}}$) containing $\mathfrak{j}$ also satisfies $\{e_k, n'\} > 0$. It follows that the factor $\Psi[e_k]^{\delta_k} C \Psi[e_k]^{-\delta_k} \in \overline{G}$ in $\mathfrak{p}_{\gamma,\overline{\mathfrak{D}}_1}$ and $\mathfrak{p}_{\gamma,\overline{\mathfrak{D}}'_1}$ coming from the region $\mathcal{H}_k^+$ coincide modulo $\overline{G}^{>1}$. Thus, we have (6.50) again, and we obtain (6.51). Therefore, we conclude that (6.49) holds for $\ell = 1$. The other case $\{e_k, n\} \leq 0$ is similar by exchanging $\mathcal{H}_k^+$ and $\mathcal{H}_k^-$.

Next, suppose that (6.49) holds up to ℓ for any general $z \in M_{\mathbb{R}}$. Let us fix $n \in (\overline{N}^+_{\mathrm{pr}})^{\ell+1}$. We take an admissible loop γ in a similar way to the above. Then, by the induction assumption, all factors of $\pi_{\ell+1}(\mathfrak{p}_{\gamma,\overline{\mathfrak{C}}_{\ell+1}})$ belongs to $\operatorname{Ker} \pi_{\ell+1,\ell}$, which is abelian as in Lemma 3.6. Then, by repeating the same argument as $\ell = 1$, we have

$$\pi_{\ell+1}(\mathfrak{p}_{\gamma,\overline{\mathfrak{C}}_{\ell+1}}) = \pi_{\ell+1}(\mathfrak{p}_{\gamma,\overline{\mathfrak{D}}'_{\ell+1}})\pi_{\ell+1}(\mathfrak{p}^{-1}_{\gamma,\overline{\mathfrak{D}}_{\ell+1}}) = \mathrm{id}. \tag{6.52}$$

Then, again by the decomposition of $\operatorname{Ker} \pi_{\ell+1,\ell}$, which is parallel to Lemma 3.6, we conclude that (6.49) holds for $\ell + 1$. □

6.4 Proof of mutation invariance of CSDs

Let us complete the proof of Theorem 6.9. We have already seen in Lemma 6.16 that $T_k(\mathfrak{D}_{\mathfrak{s}})$ is an extended scattering diagram for $(\mathfrak{s}', k)$.

We first prove the consistency of $T_k(\mathfrak{D}_{\mathfrak{s}})$ as an extended scattering diagram for $(\mathfrak{s}', k)$. Let us introduce the following algebra automorphism of $R_k = R'_k$,

$$\begin{array}{rccc} \tilde{S}_k = \tilde{S}_{k,\mathfrak{s}}: & R_k & \to & R_k \\ & x^{\tilde{m}} & \mapsto & x^{\tilde{S}_k(\tilde{m})} = x^{\tilde{m} + \langle \delta_k e_k, m\rangle \tilde{p}^*(e_k)} \quad (\tilde{m} \in \overline{P}), \\ & (1 + x^{\tilde{p}^*(e_k)})^{-1} & \mapsto & (1 + x^{\tilde{p}^*(e_k)})^{-1}. \end{array} \tag{6.53}$$

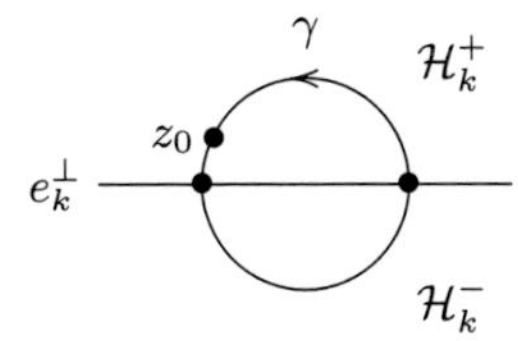

Figure 19: Admissible loop γ in the proof of Lemma 6.19.

Lemma 6.18. *The following equalities hold as an algebra automorphism of R_k:*

$$\Psi[S_k^*(n)] \circ \tilde{S}_k = \tilde{S}_k \circ \Psi[n] \quad (n \in N^+), \tag{6.54}$$

$$\Psi[e_k']^{\delta_k} = \tilde{S}_k \circ \Psi[e_k]^{-\delta_k}. \tag{6.55}$$

Proof. Here, we consider the action only on $x^{\tilde{m}}$. We have

$$\begin{aligned}(\Psi[S_k^*(n)] \circ \tilde{S}_k)(x^{\tilde{m}}) &= \Psi[S_k^*(n)](x^{\tilde{S}_k(\tilde{m})}) \\ &= x^{\tilde{S}_k(\tilde{m})} \tilde{S}_k(f[n])^{\langle S_k^*(n), \tilde{S}_k(\tilde{m})\rangle} \\ &= \tilde{S}_k(x^{\tilde{m}} f[n]^{\langle n, m\rangle}) \\ &= (\tilde{S}_k \circ \Psi[n])(x^{\tilde{m}}),\end{aligned} \tag{6.56}$$

where we used Lemma 6.6 (a) and (d). Also, we have

$$\begin{aligned}\Psi[e_k']^{\delta_k}(x^{\tilde{m}}) &= x^{\tilde{m}}(1 + x^{\tilde{p}^*(e_k')})^{\langle \delta_k e_k', m\rangle} \\ &= x^{\tilde{m}}(1 + x^{-\tilde{p}^*(e_k)})^{-\langle \delta_k e_k, m\rangle} \\ &= x^{\tilde{m} + \langle \delta_k e_k, m\rangle \tilde{p}^*(e_k)}(1 + x^{\tilde{p}^*(e_k)})^{-\langle \delta_k e_k, m\rangle} \\ &= \tilde{S}_k(x^{\tilde{m}}(1 + x^{\tilde{p}^*(e_k)})^{-\langle \delta_k e_k, m\rangle}) \\ &= (\tilde{S}_k \circ \Psi[e_k]^{-\delta_k})(x^{\tilde{m}}).\end{aligned} \tag{6.57}$$

The action on $(1 + x^{\tilde{p}^*(e_k)})^{-1}$ is similar and easier. □

Lemma 6.19. *$T_k(\mathfrak{D}_\mathfrak{s})$ is a consistent extended scattering diagram for $(\mathfrak{s}', k)$.*

Proof. It is enough to prove the consistency as an extended scattering diagram for $(\mathfrak{s}', k)$. In view of Lemma 2.14, it is enough to show that $\mathfrak{p}_{\gamma, T_k(\mathfrak{D}_\mathfrak{s})} = \mathrm{id}$ for any admissible loop γ. If γ is entirely in $\mathcal{H}_k^-$, the claim follows from the consistency of $\mathfrak{D}_\mathfrak{s}$ as an extended scattering diagram for $(\mathfrak{s}, k)$ in Lemma 6.16 (a). If γ is entirely in $\mathcal{H}_k^+$, the claim follows from the consistency of $\mathfrak{D}_\mathfrak{s}$ and (6.54). Therefore, it is enough to consider the case where γ crosses the hyperplane $e_k^\perp$ as in Figure 19, where z_0 is the base point of γ. Then, from the consistency of $\mathfrak{D}_\mathfrak{s}$, we have

$$\mathfrak{p}_{\gamma, \mathfrak{D}_\mathfrak{s}} = g_+ \circ \Psi[e_k]^{-\delta_k} \circ g_- \circ \Psi[e_k]^{\delta_k} = \mathrm{id}, \tag{6.58}$$

where $g_\pm$ is the contribution from each half space $\mathcal{H}_k^\pm$. Then, by the definition of $T_k(\mathfrak{D}_\mathfrak{s})$ and Lemma 6.18,

$$\begin{aligned}\mathfrak{p}_{\gamma,T_k(\mathfrak{D}_\mathfrak{s})} &= (\tilde{S}_k \circ g_+ \circ \tilde{S}_k^{-1}) \circ \Psi[e_k']^{\delta_k} \circ g_- \circ \Psi[e_k']^{-\delta_k} \\ &= \tilde{S}_k \circ (g_+ \circ \Psi[e_k]^{-\delta_k} \circ g_- \circ \Psi[e_k]^{\delta_k}) \circ \tilde{S}_k^{-1} \\ &= \mathrm{id}.\end{aligned} \tag{6.59}$$

□

Next we prove the following property.

Lemma 6.20. *$T_k(\mathfrak{D}_\mathfrak{s})$ is equivalent to an extended scattering diagram $\overline{\mathfrak{D}}$ for $(\mathfrak{s}', k)$ with*

$$\overline{\mathfrak{D}}_{\mathrm{in}} = \mathrm{In}_{\mathfrak{s}'}, \tag{6.60}$$

where $\mathfrak{s}' = \mu_k(\mathfrak{s})$.

Proof. Let us write $\mathbf{w}_i' := \mathbf{w}_{e_i'}$ for simplicity. For the sake of convenience, we restate here what should be proved.

Claim. $T_k(\mathfrak{D}_\mathfrak{s})$ is equivalent to an extended scattering diagram $\overline{\mathfrak{D}}$ for $(\mathfrak{s}', k)$ satisfying the following properties:

(i). $\overline{\mathfrak{D}}$ contains a unique wall $\mathbf{w}_k' = (e_k'^{\perp}, \Psi[e_k']^{\delta_k})_{e_k'}$.
(ii). For any $i \neq k$, $\overline{\mathfrak{D}}$ contains a unique wall $\mathbf{w}_i' = (e_i'^{\perp}, \Psi[e_i']^{\delta_i})_{e_i'}$.
(iii). All other walls in $\overline{\mathfrak{D}}$ are outgoing.

The property (i) holds for $T_k(\mathfrak{D}_\mathfrak{s})$ by the construction. So, let us consider the properties (ii) and (iii). We note that, by the construction of $T_k(\mathfrak{D}_\mathfrak{s})$ in Definition 6.7, a wall $S_k(\mathbf{w}_+)$ is incoming if and only if $\mathbf{w}_+$ is incoming. Thus, any incoming wall in $T_k(\mathfrak{D}_\mathfrak{s})$ other than $\mathbf{w}_k'$ only comes from $\mathbf{w}_{i,+} = ((e_i{}^{\perp})_+, \Psi[e_i]^{\delta_i})_{e_i}$ or $\mathbf{w}_{i,-} = ((e_i{}^{\perp})_-, \Psi[e_i]^{\delta_i})_{e_i}$ $(i \neq k)$. For each $i \neq k$, we consider three cases.

Case 1. $\{\delta_k e_k, e_i\} > 0$. We have $p^*(e_i) \in \mathcal{H}_k^+$. Thus, $\mathbf{w}_{i,+}$ is an incoming wall. Also, $S_k^*(e_i) = e_i + \{\delta_k e_k, e_i\} e_k = e_i'$. Therefore, we have an incoming wall in $T_k(\mathfrak{D}_\mathfrak{s})$,

$$S_k(\mathbf{w}_{i,+}) = ((e_i'^{\perp})_-, \Psi[e_i']^{\delta_i})_{e_i'} = \mathbf{w}_{i,-}', \tag{6.61}$$

where the sign $\pm$ of $(e_i'^{\perp})_\pm$ is with respect to $e_k' = -e_k$. Then, up to equivalence, it is extended to $\mathbf{w}_i'$ by adding outgoing walls.

Case 2. $\{\delta_k e_k, e_i\} < 0$. We have $p^*(e_i) \in \mathcal{H}_k^-$. Thus, $\mathbf{w}_{i,-}$ is an incoming wall. Also, $e_i = e_i'$. Therefore, we have an incoming wall in $T_k(\mathfrak{D}_\mathfrak{s})$,

$$\mathbf{w}_{i,-} = ((e_i'^{\perp})_+, \Psi[e_i']^{\delta_i})_{e_i'} = \mathbf{w}_{i,+}'. \tag{6.62}$$

Again, up to equivalence, it is extended to $\mathbf{w}'_i$ by adding outgoing walls.

Case 3. $\{\delta_k e_k, e_i\} = 0$. We have $p^*(e_i) \in \mathcal{H}_k^+ \cap \mathcal{H}_k^-$. Thus, $\mathbf{w}_{i,+}$ and $\mathbf{w}_{i,-}$ are both incoming walls, where $p^*(e_i)$ is at the boundary of both walls $\mathbf{w}_{i,\pm}$. Also, $S_k^*(e_i) = e'_i$. Therefore, we have incoming walls $\mathbf{w}_{i,+} = \mathbf{w}'_{i,-}$ and $\mathbf{w}_{i,-} = \mathbf{w}'_{i,+}$ in $T_k(\mathfrak{D}_\mathfrak{s})$. Then, joining them back again, we have $\mathbf{w}'_i$.

It is clear that the resulting extended scattering diagram $\overline{\mathfrak{D}}$ satisfies all desired properties. □

By Lemmas 6.19 and 6.20, up to equivalence, $T_k(\mathfrak{D}_\mathfrak{s})$ satisfies the assumption of $\overline{\mathfrak{D}}$ in Proposition 6.17 (by replacing $\mathfrak{s}$ therein with $\mathfrak{s}'$). Thus, we conclude that $T_k(\mathfrak{D}_\mathfrak{s})$ is equivalent to $\mathfrak{D}_{\mathfrak{s}'}$ as an extended scattering diagram for $(\mathfrak{s}', k)$. Moreover, since our choice of $\mathfrak{D}_\mathfrak{s}$ is with minimal support by Proposition 5.18, so is $T_k(\mathfrak{D}_\mathfrak{s})$. Thus, we have

$$\mathrm{Supp}(T_k(\mathfrak{D}_\mathfrak{s})) = \mathrm{Supp}(\mathfrak{D}_{\mathfrak{s}'}). \tag{6.63}$$

This implies that $T_k(\mathfrak{D}_\mathfrak{s})$ is actually an ordinary scattering diagram for $\mathfrak{s}'$, and that it is equivalent to $\mathfrak{D}_{\mathfrak{s}'}$ as an ordinary scattering diagram for $\mathfrak{s}'$. This completes the proof of Theorem 6.9.

6.5 Admissible region

In view of the remark after Theorem 6.9, we have the following corollary of Theorem 6.9.

Corollary 6.21. *Let $\mathfrak{D}_\mathfrak{s}$ be a CSD for $\mathfrak{s}$. Let $n = \sum_{i=1}^r a_i e_i \in N_{\mathrm{pr}}^+$ such that $n \neq e_1, \dots, e_r$, and let z be any general point in $n^\perp$. For each $k = 1, \dots, r$, we define*

$$b'_k = \begin{cases} -a_k + \sum_{j=1}^r [-\{\delta_k e_k, e_j\}]_+ a_j & z \in \mathcal{H}_k^+, \\ -a_k + \sum_{j=1}^r [\{\delta_k e_k, e_j\}]_+ a_j & z \in \mathcal{H}_k^-. \end{cases} \tag{6.64}$$

Then, we have

$$g_z(\mathfrak{D}_\mathfrak{s}) \neq \mathrm{id} \quad \Longrightarrow \quad b'_k \geq 0 \textit{ for any } k. \tag{6.65}$$

It is an interesting problem to give a direct proof of this property and also to extend it to other classes of consistent scattering diagrams.

Let us call the region of $z \in M_\mathbb{R}$ that satisfies the condition $b'_k \geq 0$ for any k the *admissible region* for a CSD $\mathfrak{D}_\mathfrak{s}$, where nontrivial outgoing walls are allowed.

Example 6.22. Let us explicitly write down the condition (6.65) for the rank 2 CSDs in Sections 2.2 and 3.5. For $\{e_2, e_1\} = 0$, no $n \neq e_1, e_2$ is

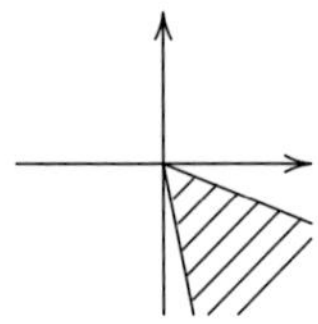

Figure 20: Admissible region for a CSD of rank 2.

allowed. This agrees with the CSD of type $A_1 \times A_1$ therein. For $\{e_2, e_1\} = 1$, the quantities in (6.64) are given as

$$b'_1 = \begin{cases} -a_1 + \delta_1 a_2 & z \in \mathcal{H}_1^+, \\ -a_1 & z \in \mathcal{H}_1^-. \end{cases} \quad b'_2 = \begin{cases} -a_2 & z \in \mathcal{H}_2^+, \\ -a_2 + \delta_2 a_1 & z \in \mathcal{H}_2^-, \end{cases} \tag{6.66}$$

where $a_1, a_2 \neq 0$. Then, the condition $b'_k \geq 0$ $(k = 1, 2)$ is equivalent to the condition

$$z \in \mathcal{H}_1^+ \cap \mathcal{H}_2^-, \quad \frac{1}{\delta_1} \leq \frac{a_2}{a_1} \leq \delta_2. \tag{6.67}$$

The first condition is equivalent to that a wall is outgoing, and the second condition follows from the formula (1.75). Therefore, we directly verified the property (6.65) in the rank 2 case. Let $\mathbf{z} = (z_1, z_2) \in \mathbb{R}^2$ be the point corresponding to $z \in M_{\mathbb{R}}$. Then, by (2.12), we have

$$\delta_1^{-1} a_1 z_1 + \delta_2^{-1} a_2 z_2 = 0. \tag{6.68}$$

Thus, the condition (6.67) is rewritten as

$$z_1 \geq 0, \quad z_2 \leq 0, \quad -\delta_2 \leq \frac{z_2}{z_1} \leq -\frac{1}{\delta_1}. \tag{6.69}$$

The admissible region is depicted in Figure 20.

6.6 Chamber structure, G-fan, and cluster complex

There is an important consequence of the mutation invariance of CSDs.

Suppose that $\mathfrak{D}_{\mathfrak{s}}$ is a CSD with minimal support. We first give few remarks:

(a). By Lemma 2.18, the interior $\mathrm{Int}(\mathcal{C}_{\mathfrak{s}}^+)$ of the positive orthant $\mathcal{C}_{\mathfrak{s}}^+$ is a *chamber* of $\mathrm{Supp}(\mathfrak{D}_{\mathfrak{s}})$, namely, it is a connected component of $M_{\mathbb{R}} \setminus \mathrm{Supp}(\mathfrak{D}_{\mathfrak{s}})$. Also, since $\mathrm{Int}(\mathcal{C}_{\mathfrak{s}}^+)$ is an open set, its boundary belongs to $\mathrm{Supp}(\mathfrak{D}_{\mathfrak{s}})$.

(b). For $\mathfrak{s}' = \mu_k(\mathfrak{s})$, we define a cone in $M_{\mathbb{R}}$ of full dimension by

$$\mathcal{C}_{\mathfrak{s}'}^{\mathfrak{s}} := T_{k,\mathfrak{s}}^{-1}(\mathcal{C}_{\mathfrak{s}'}^+). \tag{6.70}$$

Since $\mathrm{Int}(\mathcal{C}^+_{\mathfrak{s}'})$ is a chamber of $\mathrm{Supp}(\mathfrak{D}_{\mathfrak{s}'})$, $\mathrm{Int}(\mathcal{C}^{\mathfrak{s}}_{\mathfrak{s}'})$ is a chamber of $\mathrm{Supp}(\mathfrak{D}_{\mathfrak{s}})$ by (6.30). More explicitly, $\mathcal{C}^{\mathfrak{s}}_{\mathfrak{s}'}$ is the cone defined by, for $i = k$,

$$\langle e'_k, T_k(z)\rangle = \langle -e_k, z + \langle \delta_k e_k, z\rangle p^*(e_k)\rangle = -\langle e_k, z\rangle \geq 0, \tag{6.71}$$

and, for $i \neq k$, by (6.71),

$$\begin{aligned} \langle e'_i, T_k(z)\rangle &= \langle e_i + [\{\delta_k e_k, e_i\}]_+ e_k, z + [\langle \delta_k e_k, z\rangle]_+ p^*(e_k)\rangle \\ &= \langle e_i, z\rangle + [\{\delta_k e_k, e_i\}]_+ \langle e_k, z\rangle \geq 0. \end{aligned} \tag{6.72}$$

Thus, the cone $\mathcal{C}^{\mathfrak{s}}_{\mathfrak{s}'}$ is adjacent to $\mathcal{C}^+_{\mathfrak{s}}$ with a common boundary in $e_k^\perp$.

(c). Let σ be a face of $\mathcal{C}^{\mathfrak{s}}_{\mathfrak{s}'}$ of codimension 1 with the normal vector n. Let $z \in \sigma$ be general. Then, $g_z(\mathfrak{D}_{\mathfrak{s}})$ is obtained by the pull-back of $g_{T_k(z)}(\mathfrak{D}_{\mathfrak{s}'}) = \Psi[e'_i]^{\delta_i}$ by T_k. Thus, we have

$$g_z(\mathfrak{D}_{\mathfrak{s}}) = \Psi[n]^{\delta(n)}. \tag{6.73}$$

Moreover, $\delta(n) = \delta_i$ for some $i \in \{1, \dots, r\}$.

We repeat the above argument for a sequence of mutations of seeds.

Definition 6.23 (G-cone/G-chamber). Suppose that $\mathfrak{D}_{\mathfrak{s}}$ is a CSD with minimal support. We consider any sequence of mutations of seeds

$$\mathfrak{s} = \mathfrak{s}_0 \overset{\mu_{k_0}}{\mapsto} \mathfrak{s}_1 \overset{\mu_{k_2}}{\mapsto} \cdots \overset{\mu_{k_{i-1}}}{\mapsto} \mathfrak{s}_i. \tag{6.74}$$

We define a cone in $M_{\mathbb{R}}$ by

$$\mathcal{C}^{\mathfrak{s}}_{\mathfrak{s}_i} := (T^{-1}_{k_0,\mathfrak{s}_0} \circ \cdots \circ T^{-1}_{k_{i-1},\mathfrak{s}_{i-1}})(\mathcal{C}^+_{\mathfrak{s}_i}). \tag{6.75}$$

As a special case $i = 0$, we set $\mathcal{C}^{\mathfrak{s}}_{\mathfrak{s}} := \mathcal{C}^+_{\mathfrak{s}}$. Then, $\mathrm{Int}(\mathcal{C}^{\mathfrak{s}}_{\mathfrak{s}_i})$ is a chamber of $\mathrm{Supp}(\mathfrak{D}_{\mathfrak{s}})$. We call $\mathcal{C}^{\mathfrak{s}}_{\mathfrak{s}_i}$ and $\mathrm{Int}(\mathcal{C}^{\mathfrak{s}}_{\mathfrak{s}_i})$ a *G-cone* and a *G-chamber* of $\mathfrak{D}_{\mathfrak{s}}$. (In [GHKK18], $\mathcal{C}^{\mathfrak{s}}_{\mathfrak{s}_i}$ is called a *cluster chamber.*)

Let $\mathbb{T}_r$ be the *r-regular tree* wth edges labelled by 1, …, r such that all edges emanating from each vertex t of $\mathbb{T}_r$ have different labels. We abuse the notation $\mathbb{T}_r$ for the set of all vertices of $\mathbb{T}_r$. We fix the initial vertex $t_0 \in \mathbb{T}_r$ arbitrarily. Then, we attach $\mathcal{C}^+_{\mathfrak{s}}$ to t_0. Also, we attach $\mathcal{C}^{\mathfrak{s}}_{\mathfrak{s}_i}$ in (6.75) to t_i, where t_0 and t_i are sequentially connected with edges labelled by k_0, …, k_{i-1}. Thus, the G-cones of $\mathfrak{D}_{\mathfrak{s}}$ are parametrized by $\mathbb{T}_r$, though this parametrization is not necessarily one-to-one. Then, under the identification $M_{\mathbb{R}} \simeq \mathbb{R}^r$, $f_i \mapsto \mathbf{e}_i$, any G-cone $\mathcal{C}^{\mathfrak{s}}_{\mathfrak{s}_i}$ is identified with the cone $\sigma(G^{t_0}_{t_i})$ spanned by the column vectors (the *g-vectors*) of the *G-matrix* $G^{t_0}_{t_i}$ in [FZ07] for the corresponding cluster pattern. The collection of G-cones $\sigma(G^{t_0}_{t})$ ($t \in \mathbb{T}_r$) and their faces constitutes a fan $\Delta(\mathbf{G}^{t_0})$ (the *G-fan*). Moreover, as a simplicial complex, the G-fan $\Delta(\mathbf{G}^{t_0})$ is isomorphic to the *cluster complex* in [FZ03a]. See Part II for details.

By repeatedly applying the result (6.73), we have the following property.

Proposition 6.24. *Let σ be a face of a G-cone $\mathcal{C}^{\mathfrak{s}}_{\mathfrak{s}_i}$ of codimension 1 with the normal vector $n \in N^{+}_{\mathrm{pr}}$. Then, for any general $z \in \sigma$, we have*

$$g_z(\mathfrak{D}_{\mathfrak{s}}) = \Psi[n]^{\delta(n)}. \tag{6.76}$$

Moreover, $\delta(n) = \delta_i$ for some $i \in \{1, \dots, r\}$.

The above result implies that, in Theorem 5.2, for the walls that are faces of G-cones, only $s = t = 1$ occurs.

Example 6.25. (a). Let $\mathfrak{D}_{\mathfrak{s}}$ be any CSD of finite type with rank 2 or 3 in Sections 2.2 and 3.5. Then, the G-cones fill up the entire ambient space $\mathbb{R}^r$. In other words, $\mathrm{Supp}(\mathfrak{D}_{\mathfrak{s}})$ is naturally identified with the G-fan $\Delta(\mathbf{G}^{t_0})$ of the corresponding cluster pattern. In particular, the property (6.76) holds for any wall of $\mathfrak{D}_{\mathfrak{s}}$.

(b). Let $\mathfrak{D}_{\mathfrak{s}}$ be the CSD of type $A_1^{(1)}$ in Section 3.5. Then, the G-cones fill up the set $\mathbb{R}^2 \setminus \mathbb{R}_{>0}(1,-1)$. In other words, $\mathrm{Supp}(\mathfrak{D}_{\mathfrak{s}}) \setminus \mathbb{R}_{>0}(1,-1)$ in $\mathbb{R}^2 \setminus \mathbb{R}_{\geq 0}(1,-1)$ is naturally identified with the G-fan $\Delta(\mathbf{G}^{t_0})$ of the corresponding cluster pattern. In particular, the property (6.76) holds for any wall in $\mathrm{Supp}(\mathfrak{D}_{\mathfrak{s}}) \setminus \mathbb{R}_{>0}(1,-1)$. On the other hand, the wall in (3.69) is *unreachable* by *cluster mutations* (i.e., mutations in the sense of cluster algebras). The $A_2^{(2)}$ case is similar. For the non-affine type of rank 2 in Section 3.5, replace $\mathbb{R}_{>0}(1,-1)$ with the Badlands $\sigma(\mathbf{v}_+, \mathbf{v}_-) \setminus \{0\}$ therein, which is also unreachable by cluster mutations.

6.7 Rank 3 examples: infinite type

Let us continue to give examples of rank 3 CSDs based on Construction 5.14. Here we consider the CSDs of *infinite type*, which are further separated into *affine* and *non-affine* types. We use the same convention in Section 5.4 to present rank 3 CSDs.

(a). Type $A_2^{(1)}$. As a prototypical example of affine type, we consider the following case

$$B = \begin{pmatrix} 0 & -1 & -1 \\ 1 & 0 & -1 \\ 1 & 1 & 0 \end{pmatrix}, \quad \delta_1 = \delta_2 = \delta_3 = 1. \tag{6.77}$$

As a new feature compared with the finite type, we have perpendicular joints of type $A_1^{(1)}$ in Section 3.5, which are given by $\mathfrak{j}'_1 \subset (1,1,0)^{\perp} \cap \mathbf{e}_3^{\perp}$ and $\mathfrak{j}'_2 \subset (0,1,1)^{\perp} \cap \mathbf{e}_1^{\perp}$ in Figure 21. Indeed, we have

$$\{\mathbf{e}_3, (1,1,0)\} = \{(0,1,1), \mathbf{e}_1\} = 2. \tag{6.78}$$

We put this number 2 in the figure in the position of incoming direction, so that outgoing walls appear in the opposite direction. These joints $\mathfrak{j}'_1$

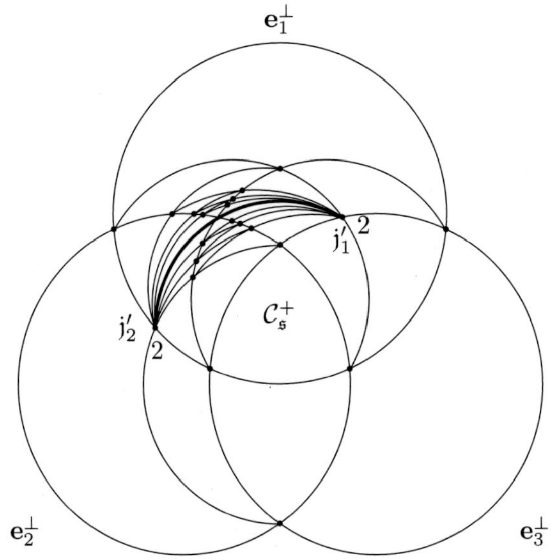

Figure 21: CSD of type $A_2^{(1)}$. The thick arc is unreachable by cluster mutations.

and j_2' are connected by walls with the common normal vector $(1,1,1)$, and their wall elements are given by $\Psi[(1,1,1)]^2$, $\Psi[(2,2,2)]$, $\Psi[(4,4,4)]^{1/2}$, ..., based on the result the CSD of type $A_1^{(1)}$ in Section 3.5. The union of their supports is the cone $\sigma(j_1',j_2')$, and it is represented by a thick arc in Figure 21. It corresponds to the unreachable wall of type $A_1^{(1)}$ with the normal vector $(1,1)$ in Example 5.14 (b). Note that j_1' and j_2' are directly connected only by $\sigma(j_1',j_2')$, because a hyperplane containing j_1' and j_2' in $\mathbb{R}^3$ is unique. Consequently, j_1' and j_2' "alternately" create new walls, and they pile up as shown in the figure. We see that those walls with normal vectors

$$(2,1,1),\ (2,2,1),\ (3,2,2),\ (3,3,2),\ \dots \tag{6.79}$$

converge to the thick arc *from above* in the figure. Similarly, those walls with normal vectors

$$(1,1,2),\ (1,2,2),\ (2,2,3),\ (3,2,2),\ \dots \tag{6.80}$$

converges to the thick arc *from below*. This agrees with the well-known results on the cluster patterns of type $A_2^{(1)}$ by [FZ02a] and [CI12]. We conclude that the G-cones fill up the region $\mathbb{R}^3 \setminus (\sigma(j_1',j_2') \setminus \{0\})$. In other words, $\mathrm{Supp}(\mathfrak{D}_{\mathfrak{s}}) \setminus (\sigma(j_1',j_2') \setminus \{0\})$ in $\mathbb{R}^3 \setminus (\sigma(j_1',j_2') \setminus \{0\})$ is naturally identified with the G-fan $\Delta(\mathbf{G}^{t_0})$ of the corresponding cluster pattern. The walls in $\sigma(j_1',j_2')$ is *unreachable* by cluster mutations.

(b). Non-affine type. As a prototypical example of non-affine type, we

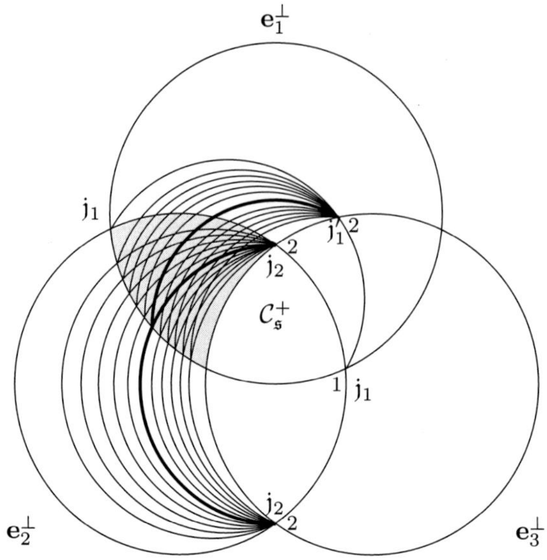

Figure 22: Example of rank 3 CSD of non-affine type. The close-up of the shaded region is given in Figure 23.

consider the following case

$$B = \begin{pmatrix} 0 & -1 & 0 \\ 1 & 0 & -2 \\ 0 & 2 & 0 \end{pmatrix}, \quad \delta_1 = \delta_2 = \delta_3 = 1. \tag{6.81}$$

The CSD has the remarkable *fractal* nature. This nature is simply originated in the fractal structure of the underlying r-regular tree $\mathbb{T}_r$ ($r \geq 3$) for the corresponding cluster complex, and it has been observed in some figures in the literature (e.g., [FG16, NC13]) for the so-called Markov quiver. This implies that it is impossible to draw a precise picture in principle. Thus, we change the strategy as follows:

- We set the goal to obtain a reasonably clear picture of the *Badlands of rank 3*.
- We abandon to faithfully follow Construction 5.14 by uniformly increasing the degree $\delta(n)$ of a newly attached wall element $\Psi[n]$. Rather, we do the mixture of Construction 5.14 and "cluster mutation approach". Namely, having Construction 5.14 in mind, we increase the region of G-cones in the scattering diagram. Since the Badlands happens to be the complement of the G-fan in this example, the shape of the Badlands will gradually emerge by this procedure.
- While proceeding, we carefully choose added walls so that both local and global structures of the Badlands are better visible. (For example,

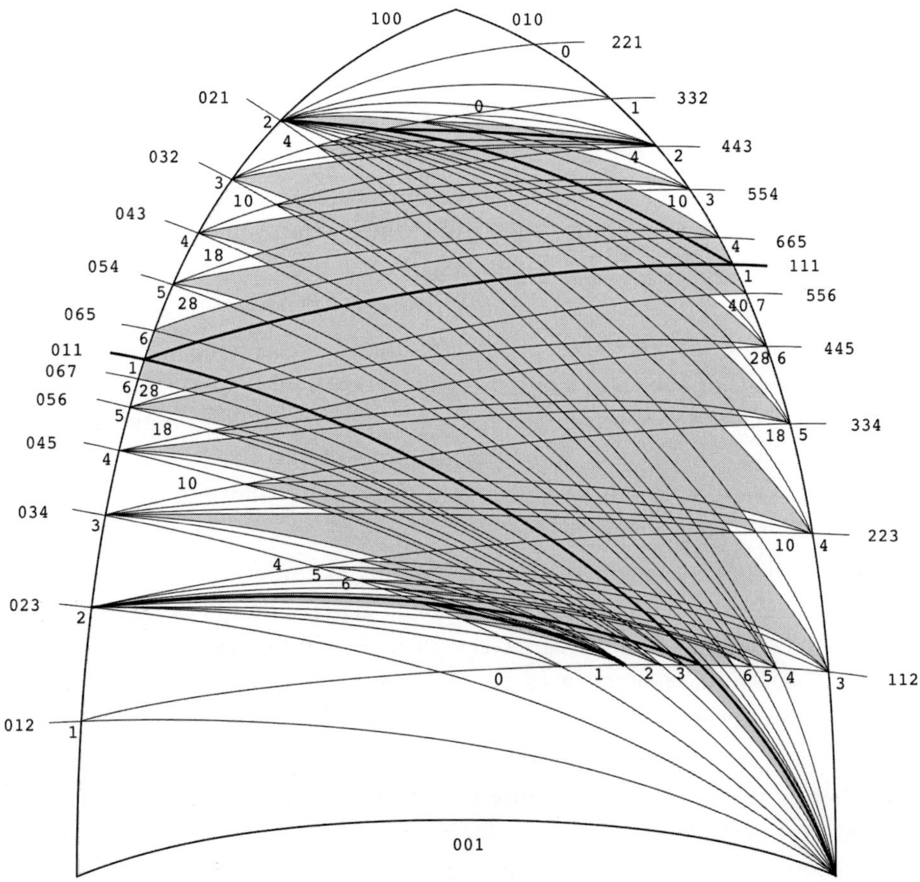

Figure 23: Uncompleted picture of the Badlands of rank 3. This is the close-up of the shaded region in Figure 22

we ignored some wall with degree 12, while we added a new wall of degree 245.)

- Due to the complexity we double-checked the result with a computer program for computing *c-vectors* of the cluster pattern. More specifically, we used SageMath 9.4 with *Cluster Algebras package* by D. Rupel and S. Stella.

The resulting CSD is depicted in Figures 22 and 23.

Figures 22 gives the big picture of the CSD. Let $\mathcal{R} = \mathcal{H}_1^{>0} \cap \mathcal{H}_2^{>0} \cap \mathcal{H}_3^{<0}$, which corresponds to the shaded region in Figure 22. Outside the region $\mathcal{R}$, the structure of the CSD is clearly presented in the figure, where there are actually infinitely many walls converging to the unreachable walls attached to joints of type $A_1^{(1)}$ represented by two thick arcs. There is no more additional wall outside $\mathcal{R}$.

The close-up of the region $\mathcal{R}$ is given in Figure 23. This is where the Badlands and its fractal nature appear. Here are remarks and observations:

- A three-digit number such as 032 represents the normal vector (0,3,2) of a wall crossing the boundary of the region $\mathcal{R}$.
- A number such as 4 at a joint in $n_1^\perp \cap n_2^\perp$ represents $|\{n_1, n_2\}| = 4$. It is put in the position of incoming direction, so that outgoing walls appear in the opposite direction as in the previous case.
- A thick arc represents the supports of walls attached to a joint of type $A_1^{(1)}$ that is unreachable by cluster mutations. They consist of the "backbone" of the total shaded region (explained below). Six thick arcs are presented here, but in the limit $\ell \to \infty$ there will be infinitely many ones added in the top and bottom parts of the back-born. However, they will be so small and tightly aligned that they are not visible in this page size anyway.
- An unshaded triangle is already completed under the algorithm, and it is a G-cone. Meanwhile, a shaded triangle or quadrilateral is not completed yet. Thus, when the degree ℓ increases, the corners are sharpened and new corners appear. In other words, the total shaded region reduces, and the complexity of the boundary increases.
- By the above comments and the picture presented in Figure 23, it is already clear that in the limit $\ell \to \infty$ the total shaded region (in $\mathbb{R}^3$) converges to some connected closed domain, say $\mathcal{B}$. It is also characterized as the unreachable region by cluster mutations. The boundary of $\mathcal{B}$ is a fractal. Moreover, if the density conjecture on the Badlands of rank 2 in Section 3.5 holds, the set $\mathcal{B}$ is also densely filled with nontrivial walls. Thus, we call $\mathcal{B}$ the *Badlands* of this CSD.

Some other examples of rank 3 CSD of infinite type are found in [Mul16, Rea]. It seems that there are several patterns for non-affine infinite type CSDs even for the rank 3 case, and to classify them is an interesting problem.

Notes

Most results are taken from [GHKK18, §1.3, Appendix C.2] with added/-modified proofs. In particular, the proof of Lemma 6.16 is significantly simplified from the one of [GHKK18, Theorem 1.28] by based on Construction 5.14 and Proposition 1.24 (b).

7 Broken lines and theta functions

In this section we introduce broken lines and theta functions for scattering diagrams. We prove the positivity of coefficients of theta functions for cluster scattering diagrams.

7.1 Broken lines and theta functions

In this section we again use the principal x-representation for wall elements. For notational simplicity, for any $\tilde{m}_i \in \tilde{M}^\circ = M^\circ \oplus N$, let m_i denote the M°-component of $\tilde{m}_i$. Namely, $\tilde{m}_i = m_i + n_i$ ($m_i \in M^\circ$, $n_i \in N$) for some n_i. We recall that $\langle n, \tilde{m}_i \rangle = \langle n, m_i \rangle$ holds for any $n \in N$ by (4.6).

Definition 7.1 (Broken line). Let $\mathfrak{D}$ be any scattering diagram for a given initial seed $\mathfrak{s}$. Let $\tilde{m}_0 \in \tilde{M}^\circ \setminus \{0\}$ and $Q \in M_{\mathbb{R}} \setminus \mathrm{Supp}(\mathfrak{D})$. A *broken line for $\tilde{m}_0$ with endpoint Q* is a piecewise-linear curve $\gamma : (-\infty, 0] \to M_{\mathbb{R}} \setminus \mathrm{Sing}(\mathfrak{D})$ satisfying the following properties:

(1). The endpoint $\gamma(0)$ is Q.

(2). The initial velocity $\gamma'(t)$ in $t \to -\infty$ is $-m_0$.

(3). The curve γ bends and change velocity only when it intersects walls of $\mathfrak{D}$. (It can cross walls also *without bending and changing velocity.*) Moreover, when γ bends, it always *crosses* walls, not *reflects* at them.

More conditions follow below. Suppose that γ bends $J \geq 0$ times. Let $t_1 < t_2 < \cdots < t_J < 0$ be the points ("time") of the parameter t for $\gamma(t)$ when γ bends. (It may cross multiple walls with a common normal vector at each time t_j.) We set $t_0 = -\infty$ and $t_{J+1} = 0$. Thus, γ is linear and with constant velocity in each interval $I_j = (t_j, t_{j+1})$ ($j = 0, \dots, J$). As an extreme case, for $J = 0$, the curve γ is a half line ending at Q with constant velocity $-m_0$. To each interval I_j, we attach a monomial $c_j x^{\tilde{m}_j} \in \mathbb{k}[\tilde{M}^\circ]$ recursively as follows:

(4). To the initial interval I_0, the monomial $x^{\tilde{m}_0}$ is attached, where $\tilde{m}_0$ is the given data.

(5). For $j \geq 1$, let γ_j be a segment of γ for the interval $(t_j - \delta, t_j + \delta)$ with sufficiently small $\delta > 0$ such that γ_j crosses walls only at t_j. Then, $c_j x^{\tilde{m}_j}$ is a monomial in $\mathfrak{p}_{\gamma_j, \mathfrak{D}}(c_{j-1} x^{\tilde{m}_{j-1}})$, where $\mathfrak{p}_{\gamma_j, \mathfrak{D}}$ acts as the principal x-representation. See (7.5) for more explicit description.

Finally, the velocity of γ in the interval I_j is given as follows:

(6). The velocity $\gamma'(t)$ in the interval I_j is given by $-m_j$, where $c_j x^{\tilde{m}_j}$ is the attached monomial. This agrees with the conditions (2) and (4).

This completes the definition of a broken line. We then attach a monomial in $\Bbbk[\tilde{M}^\circ]$ to each broken line γ by

$$c_\gamma x^{\tilde{m}_\gamma} := c_J x^{\tilde{m}_J}. \tag{7.1}$$

Several remarks are in order.

(a). At t_j, the velocity changes from $-m_j$ to $-m_{j+1} = -m_j - k_j p^*(n_j)$, where n_j is the common normal vector of the walls at t_j, and k_j is some positive integer. Since $p^*(n_j) \in n_j^\perp$, the broken line always *crosses* walls, *not reflects nor stops* at them as stated in the condition (3).

(b). In particular, we have $\tilde{m}_j = \tilde{m}_0 + \sum_{i=1}^r a_i \tilde{p}^*(e_i)$ $(a_i \in \mathbb{Z}_{\geq 0})$.

(c). For $\tilde{m}' = \tilde{m}_0 + \sum_{i=1}^r a_i \tilde{p}^*(e_i)$, we define the degree $\deg_{\tilde{m}_0}(\tilde{m}')$ by

$$\deg_{\tilde{m}_0}(\tilde{m}') := \sum_{i=1}^r a_i. \tag{7.2}$$

Then, by (a), we have

$$\deg_{\tilde{m}_0}(\tilde{m}_0) = 0 < \deg_{\tilde{m}_0}(\tilde{m}_1) < \cdots < \deg_{\tilde{m}_0}(\tilde{m}_J) = \deg_{\tilde{m}_0}(\tilde{m}_\gamma). \tag{7.3}$$

(d). For the segment γ_j in the condition (5), suppose that γ_j crosses walls $[\mathfrak{d}_\lambda, f_\lambda]_{n_j}$ $(\lambda \in \Lambda_j)$ at t_j. Note that $\gamma' = -m_{j-1}$ just before γ crosses these walls. Then, the intersection signs in (2.6) are given by

$$\epsilon_j = \begin{cases} 1 & \langle n_j, m_{j-1} \rangle > 0, \\ -1 & \langle n_j, m_{j-1} \rangle < 0. \end{cases} \tag{7.4}$$

Thus, $\epsilon_j \langle n_j, m_{j-1} \rangle = |\langle n_j, m_{j-1} \rangle|$. Therefore, by Definition 2.6, we have

$$\mathfrak{p}_{\gamma_j,\mathfrak{D}}(c_{j-1} x^{\tilde{m}_{j-1}}) = c_{j-1} x^{\tilde{m}_{j-1}} \prod_{\lambda \in \Lambda_j} f_\lambda^{|\langle \delta(n_j) n_j, m_{j-1} \rangle|}. \tag{7.5}$$

It is crucial that there is *no division* by f_λ in this expression for the forthcoming positivity of theta functions.

Definition 7.2 (General position). We say that $Q \in M_\mathbb{R} \setminus \mathrm{Supp}(\mathfrak{D})$ is *in general position* if for each $\ell > 0$ there is some neighborhood U_ℓ of Q such that, for the reduction $\mathfrak{D}_\ell$, any broken line for $Q' \in U_\ell$ converges to a broken line for Q in the limit $Q' \to Q$. (Namely, no broken line crosses $\mathrm{Sing}(\mathfrak{D}_\ell)$ in the limit.)

Below we always and implicitly assume that Q is in general position.

Definition 7.3 (Theta function). Under the same assumption and notations in Definition 7.1, the *theta function* $\vartheta_{Q,\tilde{m}_0}$ for $\tilde{m}_0$ with endpoint Q is defined by

$$\vartheta_{Q,\tilde{m}_0} := \sum_{\gamma \in B(Q,\tilde{m}_0)} c_\gamma x^{\tilde{m}_\gamma}, \tag{7.6}$$

where $B(Q, \tilde{m}_0)$ is the set of all broken lines for $\tilde{m}_0$ with endpoint Q. We also set

$$\vartheta_{Q,0} = 1. \tag{7.7}$$

The function $\vartheta_{Q,\tilde{m}_0}$ only depends on the equivalence class of $\mathfrak{D}$, though the condition for Q to be in general position depends on the choice of a representative of $\mathfrak{D}$.

Lemma 7.4 ([GHKK18, Proposition 3.4]). *For any scattering diagram $\mathfrak{D}$, we have*

$$\vartheta_{Q,\tilde{m}_0} \in x^{\tilde{m}_0}\Bbbk[[\tilde{P}]]. \tag{7.8}$$

Proof. By Remark (b), each term $c_\gamma x^{\tilde{m}_\gamma}$ belongs to $x^{\tilde{m}_0}\Bbbk[[\tilde{P}]]$. Thus, it is enough to show that, for a given $\tilde{m}' \in \tilde{m}_0 + \tilde{P}$, there are only finitely many γ such that $\tilde{m}_\gamma = \tilde{m}'$. We note the following facts:

- By the constraint (7.3), such γ bends at most $\deg_{\tilde{m}_0}(\tilde{m}')$ times.
- The walls of $\mathfrak{D}$ that contribute to $\tilde{m}'$ belongs to the reduction $\mathfrak{D}_\ell$ at $\ell = \deg_{\tilde{m}_0}(\tilde{m}')$, which has only finitely many walls.
- At each bending there are only finitely many possibilities of bending.

Therefore, there are only finitely many possibilities of such γ. □

Proposition 7.5 ([GHKK18, Proposition 3.8]). *Let $\mathfrak{D}$ be any scattering diagram for a given seed $\mathfrak{s}$, and let $\mathcal{C}^+_{\mathfrak{s}}$ be the positive orthant in* (2.40)*. For $\tilde{m}_0 \in \tilde{M}^\circ$, suppose that $m_0 \in \mathcal{C}^+_{\mathfrak{s}} \cap M^\circ$. Also, suppose that $Q \in \mathrm{Int}(\mathcal{C}^+_{\mathfrak{s}})$. Then, we have*

$$\vartheta_{Q,\tilde{m}_0} = x^{\tilde{m}_0}. \tag{7.9}$$

Proof. We may assume that $\tilde{m}_0 \neq 0$. It is clear that $Q + \mathbb{R}_{\geq 0} m_0$ is a broken line for $\tilde{m}_0$ with end point Q, and it does not intersect any walls of $\mathfrak{D}_{\mathfrak{s}}$; therefore, the associated monomial is $x^{\tilde{m}_0}$. We claim that under the condition there is no broken line γ with bending. Suppose that such γ exists. We use the notation in Definition 7.1. For each break point t_j therein, let $n_j \in N^+$ be a unique (not necessarily primitive) element such that γ intersects walls in $n_j^\perp$ at t_j and that the velocity of γ is shifted by $-p^*(n_j)$. Let L_j be the segment of γ for the interval I_j. Then, we claim the following:

$$L_j \subset H_j^- := \left\{ z \in M_{\mathbb{R}} \;\middle|\; \left\langle \sum_{i=1}^{j} n_i, z \right\rangle \leq 0 \right\} \quad (j = 1, \dots, J). \tag{7.10}$$

We show it by the induction on j. This is true for $j = 1$, because the support of the wall crossed by γ at t_1 is in $n_1^\perp$, and γ is in $\mathcal{C}^+_{\mathfrak{s}}$ for $t \to -\infty$. Now suppose that (7.10) hold up to j. Then, we have $\langle \sum_{i=1}^{j} n_i, \gamma(t_{j+1}) \rangle \leq 0$. Also, $\langle n_{j+1}, \gamma(t_{j+1}) \rangle = 0$. Thus, $\gamma(t_{j+1}) \in H_{j+1}^-$. Moreover, the new velocity vector $\gamma'_{j+1} = -m_0 - \sum_{i=1}^{j+1} p^*(n_i)$ satisfies the inequality

$$\begin{aligned} \left\langle \sum_{i=1}^{j+1} n_i, \gamma'_{j+1} \right\rangle &= \left\langle \sum_{i=1}^{j+1} n_i, -m_0 \right\rangle + \left\{ \sum_{i=1}^{j+1} n_i, \sum_{i=1}^{j+1} n_i \right\} \\ &= \left\langle \sum_{i=1}^{j+1} n_i, -m_0 \right\rangle \leq 0. \end{aligned} \tag{7.11}$$

Thus, the fact (7.10) holds for $j+1$. Since $\mathrm{Int}(\mathcal{C}_{\mathfrak{s}}^+) \cap H_j^- = \emptyset$, γ cannot reach $Q \in \mathrm{Int}(\mathcal{C}_{\mathfrak{s}}^+)$. □

7.2 Transitivity on theta functions

Now we assume that $\mathfrak{D}$ is consistent.

Proposition 7.6 ([CPS, Lemma 4.9]). *Let $\mathfrak{D}$ be a consistent scattering diagram. Let $\tilde{m}_0 \in \tilde{M}^\circ$ and Q, $Q' \in M_{\mathbb{R}} \setminus \mathrm{Supp}(\mathfrak{D})$. Then, for any admissible curve γ from Q to Q', we have*

$$\vartheta_{Q',\tilde{m}_0} = \mathfrak{p}_{\gamma,\mathfrak{D}}(\vartheta_{Q,\tilde{m}_0}). \tag{7.12}$$

The rest of this subsection is devoted to prove this proposition. We may assume that $\tilde{m}_0 \neq 0$.

Let J be the maximal ideal of $\Bbbk[[\tilde{P}]]$ generated by $\tilde{P} \setminus \{0\}$.

Lemma 7.7 ([CPS, Lemma 4.7]). *Let $\mathfrak{D}_\ell$ be the reduction of $\mathfrak{D}$ at ℓ. Suppose that Q and Q' are in the same chamber of $\mathrm{Supp}(\mathfrak{D}_\ell)$. Then, we have the equality*

$$\vartheta_{Q,\tilde{m}_0} \equiv \vartheta_{Q',\tilde{m}_0} \mod x^{\tilde{m}_0} J^{\ell+1}. \tag{7.13}$$

Proof. Since we consider $\vartheta_{Q,\tilde{m}_0}$ only modulo $x^{\tilde{m}_0} J^{\ell+1}$, one can replace $\mathfrak{D}$ with $\mathfrak{D}_\ell$ to calculate $\vartheta_{Q,\tilde{m}_0}$. First, suppose that Q and Q' are connected by a curve β in the same chamber such that any point on β is in general position for $\mathfrak{D}_\ell$. Then, each broken line with endpoint Q continuously deformed to a broken line with endpoint Q'. Therefore, the theta function is unchanged.

Suppose that there is no such curve between Q and Q'. Let us consider the special case where Q and Q' are connected by a curve β in the same chamber such that there is a unique point Q_0 on β that is not in general position. (The general case reduces to this case.) Then, there is at least one broken line γ with endpoint Q such that, in the limit $Q \to Q_0$, the broken line intersects with some boundary $\partial\mathfrak{d}$ or joint $\mathfrak{j} = \mathfrak{d}_1 \cap \mathfrak{d}_2$ of $\mathfrak{D}_\ell$.

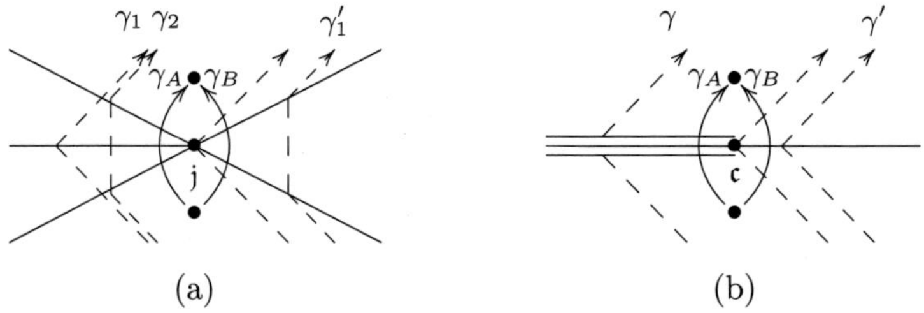

Figure 24: Broken lines in the proof of Lemma 7.7.

First, let us consider the case $\mathfrak{j}$. See Figure 24 (a). Suppose that γ crosses the walls that contain the joint $\mathfrak{j}$ with the incoming attached monomial $c_{\rm in}x^{\tilde{m}_{\rm in}}$ and the outgoing attached monomial $c_{\rm out}x^{\tilde{m}_{\rm out}}$. There may be several broken lines $\gamma_1 = \gamma, \ldots, \gamma_p$ with endpoint Q such that their bending patterns differ from γ only inside these walls. They have the common incoming attached monomial $c_{\rm in}x^{\tilde{m}_{\rm in}}$ and the common exponent $\tilde{m}_{\rm out}$ for the outgoing attached monomials. Then, all these broken lines have a common limit for $Q \to Q_0$. Let γ_A be the half of a small loop around $\mathfrak{j}$, which is on the same side of $\mathfrak{j}$ with γ_i's as in Figure 24 (a). Then, the sum of the outgoing attached monomials for $\gamma_1, \ldots, \gamma_p$ is exactly the terms of the monomials in $\mathfrak{p}_{\gamma_A,\mathfrak{D}_\ell}(c_{\rm in}x^{\tilde{m}_{\rm in}})$ with exponents $x^{\tilde{m}_{\rm out}}$. We repeat the same argument to the opposite side of $\mathfrak{j}$ with the opposite half γ_B of the small loop around $\mathfrak{j}$. Then, by the consistency of $\mathfrak{D}$, we have

$$\mathfrak{p}_{\gamma_A,\mathfrak{D}_\ell}(c_{\rm in}x^{\tilde{m}_{\rm in}}) \equiv \mathfrak{p}_{\gamma_B,\mathfrak{D}_\ell}(c_{\rm in}x^{\tilde{m}_{\rm in}}) \mod x^{\tilde{m}_0}J^{\ell+1}. \tag{7.14}$$

Thus, passing through Q_0 does not change $\vartheta_{Q,\tilde{m}_0}$ modulo $x^{\tilde{m}_0}J^{\ell+1}$. Therefore, we obtain the equality (7.13).

Next, we consider the remaining case $\partial\mathfrak{d}$. See Figure 24 (b). We may concentrate on a face $\mathfrak{c}$ of $\mathfrak{d}$ of codimension 2 such that $\mathfrak{c}$ is not a joint. Then, $\mathfrak{c}$ is contained in some walls of $\mathfrak{D}_\ell$ with the common normal vector n. In this case, for each broken line for Q, there is a unique broken line for Q' with the same attached incoming and outgoing monomial modulo $x^{\tilde{m}_0}J^{\ell+1}$, thanks to the consistency of $\mathfrak{D}$. Thus, passing through Q_0 does not change $\vartheta_{Q,\tilde{m}_0}$ modulo $x^{\tilde{m}_0}J^{\ell+1}$. □

Lemma 7.8 ([CPS, Lemma 4.8]). *Suppose that Q and Q' are in different chambers of* $\mathrm{Supp}(\mathfrak{D}_\ell)$. *Then, for any admissible curve γ from Q to Q', we have*

$$\vartheta_{Q',\tilde{m}_0} \equiv \mathfrak{p}_{\gamma,\mathfrak{D}_\ell}(\vartheta_{Q,\tilde{m}_0}) \mod x^{\tilde{m}_0}J^{\ell+1}. \tag{7.15}$$

Proof. We may assume that Q and Q' are in adjacent chambers of $\mathrm{Supp}(\mathfrak{D}_\ell)$, since the general case follows from this case by composing the automorphisms

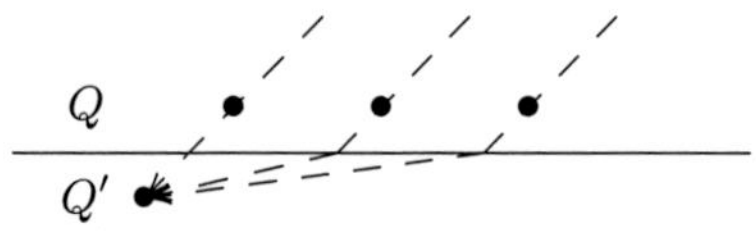

Figure 25: Broken lines in the proof of Lemma 7.8.

$\mathfrak{p}_{\gamma,\mathfrak{D}_\ell}$. Let $n \in N^+_{\mathrm{pr}}$ be the normal vector of the walls separating Q and Q'. We decompose

$$\vartheta_{Q,\tilde{m}_0} = \sum_{\tilde{m}\in\tilde{M}^\circ} c_{\tilde{m}} x^{\tilde{m}} = \vartheta_+ + \vartheta_0 + \vartheta_-, \tag{7.16}$$

$$\vartheta_+ = \sum_{\substack{\tilde{m}\in\tilde{M}^\circ\\ \langle n,m\rangle>0}} c_{\tilde{m}} x^{\tilde{m}}, \quad \vartheta_0 = \sum_{\substack{\tilde{m}\in\tilde{M}^\circ\\ \langle n,m\rangle=0}} c_{\tilde{m}} x^{\tilde{m}}, \quad \vartheta_- = \sum_{\substack{\tilde{m}\in\tilde{M}^\circ\\ \langle n,m\rangle<0}} c_{\tilde{m}} x^{\tilde{m}}. \tag{7.17}$$

Also, we decompose $\vartheta_{Q',\tilde{m}_0} = \vartheta'_+ + \vartheta'_0 + \vartheta'_-$ in the same way. Thanks to Lemma 7.7, we may take Q and Q' to be very close to the wall and also move parallel to the wall inside each chamber as we like. Without loosing generality, one can assume that $\langle n, Q\rangle > 0$ and $\langle n, Q'\rangle < 0$.

(i). ϑ_+ and ϑ'_+. For any exponent $\tilde{m}$ of ϑ'_+, $\langle n, m\rangle > 0$ holds. Then, any broken line for ϑ'_+ is obtained by extending a broken line for ϑ_+. See Figure 25. Thus, we have

$$\vartheta'_+ \equiv \mathfrak{p}_{\gamma,\mathfrak{D}_\ell}(\vartheta_+) \mod x^{\tilde{m}_0} J^{\ell+1}. \tag{7.18}$$

(ii). ϑ_- and ϑ'_-. The situation is opposite, and we have

$$\vartheta_- \equiv \mathfrak{p}^{-1}_{\gamma,\mathfrak{D}_\ell}(\vartheta'_-) \mod x^{\tilde{m}_0} J^{\ell+1}. \tag{7.19}$$

(iii). ϑ_0 and ϑ'_0. The broken lines for both do not cross the walls separating Q and Q' in the final step. Then, the proof of Lemma 7.7 applies to ϑ_0 and ϑ'_0, and we have

$$\vartheta'_0 \equiv \vartheta_0 \equiv \mathfrak{p}_{\gamma,\mathfrak{D}_\ell}(\vartheta_0) \mod x^{\tilde{m}_0} J^{\ell+1}. \tag{7.20}$$

We can unify the above three equalities into a single one

$$\vartheta_{Q',\tilde{m}_0} \equiv \mathfrak{p}_{\gamma,\mathfrak{D}_\ell}(\vartheta_{Q,\tilde{m}_0}) \mod x^{\tilde{m}_0} J^{\ell+1} \tag{7.21}$$

as desired. □

Since ℓ is arbitrary in (7.15), we have the desired equality (7.12). This completes the proof of Proposition 7.6.

7.3 Positivity and mutation invariance of theta functions

Now let us specialize to a CSD $\mathfrak{D}_{\mathfrak{s}}$ for a given seed $\mathfrak{s}$.

We have the following fundamental result on the positivity of theta functions for a CSD.

Theorem 7.9 (Positivity of theta functions [GHKK18, Theorem 1.13 & Remark 3.2]). *For a CSD $\mathfrak{D}_{\mathfrak{s}}$, every theta function $\vartheta_{Q,\tilde{m}_0} \in x^{\tilde{m}_0}\Bbbk[[\tilde{P}]]$ has only positive integer coefficients.*

Proof. We choose a positive realization of $\mathfrak{D}_{\mathfrak{s}}$ in Theorem 5.2, so that every wall function $f[tn]^s$ of $\mathfrak{D}_{\mathfrak{s}}$ has only positive integer coefficients. Then, the product in the right hand side of (7.5) has also only positive integer coefficients. Thus, c_γ in (7.1) is a positive integer. Therefore, the theta function defined by (7.6) has only positive integer coefficients. □

In the rest, let us present the mutation invariance of theta functions of a CSD and its consequences. Let us define a piecewise-linear transformation in parallel to T_k in (6.25),

$$\begin{aligned} \tilde{T}_k = \tilde{T}_{k,\mathfrak{s}}:\ \tilde{M}^\circ &\to \tilde{M}^\circ \\ \tilde{m} &\mapsto \tilde{m} + [\langle \delta_k e_k, m\rangle]_+ \tilde{p}^*(e_k) = \begin{cases} \tilde{S}_k(\tilde{m}) & m \in \mathcal{H}_k^+, \\ \tilde{m} & m \in \mathcal{H}_k^-, \end{cases} \end{aligned} \tag{7.22}$$

where $\tilde{S}_k(\tilde{m})$ is the one in (6.10).

Proposition 7.10 ([GHKK18, Proposition 3.6]). *For a given seed $\mathfrak{s}$, let $\mathfrak{s}' = \mu_k(\mathfrak{s})$ be the mutation of $\mathfrak{s}$ in direction k.*

(a). There is a one-to-one correspondence between the broken lines for $\tilde{m}_0$ with endpoint Q with respect to a CSD $\mathfrak{D}_{\mathfrak{s}}$ and the broken lines for $\tilde{T}_k(\tilde{m}_0)$ with endpoint $T_k(Q)$ with respect to a CSD $\mathfrak{D}_{\mathfrak{s}'} = T_k(\mathfrak{D}_{\mathfrak{s}})$. The locus of the line is transformed by the piecewise-linear map T_k in (6.25).

(b). (Mutation invariance of theta functions.) The theta function $\vartheta^{\mathfrak{s}}_{Q,\tilde{m}_0}$ with respect to $\mathfrak{D}_{\mathfrak{s}}$ and the one $\vartheta^{\mathfrak{s}'}_{Q,\tilde{m}_0}$ with respect to $\mathfrak{D}_{\mathfrak{s}'}$ are related by

$$\vartheta^{\mathfrak{s}'}_{T_k(Q),\tilde{T}_k(\tilde{m}_0)} = \begin{cases} \tilde{S}_k(\vartheta^{\mathfrak{s}}_{Q,\tilde{m}_0}) & Q \in \mathcal{H}_k^+, \\ \vartheta^{\mathfrak{s}}_{Q,\tilde{m}_0} & Q \in \mathcal{H}_k^-. \end{cases} \tag{7.23}$$

where $\mathcal{H}_k^\pm$ is the one in (6.23), and $\tilde{S}_k$ acts on the exponents of x in $\vartheta^{\mathfrak{s}}_{Q,\tilde{m}_0}$.

Proof. (a). In $\mathcal{H}_k^-$ the walls and broken lines are unchanged, while in $\mathcal{H}_k^+$ the walls and broken lines are transformed by the linear map S_k. Thus, the rule for broken lines are preserved in each $\mathcal{H}_k^\pm$. Therefore, we only need to check the rule when a broken line crosses the wall $\mathbf{w}_{e_k} = [e_k^\perp, 1 + x^{\tilde{p}^*(e_k)}]_{e_k}$.

Case (1). Suppose that a broken line γ with respect to $\mathfrak{D}_{\mathfrak{s}}$ crosses $e_k^{\perp}$ from $\mathcal{H}_k^-$ with attached monomial $cx^{\tilde{m}}$. Then, $\langle e_k, -m\rangle > 0$. After crossing $e_k^{\perp}$, the new attached monomial is a term in $cx^{\tilde{m}}(1+x^{\tilde{p}^*(e_k)})^{|\langle \delta_k e_k, m\rangle|}$. Applying $\tilde{S}_k$, it transforms to a term in

$$\begin{aligned}
&cx^{\tilde{m}+\langle \delta_k e_k, m\rangle \tilde{p}^*(e_k)}(1+x^{\tilde{p}^*(e_k)})^{|\langle \delta_k e_k, m\rangle|}\\
&= cx^{\tilde{m}-|\langle \delta_k e_k, m\rangle|\tilde{p}^*(e_k)}(1+x^{\tilde{p}^*(e_k)})^{|\langle \delta_k e_k, m\rangle|}\\
&= cx^{\tilde{m}}(1+x^{\tilde{p}^*(e'_k)})^{|\langle \delta_k e'_k, m\rangle|}.
\end{aligned} \tag{7.24}$$

This is the rule for a broken line with respect to $T_k(\mathfrak{D}_{\mathfrak{s}})$.

Case (2). Suppose that a broken line γ with respect to $\mathfrak{D}_{\mathfrak{s}}$ crosses $e_k^{\perp}$ from $\mathcal{H}_k^+$ with attached monomial $cx^{\tilde{m}}$. Then, $\langle e_k, -m\rangle < 0$. After crossing $e_k^{\perp}$, the new attached monomial is a term in $cx^{\tilde{m}}(1+x^{\tilde{p}^*(e_k)})^{|\langle \delta_k e_k, m\rangle|}$, which is rewritten as

$$cx^{\tilde{m}+\langle \delta_k e_k, m\rangle \tilde{p}^*(e_k)}(1+x^{\tilde{p}^*(e'_k)})^{|\langle \delta_k e'_k, m\rangle|}. \tag{7.25}$$

Since the attached monomial for the incoming segment is transformed from $cx^{\tilde{m}}$ to $cx^{\tilde{m}+\langle \delta_k e_k, m\rangle \tilde{p}^*(e_k)}$, this is the correct rule for a broken line with respect to $T_k(\mathfrak{D}_{\mathfrak{s}})$.

(b). Let γ be a broken line with respect to $\mathfrak{D}_{\mathfrak{s}}$, and let $\gamma' = T_k(\gamma)$ be the corresponding broken line with respect to $\mathfrak{D}_{\mathfrak{s}'}$. Let $c_\gamma x^{\tilde{m}_\gamma}$ and $c'_\gamma x^{\tilde{m}'_\gamma}$ be the monomials attached with γ and γ', respectively. Then, by (a), we have

$$c'_\gamma x^{\tilde{m}'_\gamma} = \begin{cases} c_\gamma x^{\tilde{S}_k(\tilde{m}_\gamma)} & Q \in \mathcal{H}_k^+, \\ c_\gamma x^{\tilde{m}_\gamma} & Q \in \mathcal{H}_k^-. \end{cases} \tag{7.26}$$

Thus, we have (7.23). □

We have the following corollary of Propositions 7.5 and 7.10.

Corollary 7.11 ([GHKK18, Corollary 3.9]). *Let $\mathfrak{D}_{\mathfrak{s}}$ be a CSD, and let $\mathcal{C}^{\mathfrak{s}}_{\mathfrak{s}_i}$ be the G-cone in (6.75). For $\tilde{m}_0 \in \tilde{M}^\circ$, suppose that $m_0 \in \mathcal{C}^{\mathfrak{s}}_{\mathfrak{s}_i} \cap M^\circ$. Also, suppose that $Q \in \mathrm{Int}(\mathcal{C}^{\mathfrak{s}}_{\mathfrak{s}_i})$. Then, we have*

$$\vartheta_{Q,\tilde{m}_0} = x^{\tilde{m}_0}. \tag{7.27}$$

Proof. By (6.75),

$$\mathcal{C}^+_{\mathfrak{s}_i} = (T_{k_{i-1},\mathfrak{s}_{i-1}} \circ \cdots \circ T_{k_0,\mathfrak{s}_0})(\mathcal{C}^{\mathfrak{s}}_{\mathfrak{s}_i}). \tag{7.28}$$

Let $T := T_{k_{i-1},\mathfrak{s}_{i-1}} \circ \cdots \circ T_{k_0,\mathfrak{s}_0}$ and $\tilde{T} := \tilde{T}_{k_{i-1},\mathfrak{s}_{i-1}} \circ \cdots \circ \tilde{T}_{k_0,\mathfrak{s}_0}$. Let $S : M_{\mathbb{R}} \to M_{\mathbb{R}}$ be the linear map that locally coincides with T on the cone $\mathcal{C}^{\mathfrak{s}}_{\mathfrak{s}_i}$. Let $\tilde{T}$ and $\tilde{S}$ be the corresponding maps for $\tilde{M}^\circ \to \tilde{M}^\circ$. Then, applying (7.23) repeatedly, we obtain

$$\vartheta^{\mathfrak{s}_i}_{T(Q),\tilde{T}(\tilde{m}_0)} = \tilde{S}(\vartheta_{Q,\tilde{m}_0}), \tag{7.29}$$

where $\tilde{S}$ acts on the exponents of x. Meanwhile, $\vartheta^{\mathfrak{s}_i}_{T(Q),\tilde{T}(\tilde{m}_0)} = x^{\tilde{T}(\tilde{m}_0)}$ by Proposition 7.5. Thus, we have $\vartheta_{Q,\tilde{m}_0} = x^{\tilde{m}_0}$. □

Combining Proposition 7.6 and Corollary 7.11, we have another fundamental theorem on theta functions for a CSD.

Theorem 7.12 ([GHKK18, Theorem 4.9]). *Let $\mathfrak{D}_{\mathfrak{s}}$ be a CSD, and let $\mathcal{C}^{\mathfrak{s}}_{\mathfrak{s}_i}$ be the G-cone in* (6.75)*. For $\tilde{m}_0 \in \tilde{M}^\circ$, suppose that $m_0 \in \mathcal{C}^{\mathfrak{s}}_{\mathfrak{s}_i} \cap M^\circ$. Also, suppose that $Q \in \mathrm{Int}(\mathcal{C}^+_{\mathfrak{s}})$. Then, for any admissible curve γ from any point in $\mathrm{Int}(\mathcal{C}^{\mathfrak{s}}_{\mathfrak{s}_i})$ to Q, we have*

$$\vartheta_{Q,\tilde{m}_0} = \mathfrak{p}_{\gamma,\mathfrak{D}_{\mathfrak{s}}}(x^{\tilde{m}_0}). \tag{7.30}$$

To conclude, let us describe the relation between theta functions and cluster variables (x-variables) for the corresponding cluster pattern with principal coefficients in [FZ07]. See Section 6 of Part II for a detailed account.

As explained briefly in Section 6.6, each G-cone $\mathcal{C}^{\mathfrak{s}}_{\mathfrak{s}_i}$ in (6.75) is identified with a cone spanned by the g-vectors $\mathbf{g}_{1;t}, \dots, \mathbf{g}_{r;t}$ of the G-matrix $G^{t_0}_t$ for some $t \in \mathbb{T}_r$. Let $\mathbf{x}_t = (x_{1;t}, \dots, x_{r;t})$ be the *cluster* at t for the corresponding cluster pattern with principal coefficients. By the celebrated *Laurent phenomenon* by [FZ02a], any cluster variables $x_{i;t}$ ($i = 1, \dots, r$) are expressed as a Laurent polynomial in the initial cluster variables $\mathbf{x}_{t_0} = \mathbf{x} = (x_1, \dots, x_r)$ and the initial principal coefficients $\mathbf{y}_{t_0} = \mathbf{y} = (y_1, \dots, y_r)$, where negative exponents appear only for variables x_i. We identify $x^{\tilde{m}}$ ($\tilde{m} \in \tilde{M}^\circ$) with the above variables as $x^{f_i} = x_i$ and $x^{e_i} = y_i$. See (1.7). We note that

$$x^{\tilde{p}^*(e_i)} = y_i \prod_{j=1}^{r} x_j^{b_{ji}} \tag{7.31}$$

are the *$\hat{y}$-variables* $\hat{y}_i$ in [FZ07]. Under this identification, it follows from Theorem 7.12 that the cluster variables $x_{i;t}$ are identified with (a part of) theta functions as

$$x_{i;t} = \vartheta_{Q,\mathbf{g}_{i;t}}, \tag{7.32}$$

where the RHS is the one given in (7.30).

Notes

All results are taken from [GHKK18, §3, §4] and [CPS, §4] adapted to the present formulation of scattering diagrams.

Appendix A Source code of Ordering Algorithm

For the reader's convenience we present a source code of Ordering Algorithm (Algorithm 5.7) written for SageMath 9.4 [1]. It is an almost direct translation of Algorithm 5.7, and no effort is made for the efficiency and sophistication. (If any minor or major bug is found, please let us know.)

The following examples explain how to use it on Sage Notebook.

Example A.1. (a). Let us order the product $\Psi[e_2]^5\Psi[e_1]^3$ modulo $G^{>5}$.

```
In:
order(5,[[0,1,5],[1,0,3]])
Out:
[[1, 0, 3],
 [3, 1, 5],
 [2, 1, 15],
 [3, 2, 125],
 [1, 1, 15],
 [2, 2, 60],
 [2, 3, 270],
 [1, 2, 30],
 [1, 3, 30],
 [1, 4, 15],
 [0, 1, 5]]
```

(b). One can order any (not necessarily anti-ordered) product with factors of the form $\Psi[tn]^{s/t}$ ($n \in N_{\mathrm{pr}}^{+},\ s,t \in \mathbb{Z}_{>0}$) as specified in Algorithm 5.7.

```
In:
order(6,[[0,1,1],[1,2,1],[2,1,1],[2,2,1/2],[1,0,2]])
Out:
[[1, 0, 2],
 [4, 1, 1],
 [3, 1, 2],
 [2, 1, 2],
 [4, 2, 19],
 [3, 2, 16],
 [1, 1, 2],
 [2, 2, 13/2],
 [3, 3, 33],
 [2, 3, 10],
 [1, 2, 1],
 [2, 4, 9/2],
 [1, 3, 1],
 [0, 1, 1]]
```

Enjoy exploring the Badlands!

[1] Sage Mathematics Software, The Sage Development Team, https://www.sagemath.org.

Source code of Ordering Algorithm for SageMath 9.4

```
# Ordering Algorithm for dilogarithm products
# written by Tomoki Nakanishi for SageMath 9.4
# ver.2021.10.30.b
# usage: order(L,C)

def bf(a,b):
    # bilinear form {a,b} such that {e_2,e_1}=1
    # a, b: rank 2 integer vectors
    res = a[1]*b[0]-a[0]*b[1]
    return res

def check_ordered(C):
    # check dilog product C is ordered completely
    # C has factors of [n]^{c}, c>0 any, len(C)>1
    l = len(C) # l>1
    for i in range(l-1):
        if bf(C[i][:2],C[i+1][:2]) > 0: # if {n',n}>0
            return False # not ordered
        elif bf(C[i][:2],C[i+1][:2]) == 0: # {n',n}=0
            if C[i][0]+C[i][1] > C[i+1][0]+C[i+1][1]: # deg(n')>deg(n)
                return False # not ordered
    return True # ordered

def decompose_initial(C):
    # decompose dilog product C for the main routine (1)
    # C has factors [tn]^{s/t}
    res = []
    l = len(C)
    for i in range(l):
        if C[i][2].is_integer() and C[i][2] > 1: # if the power is integer >1
            for j in range(C[i][2]):
                res = res+[[C[i][0],C[i][1],1]] # decompose to [tn]'s
        else:
            res = res+[C[i]]
    return res

def decompose(p,C):
    # decompose dilog product C for subroutine (p), p>1
    # p>0: integer
    # C=[n']^{c'}[n]^{c}) is an p-exchangeable pair
    res = []
    c0 = C[0][2] # multiple of 1/p
    c1 = C[1][2] # multiple of 1/p
    for i in range(p*c0):
        res = res+[C[0][:2]+[1/p]] # decompose to [n]^{1/p}
    for i in range(p*c1):
```

```
        res = res+[C[1][:2]+[1/p]] # decompose to [n]^{1/p}
    return res

def join(C):
    # join factors with common [n] in dilog product C
    # C has factors of [n]^{c}, len(C)>1
    res = C
    l = len(res) # l>1
    flag_joined = False
    while not flag_joined:
        l = len(res)
        for i in range(l-1):
            if  res[i][:2] == res[i+1][:2]:
                res = res[:i]+[res[i][:2]+[res[i][2]+res[i+1][2]]]+res[i+2:]
                break
            if i == l-2: # if this is the last pair, it is joined
                    flag_joined = True
    return res

def order_p_partial(L,p,C):
    # order dilog product C up to G^{>L} by p-pentagon (i.e.,{n',n}=p)
    # L,p>0: integer
    # C has factors of [n]^{1/p}, len(C)>1
    res = C
    flag_ordered_p = False
    while not flag_ordered_p: # finish if p-ordered
        l = len(res) # l>1
        for i in range(l-1):
            if bf(res[i][:2],res[i+1][:2]) < 0: # if ordered, do nothing
                pass
            elif bf(res[i][:2],res[i+1][:2]) == 0: # if parallel
                if res[i][0]+res[i][1] > res[i+1][0]+res[i+1][1]:
                    # if deg(left)>deg(right), commute it
                    res = res[:i]+[res[i+1]]+[res[i]]+res[i+2:]
                    break
            elif bf(res[i][:2],res[i+1][:2]) > 0: # if anti-ordered
                if res[i][0]+res[i][1]+res[i+1][0]+res[i+1][1] > L:
                    # if deg(n+n')>L, commute it
                    res = res[:i]+[res[i+1]]+[res[i]]+res[i+2:]
                    break
                elif bf(res[i][:2],res[i+1][:2]) == p:
                    # else if {n',n}=p, apply p-pentagon relation
                    if res[i][2] == 1/p and res[i+1][2]== 1/p:
                        res = res[:i]+[res[i+1]]+[[res[i][0]+res[i+1][0],\
                            res[i][1]+res[i+1][1],1/p],res[i]]+res[i+2:]
                        break
            if i == l-2: # if we reach the last pair, it is p-ordered
                flag_ordered_p = True
```

```
    return res

def order_p(L,p,C):
    # subroutine (p); recursively defined
    # L,p>0: integer
    # for p=1, C has factors [tn]^{s/t} (n:primitive vector, s,t>0 integers)
    # for p>1, C has factors of [n]^{1/p}
    # len(C)>1
    res = C
    while not check_ordered(res): # do until completely ordered
        if p == 1: # only in the main routine (1)
            res=decompose_initial(res) # decompose to [n]
        res = order_p_partial(L,p,res) # p-order the product
        l = len(res) # l>1
        for i in range(l-1):
            if bf(res[i][:2],res[i+1][:2]) > 0: # find q-admissible pair
                q = bf(res[i][:2],res[i+1][:2])
                Cin = decompose(q,[res[i],res[i+1]]) # decompose to [n]^{1/q}
                Cout = order_p(L,q,Cin) # go to subroutine (q) (recursively)
                res = res[:i]+Cout+res[i+2:]
                break
    res = join(res)
    return res

def order(L,C):
    """    L: integer > 0, C: list of [n1,n2,c]
    n1,n2: integers>=0, c: rational>0 such that c*gcd(n1,n2)=integer>0 """
    # main routine (1)
    # L>0: integer
    # C has factors [tn]^{s/t} (n:primitive positive vector, s,t>0 integers)
    if not (L > 0 and L.is_integer()): # check L
        return print("the␣first␣argument␣is␣illegal")
    l = len(C) # l>=1
    for i in range(l):
        n1 = C[i][0]
        n2 = C[i][1]
        if not (n1.is_integer() and n2.is_integer()\
                and n1 >= 0 and n2 >= 0 and n1+n2 > 0): # check (n1,n2)
            return print("the␣second␣argument␣is␣illegal")
        s = C[i][2]*math.gcd(n1,n2)
        if not (s > 0 and s.is_integer()): # check c
            return print("the␣second␣argument␣is␣illegal")
    if l == 1: # for a single factor, just return
        return C
    res = order_p(L,1,C) # enter the main routine (1)
    return res
```

References

[BFZ05] A. Berenstein, S. Fomin, and A. Zelevinsky, *Cluster algebras III: Upper bounds and double Bruhat cells*, Duke Math. J. **126** (2005), 1–52; arXiv:math/035434 [math.RT].

[Bou68] N. Bourbaki, *Groupes et algèbres de Lie, Ch, 4–6*, Hermann, Paris, 1968.

[Bri17] T. Bridgeland, *Scattering diagrams, Hall algebras and stability conditions*, Algebraic Geometry **4** (2017), 523–561; arXiv:1603.00416 [math.AG].

[Car06] R. W. Carter, *Cluster algebras*, Textos de Matemática Vol. 37, Department de Matemática da Universidade de Coimbra, 2006.

[CGM+17] M.-W. Cheung, M. Gross, M. Muller, G. Musiker, D. Rupel, S. Stella, and H. Williams, *The greedy basis equals the theta basis: A rank two haiku*, J. Combin. Theory, Ser. A **145** (2017), 150–171; arXiv:1508.01404 [math.QA].

[CHL20] P. Cao, M. Huang, and F. Li, *A conjecture on C-matrices of cluster algebras*, Nagoya Math. J. **238** (2020), 37–46; arXiv:1702.01221 [math.RA].

[CI12] G Cerulli Irelli, *Cluster algebras of type $A_2^{(1)}$*, Algebras and Representation Theory **15** (2012), 977–1021; arXiv:0904.2543 [math.RA].

[CIKLFP13] G. Cerulli Irelli, B. Keller, D. Labardini-Fragoso, and P. Plamondon, *Linear independence of cluster monomials for skew-symmetric cluster algebras*, Compos. Math. **149** (2013), 1753–1764; arXiv:1203.1307 [math.RT].

[CILF11] G. Cerulli Irelli and D Labardini-Fragoso, *Quivers with potentials associated to triangular surfaces, part IIIs: Tagged triangulations and cluster monomials*, Compos. Math. **148** (2011), 1833–1866; arXiv:1108.1774 [math.RT].

[CK08] P. Caldero and B. Keller, *From triangulated categories to cluster algebras*, Invent. Math. **172** (2008), 169–211; arXiv:math/0506018 [math.RT].

[CL20] P. Cao and F. Li, *The enough g-pairs property and denominator vectors of cluster algebras*, Mathematische Annalen **377** (2020), 1547–1572; arXiv:1803.05281 [math.RT].

[CPS] M. Carl, M. Pumperla, and B. Siebert, *A tropical view of Landau-Ginzburg models*, preprint, 2010, available at https://www.math.uni-hamburg.de/home/siebert/preprints/LGtrop.pdf.

[CS14] L. Chekhov and M. Shapiro, *Teichmüller spaces of Riemann surfaces with orbifold points of arbitrary order and cluster variables*, Int. Math. Res. Notices **2014** (2014), 2746–2772; arXiv:1111.3963 [math–ph].

[Dav18] B. Davison, *Positivity for quantum cluster algebras*, Ann. of Math. **187** (2018), 157–219; arXiv:1601.07918 [math.RT].

[DM21] B. Davison and T. Mandel, *Strong positivity for quantum theta bases of quantum cluster algebras*, Invent. Math. (2021), published online; arXiv:1910.12915 [math.RT].

[DWZ10] H. Derksen, J. Weyman, and A. Zelevinsky, *Quivers with potentials and their representations II: Applications to cluster algebras*, J. Amer. Math. Soc. **23** (2010), 749–790; arXiv:0904.0676 [math.RA].

[FG07] V. V. Fock and A. B. Goncharov, *Dual Teichmüller and lamination spaces*, Handbook of Teichmüller theory, Vol. I, Eur. Math. Soc., 2007, pp. 647–684; arXiv:math/0510312 [math.DG].

[FG09a] ______, *Cluster ensembles, quantization and the dilogarithm*, Annales Sci. de l'École Norm. Sup. **42** (2009), 865–930; arXiv:math/0311245 [math.AG].

[FG09b] ______, *The quantum dilogarithm and representations of quantum cluster varieties*, Invent. Math. **172** (2009), 223–286; arXiv:math/0702397 [math.QA].

[FG16] V. Fock and A. Goncharov, *Cluster poisson varieties at infinity*, Selecta Math. (N.S.) **22** (2016), 2569–2589; arXiv:1104.0407 [math.AG].

[FG19] S. Fujiwara and Y. Gyoda, *Duality between final-seed and initial-seed mutations in cluster algebras*, SIGMA **15** (2019), 040, 24 pages; arXiv:1808.02156.

[FST08] S. Fomin, M. Shapiro, and D. Thurston, *Cluster algebras and triangulated surfaces. Part I: Cluster complexes*, Acta Math. **201** (2008), 83–146; arXiv:math/0608367 [math.RA].

[Ful97] W. Fulton, *Young tableaux*, London Mathematical Society Student Texts 35, Cambridge University Press, 1997.

[FZ02a] S. Fomin and A. Zelevinsky, *Cluster algebras I. Foundations*, J. Amer. Math. Soc. **15** (2002), 497–529 (electronic); arXiv:math/0104151 [math.RT].

[FZ02b] ______, *The Laurent phenomenon*, Adv. in Appl. Math. **28** (2002), 119–144; arXiv:math/0104241 [math.CO].

[FZ03a] ______, *Cluster algebras II. Finite type classification*, Invent. Math. **154** (2003), 63–121; arXiv:math/0208229 [math.RA].

[FZ03b] ______, *Cluster algebras: Notes for the CDM-03 Conference*, Current developments in mathematics (Somerville, MA), Int. Press., 2003, pp. 1–34.

[FZ03c] ______, *Y-systems and generalized associahedra*, Ann. of Math. **158** (2003), 977–1018; arXiv:hep–th/0111053.

[FZ07] ______, *Cluster algebras IV. Coefficients*, Compositio Mathematica **143** (2007), 112–164; arXiv:math/0602259 [math.RT].

[FZW16] S. Fomin, A. Zelevinsky, and L. Williams, *Introduction to cluster algebras, Chapter 1–3*, 2016, arXiv:1608.05735 [math.CO].

[FZW17] ______, *Introduction to cluster algebras, Chapter 4–5*, 2017, arXiv:1707.07190 [math.CO].

[FZW20] ______, *Introduction to cluster algebras, Chapter 6*, 2020, arXiv:2008.09189 [math.AC].

[FZW21] ______, *Introduction to cluster algebras, Chapter 7*, 2021, arXiv:2106.02160 [math.CO].

[GHK15] M. Gross, P. Hacking, and S. Keel, *Birational geometry of cluster algebras*, Algebr. Geom. **2** (2015), 137–175; arXiv:math/1309.2573.

[GHKK18] M. Gross, P. Hacking, S. Keel, and M. Kontsevich, *Canonical bases for cluster algebras*, J. Amer. Math. Soc. **31** (2018), 497–608; arXiv:1411.1394 [math.AG].

[GLS12] C. Geiss, B. Leclerc, and J. Schröer, *Generic bases for cluster algebras and the Chamber Ansatz*, J. Amer. Math. Soc. **25** (2012), 21–76; arXiv:1004.2781 [math.RT].

[GNR17] M. Gekhtman, T. Nakanishi, and D. Rupel, *Hamiltonian and Lagrangian formalisms of mutations in cluster algebras and application to dilogarithm identities*, J. Integrable Syst. **2** (2017), 1–35; arXiv: 1611.02813.

[GPS10] M. Gross, R. Pandharipande, and B. Siebert, *The tropical vertex*, Duke Math. J. (2010), no. 153, 297 – 362; arXiv:0902.0779 [math.AG].

[GR17] M. Glick and D. Rupel, *Introduction to cluster algebras*, Symmetries and Integrability of Difference Equations (Levi D., Rebelo R., and P. Winternitz, eds.), CRM Series in Mathematical Physics, Springer, 2017, pp. 325–358; arXiv:1803.08960 [math.CO].

[Gro11] M. Gross, *Tropical geometry and mirror symmetry*, CBMS Regional Conf. Ser. in Math., no. 114, Amer. Math. Soc., 2011.

[GS11] M. Gross and B. Siebert, *From affine geometry to complex geometry*, Annals of Math. **174** (2011), 95–138; arXiv:math/0703822.

[GSV08] M. Gekhtman, M. Shapiro, and A Vainshtein, *On the properties of the exchange graph of a cluster algebra*, Math. Res. Lett. (2008), 321–330; arXiv:math/0703151.

[GSV10] M. Gekhtman, M. Shapiro, and A. Vainshtein, *Cluster algebras and Poisson geometry*, Mathematical Surveys and Monographs, no. 167, American Mathematical Society, 2010.

[Hum90] J. E. Humphreys, *Reflection groups and Coxeter groups*, Cambridge Univ. Press, Cambridge, 1990.

[IN14] K. Iwaki and T. Nakanishi, *Exact WKB analysis and cluster algebras*, J. Phys. A: Math. Theor. **47** (2014), 474009; arXiv:1401.7094 [math.CA].

[IOTW] K. Igusa, K. Orr, G. Todorov, and J. Weyman, *Modulated semi-invariants*, arXiv:1507.03051 [math.RT].

[IT21] K Igusa and G. Todorov, *Picture groups and maximal green sequences*, Electronic Research Archive **29** (2021), 3031–3068; arXiv:2007.14584 [math.RT].

[ITW] K. Igusa, G. Todorov, and J. Weyman, *Picture groups of finite type and cohomology in type A_n*, arXiv:1609.02636 [math.RT].

[Jac79] N. Jacobson, *Lie algebras*, Dover Publications, New York, 1979.

[Kac90] V. G. Kac, *Infinite dimensional Lie algebras*, 3rd ed., Cambridge University Press, 1990.

[Kel10] B. Keller, *Cluster algebras, quiver representations and triangulated categories*, Triangulated categories (T. Holm, P. Jørgensen, and R. Rouquier, eds.), Lecture Note Series, vol. 375, London Mathematical Society, Cambridge University Press, 2010, pp. 76–160; arXiv:0807.1960 [math.RT].

[Kel13] B. Keller, *Cluster algebras and derived categories*, Derived Categories in Algebraic Geometry (Y. Kawamata, ed.), EMS Ser. Congr. Rep., Eur. Math. Soc., 2013, pp. 123–183; arXiv:1202.4161 [math.RT].

[KQ14] Y. Kimura and F. Qin, *Graded quiver varieties, quantum cluster algebras and dual canonical basis*, Adv. in Math. **262** (2014), 261–312; arXiv:1205.2066.

[KS06] M. Kontsevich and Y. Soibelman, *Affine structures and non-Archimedean analytic spaces*, Prog. Math. **244** (2006), 321–385; arXiv:math/0406564.

[KS08] ______, *Stability structures, motivic Donaldson-Thomas invariants and cluster transformations*, 2008, arXiv:0811.2435 [math.AG].

[KS14] ______, *Wall-crossing structures in Donaldson-Thomas invariants, integrable systems and mirror symmetry*, Homological mirror symmetry and tropical geometry, Lect. Notes Unione Ital., vol. 15, Springer, 2014, pp. 197–308; arXiv:1303.3253 [math.AG].

[Lew81] L. Lewin, *Polylogarithms and associated functions*, North-Holland, Amsterdam, 1981.

[LS15] K. Lee and R. Schiffler, *Positivity for cluster algebras*, Ann. of Math. **182** (2015), 72–125; arXiv:1306.2416.

[Mag54] W. Magnus, *On the exponential solution of differential equations for a linear operator*, Commun. Pure Appl. Math. **7** (1954), 649–673.

[Mar13] R. J. Marsh, *Lecture notes on cluster algebras*, Zurich Lectures in Advanced Mathematics, European Mathematical Society, Zürich, 2013.

[Mat21] K. Matsushita, *Consistency relations of rank 2 cluster scattering diagrams of affine type and the pentagon relation*, 2021, arXiv:2112.04743 [math.QA].

[May69] W. May, *Commutative group algebras*, Trans. Amer. Math. Soc. **136** (1969), 139–149.

[Mou19] L. Mou, *Scattering diagrams of quivers with potentials and mutations*, arXiv:1910.1371 [math.RT], 2019.

[MSW11] G. Musiker, R. Schiffler, and L. Williams, *Positivity for cluster algebras from surfaces*, Adv. in Math. **227** (2011), 2241–2308; arXiv:0906.0748 [math.CO].

[Mul16] G. Muller, *The existence of a maximal green sequence is not invariant under quiver mutation*, Electron. J. Combin. **23** (2016), #P2.47, 23 pages; arXiv:1503.04675.

[Nag13] K. Nagao, *Donaldson-Thomas theory and cluster algebras*, Duke Math. J. **7** (2013), 1313–1367; arXiv:1002.4884 [math.AG].

[Nak12] T. Nakanishi, *Tropicalization method in cluster algebras*, Contemp. Math. **580** (2012), 95–115; arXiv:1110.5472 [math.QA].

[Nak15] ______, *Structure of seeds in generalized cluster algebras*, Pacific J. Math. **277** (2015), 201–218; arXiv:1409.5967 [math.RA].

[Nak21] ______, *Synchronicity phenomenon in cluster patterns*, J. London Math. Soc. **103** (2021), 1120–1152; arXiv:1906.12036 [math.RA].

[NC13] A. Nájera Chávez, *On the c-vectors and g-vectors of the Markov cluster algebra*, Séminaire Lontharingien de Combinatoire **69** (2013), B29d;arXiv:1112.2357.

[NR16] T. Nakanishi and D. Rupel, *Companion cluster algebras to a generalized cluster algebra*, Travaux Mathématiques (2016), 129–149; arXiv:1504.06758 [math.RA].

[NZ12] T. Nakanishi and A. Zelevinsky, *On tropical dualities in cluster algebras*, Contemp. Math. **565** (2012), 217–226, arXiv:1101.3736 [math.RA].

[Pla11] P. Plamondon, *Cluster algebras via cluster categories with infinite-dimensional morphism spaces*, Compos. Math. **147** (2011), 1921–1954; arXiv:1004.0830 [math.RT].

[Pla18] P. Plamondon, *Cluster characters*, Homological Methods, Representation Theory, and Cluster Algebras, CRM Short Courses, Springer, 2018, pp. 101–125; arXiv:1610.07546 [math.RT].

[Qin19] F. Qin, *Bases for upper cluster algebras and tropical points*, 2019, arXiv:1902.09507 [math.RT].

[Rea] N. Reading, *Dominance phenomena: mutation, scattering, and cluster algebras*, arXiv:1802.10107 [math.CO].

[Rea14] ———, *Universal geometric cluster algebras*, Mathematische Zeitschrift **277** (2014), 499–547; arXiv:1209.3987.

[Rea20a] ———, *A combinatorial approach to scattering diagrams*, Algebraic Combinatorics **3** (2020), 603–636; arXiv:1806.05094 [math.CO].

[Rea20b] ———, *Scattering fans*, Int. Math. Res. Notices **23** (2020), 9640–9673; arXiv:1712.06968.

[Rei09] M. Reineke, *Poisson automorphisms and quiver moduli*, J. Inst. Math. Jussieu **9** (2009), 653–667, arXiv:0804.3214 [math.RT].

[RS16] N. Reading and D. Speyer, *Combinatorial frameworks for cluster algebras*, Int. Math. Res. Notices **2016** (2016), 109–173; arXiv:1111.2652 [math.CO].

[Sco06] J. Scott, *Grassmannians and cluster algebras*, Proc. London Math. Soc. **92** (2006), 345–380; arXiv:0311148 [math.CO].

[Wil14] L. Williams, *Cluster algebras: an introduction*, Bull. Amer. Math. Soc. **51** (2014), 1–26; arXiv:1212.6263 [math.RA].

[Zho20] Y. Zhou, *Cluster structures and subfans in scattering diagrams*, SIGMA **16** (2020), 013, 35 pages; arXiv:1901.04166 [math.AG].

Index